Applied Mathematical Sciences
Volume 120

Editors
J.E. Marsden L. Sirovich F. John (deceased)

Advisors
M. Ghil J.K. Hale T. Kambe
J. Keller K. Kirchgässner
B.J. Matkowsky C.S. Peskin
J.T. Stuart

Springer
New York
Berlin
Heidelberg
Barcelona
Budapest
Hong Kong
London
Milan
Paris
Santa Clara
Singapore
Tokyo

Applied Mathematical Sciences

(continued following index)

Andreas Kirsch

An Introduction to the Mathematical Theory of Inverse Problems

With 12 Illustrations

 Springer

Andreas Kirsch
Mathematisches Institut II
Universität (TH) Karlsruhe
Englerstrasse 2
D-76128 Karlsruhe, Germany

Editors

J.E. Marsden
Control and Dynamical Systems, 104-44
California Institute of Technology
Pasadena, CA 91125
USA

L. Sirovich
Division of Applied Mathematics
Brown University
Providence, RI 02912
USA

Mathematics Subject Classification (1991): 35R30, 35R25, 65R30, 78A45, 35Q60, 35P25, 45A05, 65M30

Library of Congress Cataloging-in-Publication Data
Kirsch, Andreas, 1953–
    An introduction to the mathematical theory of inverse problems/
Andreas Kirsch
        p.   cm. — (Applied mathematical sciences;120)
    Includes bibliographical references and index.
    ISBN 0-387-94530-X (hardcover: alk. paper)
    1. Inverse problems (Differential equations)   I. Title.
II. Series: Applied mathematical sciences (Springer-Verlag New York
Inc.) ; v. 120.
QA1.A647   vol. 120
[QA371]
515′.35 — dc20   96-17232

Printed on acid-free paper.

Production managed by Hal Henglein; manufacturing supervised by Jacqui Ashri.
Camera-ready copy prepared from the author's TeX file.
Printed and bound by Braun-Brumfield, Inc., Ann Arbor, MI.
Printed in the United States of America.

9 8 7 6 5 4 3 2 1

ISBN 0-387-94530-X Springer-Verlag New York Berlin Heidelberg   SPIN 10501642

Preface

Following Keller [119] we call two problems *inverse* to each other if the formulation of each of them requires full or partial knowledge of the other. By this definition, it is obviously arbitrary which of the two problems we call the direct and which we call the inverse problem. But usually, one of the problems has been studied earlier and, perhaps, in more detail. This one is usually called the *direct* problem, whereas the other is the *inverse* problem. However, there is often another, more important difference between these two problems. Hadamard (see [91]) introduced the concept of a *well-posed problem*, originating from the philosophy that the mathematical model of a physical problem has to have the properties of uniqueness, existence, and stability of the solution. If one of the properties fails to hold, he called the problem *ill-posed*. It turns out that many interesting and important inverse problems in science lead to ill-posed problems, while the corresponding direct problems are well-posed. Often, existence and uniqueness can be forced by enlarging or reducing the solution space (the space of "models"). For restoring stability, however, one has to change the topology of the spaces, which is in many cases impossible because of the presence of measurement errors. At first glance, it seems to be impossible to compute the solution of a problem numerically if the solution of the problem does not depend continuously on the data, i.e., for the case of ill-posed problems. Under additional a priori information about the solution, such as smoothness and bounds on the derivatives, however, it is possible to restore stability and construct efficient numerical algorithms.

We make no claim to cover all of the topics in the theory of inverse problems. Indeed, with the rapid growth of this field and its relationship to many fields of natural and technical sciences, such a task would certainly be impossible for a single author in a single volume. The aim of this book is twofold: First, we will introduce the reader to the basic notions and difficulties encountered with ill-posed problems. We will then study the basic properties of regularization methods for *linear* ill-posed problems. These methods can roughly be classified into two groups, namely, whether the regularization parameter is chosen a priori or a posteriori. We will study some of the most important regularization schemes in detail.

The second aim of this book is to give a first insight into two special *nonlinear* inverse problems that are of vital importance in many areas of the applied sciences. In both inverse spectral theory and inverse scattering theory, one tries to determine a coefficient in a differential equation from measurements of either the eigenvalues of the problem or the field "far away" from the scatterer. We hope that these two examples clearly show that a successful treatment of nonlinear inverse problems requires a solid knowledge of characteristic features of the corresponding direct problem. The combination of classical analysis and modern areas of applied and numerical analysis is, in the author's opinion, one of the fascinating features of this relatively new area of applied mathematics.

This book arose from a number of graduate courses, lectures, and survey talks during my time at the universities of Göttingen and Erlangen/Nürnberg. It was my intention to present a fairly elementary and complete introduction to the field of inverse problems, accessible not only to mathematicians but also to physicists and engineers. I tried to include as many proofs as possible as long as they required knowledge only of classical differential and integral calculus. The notions of functional analysis make it possible to treat different kinds of inverse problems in a common language and extract its basic features. For the convenience of the reader, I have collected the basic definitions and theorems from linear and nonlinear functional analysis at the end of the book in an appendix. Results on nonlinear mappings, in particular for the Fréchet derivative, are only needed in Chapters 4 and 5.

The book is organized as follows: In Chapter 1, we begin with a list of pairs of direct and inverse problems. Many of them are quite elementary and should be well-known. We formulate them from the point of view of inverse theory to demonstrate that the study of particular inverse problems has a long history. Sections 1.3 and 1.4 introduce the notions of ill-posedness and the worst-case error. While ill-posedness of a problem (roughly speaking) implies that the solution cannot be computed numerically – which is a very pessimistic point of view – the notion of the worst-case error leads to the possibility that stability can be recovered if additional information is available. We illustrate these notions with several elementary examples.

In Chapter 2, we study the general regularization theory for linear ill-posed equations in Hilbert spaces. The general concept in Section 2.1 is followed by the most important special examples: Tikhonov regularization in Section 2.2, Landweber iteration in Section 2.3, and spectral cutoff in Section 2.4. These regularization methods are applied to a test example in Section 2.5. While in Sections 2.1–2.5 the regularization parameter has been chosen a priori, i.e., before starting the actual computation, Sections 2.6–2.8 are devoted to regularization methods in which the regularization parameter is chosen implicitly by the stopping rule of the algorithm. In Sections 2.6 and 2.7, we study Morozov's discrepancy principle and, again, Landweber's iteration method. In contrast to these *linear* regularization schemes, we will investigate the conjugate gradient method in Section 2.8. This algorithm can be interpreted as a *nonlinear* regularization method and is much more difficult to analyze.

Chapter 2 deals with ill-posed problems in infinite-dimensional spaces. However, in practical situations, these problems are first discretized. The discretization of linear ill-posed problems leads to badly conditioned finite linear systems. This subject will be treated in Chapter 3. In Section 3.1, we recall basic facts about general projection methods. In Section 3.2, we will study several Galerkin methods as special cases and apply the results to Symm's integral equation in Section 3.3. This equation serves as a popular model equation in many papers on the numerical treatment of integral equations of the first kind with weakly singular kernels. We will present a complete and elementary existence and uniqueness theory of this equation in Sobolev spaces and apply the results about Galerkin methods to this equation. In Section 3.4, we study collocation methods. Here, we restrict ourselves to two examples: the moment collocation and the collocation of Symm's integral equation with trigonometric polynomials or piecewise constant functions as basis functions. In Section 3.5, we compare the different regularization techniques for a concrete numerical example of Symm's integral equation. Chapter 3 is completed by an investigation of the Backus–Gilbert method. Although this method does not quite fit into the general regularization theory, it is nevertheless widely used in the applied sciences to solve moment problems.

In Chapter 4, we study an *inverse eigenvalue problem* for a linear ordinary differential equation of second order. In Sections 4.2 and 4.3, we develop a careful analysis of the direct problem, which includes the asymptotic behavior of the eigenvalues and eigenfunctions. Section 4.4 is devoted to the question of uniqueness of the inverse problem, i.e., the problem of recovering the coefficient in the differential equation from the knowledge of one or two spectra. In Section 4.5, we show that this inverse problem is closely related to a parameter identification problem for parabolic equations. Section 4.6 describes some numerical reconstruction techniques for the inverse spectral problem.

In Chapter 5, we introduce the reader to the field of *inverse scattering theory*. Inverse scattering problems occur in several areas of science and technology, such as medical imaging, nondestructive testing of material, and geological prospecting. In Section 5.2, we study the direct problem and prove uniqueness, existence, and continuous dependence on the data. In Section 5.3, we study the asymptotic form of the scattered field as $r \to \infty$ and introduce the *far field pattern*. The corresponding inverse scattering problem is to recover the *index of refraction* from a knowledge of the far field pattern. We give a complete proof of uniqueness of this inverse problem in Section 5.4. Finally, Section 5.5 is devoted to the study of some recent reconstruction techniques for the inverse scattering problem.

Chapter 5 differs from previous ones in the unavoidable fact that we have to use some results from scattering theory without giving proofs. We will only formulate these results, and for the proofs we refer to easily accessible standard literature.

There exists a tremendous amount of literature on several aspects of inverse theory ranging from abstract regularization concepts to very concrete applications. Instead of trying to give a complete list of all relevant contributions, I mention only the monographs [15, 81, 86, 109, 130, 136, 137, 138, 144, 157, 158, 174, 215, 216], the proceedings, [5, 29, 53, 70, 93, 172, 192, 212], and survey articles [67, 116, 119, 122, 173] and refer to the references therein.

This book would not have been possible without the direct or indirect contributions of numerous colleagues and students. But, first of all, I would like to thank my father for his ability to stimulate my interest and love of mathematics during all the years. Also, I am deeply indebted to my friends and teachers, Professor Dr. Rainer Kress and Professor David Colton, who introduced me to the field of scattering theory and influenced my mathematical life in an essential way. This book is dedicated to my long friendship with them!

Particular thanks are given to Dr. Frank Hettlich, Dr. Stefan Ritter, and Dipl.-Math. Markus Wartha for carefully reading the manuscript. Furthermore, I would like to thank Professor William Rundell and Dr. Martin Hanke for their manuscripts on inverse Sturm–Liouville problems and conjugate gradient methods, respectively, on which parts of Chapters 4 and 2 are based.

Karlsruhe, April 1996   Andreas Kirsch

Contents

1

Introduction and Basic Concepts

1.1 Examples of Inverse Problems

In this section, we present some examples of pairs of problems that are inverse to each other. We start with some simple examples that are normally not even recognized as inverse problems. Most of them are taken from the survey article [119] and the monograph [87].

Example 1.1
Find a polynomial p of degree n with given zeros $x_1, \ldots, x_n$. This problem is inverse to the direct problem: Find the zeros $x_1, \ldots, x_n$ of a given polynomial p. In this example, the inverse problem is easier to solve. Its solution is $p(x) = c(x - x_1) \ldots (x - x_n)$ with an arbitrary constant c.

Example 1.2
Find a polynomial p of degree n that assumes given values $y_1, \ldots, y_n \in \mathbb{R}$ at given points $x_1, \ldots, x_n \in \mathbb{R}$. This problem is inverse to the direct problem of calculating the given polynomial at given $x_1, \ldots, x_n$. The inverse problem is the *Lagrange interpolation problem.*

Example 1.3
Given a real symmetric $n \times n$ matrix A and n real numbers $\lambda_1, \ldots, \lambda_n$, find a diagonal matrix D such that $A + D$ has the eigenvalues $\lambda_1, \ldots, \lambda_n$. This problem is inverse to the direct problem of computing the eigenvalues of the given matrix $A + D$.

Example 1.4

This inverse problem is used on intelligence tests: Given the first few terms $a_1, a_2, \ldots, a_k$ of a sequence, find the law of formation of the sequence, i.e., find a_n for all n! Usually, only the next two or three terms are asked for to show that the law of formation has been found. The corresponding direct problem is to evaluate the sequence (a_n) given the law of formation. It is clear that such inverse problems always have many solutions (from the mathematical point of view), and for this reason their use on intelligence tests has been criticized.

Example 1.5 *(Geological prospecting)*

In general, this is the problem of determining the location, shape, and/or some parameters (such as conductivity) of geological anomalies in Earth's interior from measurements at its surface. We consider a simple one-dimensional example and describe the following inverse problem.

Determine changes $\rho = \rho(x)$, $0 \leq x \leq 1$, of the mass density of an anomalous region at depth h from measurements of the vertical component $f_v(x)$ of the change of force at x. $\rho(x')\Delta x'$ is the mass of a "volume element" at x' and $\sqrt{(x - x')^2 + h^2}$ is its distance from the instrument. The change of gravity is described by Newton's law of gravity $f = \gamma \frac{m}{r^2}$ with gravitational constant γ. For the vertical component, we have

$$\Delta f_v(x) = \gamma \frac{\rho(x')\Delta x'}{(x - x')^2 + h^2} \cos \theta = \gamma \frac{h\,\rho(x')\Delta x'}{\left[(x - x')^2 + h^2\right]^{3/2}}.$$

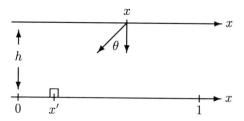

This yields the following integral equation for the determination of ρ:

$$f_v(x) = \gamma h \int_0^1 \frac{\rho(x')}{\left[(x - x')^2 + h^2\right]^{3/2}}\, dx' \quad \text{for } 0 \leq x \leq 1. \qquad (1.1)$$

We refer to [4, 81, 229] for further reading on this and related inverse problems in geological prospecting.

Example 1.6 *(Inverse scattering problem)*

Find the shape of a scattering object, given the intensity (and phase) of sound or electromagnetic waves scattered by this object. The corresponding direct problem is that of calculating the scattered wave for a given object.

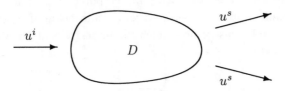

More precisely, the *direct problem* can be described as follows. Let a bounded region $D \subset \mathbb{R}^N$ ($N = 2$ or 3) be given with smooth boundary ∂D (the scattering object) and a plane *incident* wave $u^i(x) = e^{ik\hat{\theta}\cdot x}$, where $k > 0$ denotes the wave number and $\hat{\theta}$ is a unit vector that describes the direction of the incident wave. The direct problem is to find the *total field* $u = u^i + u^s$ as the sum of the incident field u^i and the *scattered field* u^s such that

$$\triangle u + k^2 u = 0 \quad \text{in } \mathbb{R}^N \setminus \overline{D}, \qquad u = 0 \quad \text{on } \partial D, \qquad (1.2a)$$

$$\frac{\partial u^s}{\partial r} - iku^s = \mathcal{O}\left(r^{-(N+1)/2}\right) \quad \text{for } r = |x| \to \infty \text{ uniformly in } \frac{x}{|x|}. \quad (1.2b)$$

For *acoustic* scattering problems, $v(x, t) = u(x)e^{-i\omega t}$ describes the pressure and $k = \omega/c$ is the wave number with speed of sound c. For suitably polarized time harmonic *electromagnetic* scattering problems, Maxwell's equations reduce to the *two-dimensional Helmholtz equation* $\triangle u + k^2 u = 0$ for the components of the electric (or magnetic) field u. The wave number k is given in terms of the dielectric constant ε and permeability μ by $k = \sqrt{\varepsilon\mu}\,\omega$.

In both cases, the radiation condition (1.2b) yields the following asymptotic behavior:

$$u^s(x) = \frac{\exp(ik\,|x|)}{|x|^{(N-1)/2}} u_\infty(\hat{x}) + \mathcal{O}\left(|x|^{-(N+1)/2}\right) \quad \text{as } |x| \to \infty,$$

where $\hat{x} = x/\,|x|$. The *inverse problem* is to determine the shape of D when the *far field pattern* $u_\infty(\hat{x})$ is measured for all $\hat{x}$ on the unit sphere in $\mathbb{R}^N$.

These and related inverse scattering problems have various applications in computer tomography, seismic and electromagnetic exploration in geophysics, and nondestructive testing of materials, for example. An inverse scattering problem of this type will be treated in detail in Chapter 5.

Standard literature on these direct and inverse scattering problems are the monographs [37, 38, 139] and the survey articles [34, 203].

Example 1.7 *(Computer tomography)*

The most spectacular application of the Radon transform is in medical imaging. For example, consider a fixed plane through a human body. Let $\rho(x, y)$ denote the change of density at the point (x, y), and let L be any line in the plane. Suppose that we direct a thin beam of X–rays into the body along L and measure how much the intensity is attenuated by going through the body.

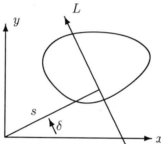

Let L be parametrized by (s, δ), where $s \in \mathbb{R}$ and $\delta \in [0, \pi)$. The ray $L_{s,\delta}$ has the coordinates

$$se^{i\delta} + iue^{i\delta} \in \mathbb{C}, \quad u \in \mathbb{R},$$

where we have identified $\mathbb{C}$ with $\mathbb{R}^2$. The attenuation of the intensity I is approximately described by $dI = -\gamma \rho I\, du$ with some constant γ. Integration along the ray yields

$$\ln I(u) = -\gamma \int_{u_0}^{u} \rho\left(se^{i\delta} + iue^{i\delta}\right) du$$

or, assuming that ρ is of compact support, the relative intensity loss is given by

$$\ln I(\infty) = -\gamma \int_{-\infty}^{\infty} \rho\left(se^{i\delta} + iue^{i\delta}\right) du.$$

In principle, from the attenuation factors we can compute all line integrals

$$(R\rho)(s, \delta) := \int_{-\infty}^{\infty} \rho\left(se^{i\delta} + iue^{i\delta}\right) du, \quad s \in \mathbb{R},\ \delta \in [0, \pi). \tag{1.3}$$

$R\rho$ is called the *Radon transform* of ρ. The *direct problem* is to compute the Radon transform $R\rho$ when ρ is given. The *inverse problem* is to determine the density ρ for a given Radon transform $R\rho$ (i.e., measurements of all line integrals).

The problem simplifies in the following special case, where we assume that ρ is radially symmetric and we choose only vertical rays. Then $\rho = \rho(r)$, $r = \sqrt{x^2 + y^2}$, and the ray L_x passing through $(x, 0)$ can be parametrized by (x, u), $u \in \mathbb{R}$. This leads to (the factor 2 is due to symmetry)

$$V(x) := \ln I(\infty) = -2\gamma \int_0^\infty \rho\left(\sqrt{x^2 + u^2}\right) du.$$

Again, we assume that ρ is of compact support in $\{x : |x| \leq R\}$. The change of variables $u = \sqrt{r^2 - x^2}$ leads to

$$V(x) = -2\gamma \int_x^\infty \frac{r}{\sqrt{r^2 - x^2}}\, \rho(r)\, dr = -2\gamma \int_x^R \frac{r}{\sqrt{r^2 - x^2}}\, \rho(r)\, dr. \quad (1.4)$$

A further change of variables $z = R^2 - r^2$ and $y = R^2 - x^2$ transforms this equation into the following *Abel's integral equation* for the function $z \mapsto \rho\left(\sqrt{R^2 - z}\right)$:

$$V\left(\sqrt{R^2 - y}\right) = -\gamma \int_0^y \frac{\rho\left(\sqrt{R^2 - z}\right)}{\sqrt{y - z}}\, dz, \quad 0 \leq y \leq R. \quad (1.5)$$

The standard mathematical literature on the Radon transform and its applications are the monographs [102, 104, 166]. We refer also to the survey articles [105, 145, 147, 152].

The following example is due to Abel himself.

Example 1.8 *(Abel's integral equation)*
Let a mass element move along a curve Γ from a point p_1 on level $h > 0$ to a point p_0 on level $h = 0$. The only force acting on this mass element is the gravitational force mg.

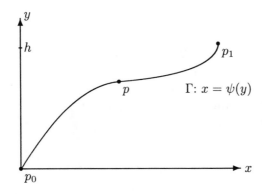

The *direct problem* is to determine the time T in which the element moves from p_1 to p_0 when the curve Γ is given. In the *inverse problem*, one measures the time $T = T(h)$ for several values of h and tries to determine the curve Γ. Let the curve be parametrized by $x = \psi(y)$. Let p have the coordinates $(\psi(y), y)$.

By conservation of energy, i.e.,

$$E + U = \frac{m}{2}v^2 + mgy = \text{const} = mgh,$$

we conclude for the velocity that

$$\frac{ds}{dt} = v = \sqrt{2g(h-y)}.$$

The total time T from p_1 to p_0 is

$$T = T(h) = \int_{p_0}^{p_1} \frac{ds}{v} = \int_0^h \sqrt{\frac{1 + \psi'(y)^2}{2g\,(h-y)}}\,dy \quad \text{for } h > 0.$$

Set $\varphi(y) = \sqrt{1 + \psi'(y)^2}$ and let $f(h) := T(h)\sqrt{2g}$ be known (measured). Then we have to determine the unknown function φ from Abel's integral equation

$$\int_0^h \frac{\varphi(y)}{\sqrt{h-y}}\,dy = f(h) \quad \text{for } h > 0. \tag{1.6}$$

A similar – but more important – problem occurs in seismology. One studies the problem to determine the velocity distribution c of Earth from measurements of the travel times of seismic waves (see [22]).

For further examples of inverse problems leading to Abel's integral equations, we refer to the lecture notes by R. Gorenflo and S. Vessella [84], the monograph [158], and the papers [141, 222].

Example 1.9 *(Backwards heat equation)*
Consider the one-dimensional heat equation

$$\frac{\partial u(x,t)}{\partial t} = \frac{\partial^2 u(x,t)}{\partial x^2} \tag{1.7a}$$

with boundary conditions

$$u(0,t) = u(\pi,t) = 0, \quad t \geq 0, \tag{1.7b}$$

and initial condition

$$u(x,0) = u_0(x), \quad 0 \leq x \leq \pi. \tag{1.7c}$$

Separation of variables leads to the solution

$$u(x,t) = \sum_{n=1}^{\infty} a_n e^{-n^2 t} \sin(nx) \quad \text{with} \quad a_n = \frac{2}{\pi} \int_0^{\pi} u_0(y) \sin(ny) dy. \quad (1.8)$$

The *direct problem* is to solve the classical initial boundary value problem: Given the initial temperature distribution u_0 and the final time T, determine $u(\cdot, T)$. In the *inverse problem*, one measures the final temperature distribution $u(\cdot, T)$ and tries to determine the temperature at earlier times $t < T$, e.g., the initial temperature $u(\cdot, 0)$.

From solution formula (1.8), we see that we have to determine $u_0 := u(\cdot, 0)$ from the following integral equation:

$$u(x,T) = \frac{2}{\pi} \int_0^{\pi} k(x,y) u_0(y) dy, \quad 0 \le x \le \pi, \quad (1.9)$$

where

$$k(x,y) := \sum_{n=1}^{\infty} e^{-n^2 T} \sin(nx) \sin(ny). \quad (1.10)$$

We refer to the monographs [15, 138, 158] and papers [24, 31, 33, 59, 60, 72, 153, 202] for further reading on this subject.

Example 1.10 *(Diffusion in inhomogeneous medium)*
The equation of diffusion in an inhomogeneous medium (now in two dimensions) is described by the equation

$$\frac{\partial u(x,t)}{\partial t} = \frac{1}{c} \text{div}(\kappa \, \text{grad} \, u(x,t)), \quad x \in D, \, t > 0, \quad (1.11)$$

where c is a constant and $\kappa = \kappa(x)$ is a parameter describing the medium. In the stationary case, this reduces to

$$\text{div}(\kappa \, \text{grad} \, u) = 0 \quad \text{in } D. \quad (1.12)$$

The *direct problem* is to solve the boundary value problem for this equation for given boundary values $u|_{\partial D}$ and given function κ. In the *inverse problem*, one measures u and the flux $\frac{\partial u}{\partial \nu}$ on the boundary ∂D and tries to determine the unknown function κ in D.

This is an example of a *parameter identification problem* for a partial differential equation. Among the extensive literature on parameter identification problems, we only mention the classical papers [128, 183, 182], the monographs [13, 15, 158], and the survey article [160].

Example 1.11 *(Sturm–Liouville eigenvalue problem)*
Let a string of length L and mass density $\rho = \rho(x) > 0$, $0 \leq x \leq L$, be fixed at the endpoints $x = 0$ and $x = L$. Plucking the string produces tones due to vibrations. Let $v(x,t)$, $0 \leq x \leq L$, $t > 0$, be the displacement at x and time t. It satisfies the *wave equation*

$$\rho(x)\frac{\partial^2 v(x,t)}{\partial t^2} = \frac{\partial^2 v(x,t)}{\partial x^2}, \quad 0 < x < L,\ t > 0, \tag{1.13}$$

subject to boundary conditions $v(0,t) = v(L,t) = 0$ for $t > 0$.

A periodic displacement of the form

$$v(x,t) = w(x)\big[a\cos\omega t + b\sin\omega t\big]$$

with frequency $\omega > 0$ is called a *pure tone*. This form of v solves the boundary value problem (1.13) if and only if w and ω satisfy the Sturm–Liouville eigenvalue problem

$$w''(x) + \omega^2 \rho(x)\,w(x) = 0,\ 0 < x < L, \quad w(0) = w(L) = 0. \tag{1.14}$$

The *direct problem* is to compute the eigenfrequencies ω and the corresponding eigenfunctions for known function ρ. In the *inverse problem*, one tries to determine the mass density ρ from a number of measured frequencies ω.

We will see in Chapter 4 that parameter estimation problems for parabolic and hyperbolic initial boundary value problems are closely related to inverse spectral problems.

Example 1.12 *(Inverse Stefan problem)*
The physicist Stefan (see [207]) modeled the melting of arctic ice in the summer by a simple one-dimensional model. In particular, consider a homogeneous block of ice filling the region $x \geq \ell$ at time $t = 0$. The ice starts to melt by heating the block at the left end. Thus, at time $t > 0$ the region between $x = 0$ and $x = s(t)$ for some $s(t) > 0$ is filled with water and the region $x \geq s(t)$ is filled with ice.

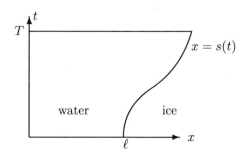

Let $u(x, t)$ be the temperature at $0 < x < s(t)$ and time t. Then u satisfies the one-dimensional heat equation

$$\frac{\partial u(x, t)}{\partial t} = \frac{\partial^2 u(x, t)}{\partial x^2} \quad \text{in } D := \{(x, t) \in \mathbb{R}^2 : 0 < x < s(t), \ t > 0\}$$

(1.15)

subject to *boundary conditions* $\frac{\partial}{\partial x} u(0, t) = f(t)$ and $u(s(t), t) = 0$ for $t \in [0, T]$ and *initial condition* $u(x, 0) = u_0(x)$ for $0 \leq x \leq \ell$.

Here, u_0 describes the initial temperature and $f(t)$ the heat flux at the left boundary $x = 0$. The speed at which the interface between water and ice moves is proportional to the heat flux. This is described by the following *Stefan condition*:

$$\frac{ds(t)}{dt} = -\frac{\partial u(s(t), t)}{\partial x} \quad \text{for } t \in [0, T].$$

(1.16)

The *direct problem* is to compute the curve s when the boundary data f and u_0 are given. In the *inverse problem*, one has given a desired curve s and tries to reconstruct u and f (or u_0).

We refer to the monographs [27, 158] and the classical papers [28, 73] for a detailed introduction to Stefan problems.

In all of these examples, we can formulate the direct problem as the evaluation of an operator K acting on a known "model" x in a model space X and the inverse problem as the solution of the equation $K(x) = y$:

Direct problem: given x (and K), evaluate $K(x)$.
Inverse problem: given y (and K), solve $K(x) = y$ for x.

In order to formulate an inverse problem, the definition of the operator K, including its domain and range, has to be given. The formulation as an operator equation allows us to distinguish among finite, semifinite, and infinite-dimensional, linear and nonlinear problems.

In general, the evaluation of $K(x)$ means solving a boundary value problem for a differential equation or evaluating an integral.

For more general and "philosophical" aspects of inverse theory, we refer to [6, 173].

1.2 Ill-Posed Problems

For all of the pairs of problems presented in the last section, there is a fundamental difference between the direct and the inverse problems. In all cases, the inverse problem is *ill-posed* or *improperly posed* in the sense of Hadamard, while the direct problem is well-posed. In his lectures published in [91], Hadamard claims that a mathematical model for a physical problem (he was thinking in terms of a boundary value problem for a partial differential equation) has to be *properly posed* or *well-posed* in the sense that it has the following three properties:

1. There exists a solution of the problem (existence).

2. There is at most one solution of the problem (uniqueness).

3. The solution depends continuously on the data (stability).

Mathematically, the existence of a solution can be enforced by enlarging the solution space. The concept of distributional solutions of a differential equation is an example. If a problem has more than one solution, then information about the model is missing. In this case, additional properties, such as sign conditions, can be built into the model. The requirement of stability is the most important one. If a problem lacks the property of stability, then its solution is practically impossible to compute because any measurement or numerical computation is polluted by unavoidable errors: thus the data of a problem are always perturbed by noise! If the solution of a problem does not depend continuously on the data, then in general the computed solution has nothing to do with the true solution. Indeed, there is no way to overcome this difficulty unless additional information about the solution is available. Here, we remind the reader of the following statement (see Lanczos [134]):

A lack of information cannot be remedied by any mathematical trickery!

Mathematically, we formulate the notion of well-posedness in the following way.

Definition 1.13 *(well-posedness)*
Let X and Y be normed spaces, $K : X \rightarrow Y$ a (linear or nonlinear) mapping. The equation $Kx = y$ is called properly posed *or* well-posed *if the following holds:*

1. Existence: *For every $y \in Y$ there is (at least one) $x \in X$ such that $Kx = y$.*

2. Uniqueness: *For every $y \in Y$ there is at most one $x \in X$ with $Kx = y$.*

3. Stability: *The solution x depends continuously on y, i.e., for every sequence $(x_n) \subset X$ with $Kx_n \rightarrow Kx$ $(n \rightarrow \infty)$, it follows that $x_n \rightarrow x$ $(n \rightarrow \infty)$.*

Equations for which (at least) one of these properties does not hold are called improperly posed *or* ill-posed.

It is important to specify the full triple (X, Y, K) and their norms. Existence and uniqueness depend only on the algebraic nature of the spaces and the operator, i.e., whether the operator is onto or one-to-one. Stability, however, depends also on the topologies of the spaces, i.e., whether the inverse operator $K^{-1} : Y \to X$ is continuous.

These requirements are not independent of each other. For example, due to the open mapping theorem (see Theorem A.25 of Appendix A), the inverse operator K^{-1} is automatically continuous if K is linear and continuous and X and Y are Banach spaces.

As an example for an ill-posed problem, we study the classical example given by Hadamard in his famous paper [91].

Example 1.14 *(Cauchy's problem for the Laplace equation)*
Find a solution u of the Laplace equation

$$\Delta u(x, y) := \frac{\partial^2 u(x, y)}{\partial x^2} + \frac{\partial^2 u(x, y)}{\partial y^2} = 0 \quad \text{in } \mathbb{R} \times [0, \infty) \qquad (1.17a)$$

that satisfies the "initial conditions"

$$u(x, 0) = f(x), \quad \frac{\partial}{\partial y} u(x, 0) = g(x), \quad x \in \mathbb{R}, \qquad (1.17b)$$

where f and g are given functions. Obviously, the (unique) solution for $f(x) = 0$ and $g(x) = \frac{1}{n} \sin(nx)$ is given by

$$u(x, y) = \frac{1}{n^2} \sin(nx) \sinh(ny), \quad x \in \mathbb{R}, \ y \geq 0.$$

Therefore, we have

$$\sup_{x \in \mathbb{R}} \{|f(x)| + |g(x)|\} = \frac{1}{n} \longrightarrow 0, \quad n \to \infty,$$

but

$$\sup_{x \in \mathbb{R}} |u(x, y)| = \frac{1}{n^2} \sinh(ny) \longrightarrow \infty, \quad n \to \infty$$

for all $y > 0$. The error in the data tends to zero while the error in the solution u tends to infinity!

Many inverse problems and some of the examples of the last section (for further examples, we refer to [87]) lead to integral equations of the first kind with continuous or weakly singular kernels. Such integral operators are *compact* with respect to any reasonable topology. The following example will often serve as a model case in these lectures.

Example 1.15 *(Differentiation)*
The direct problem is to find the antiderivative y with $y(0) = 0$ of a given
continuous function x on $[0, 1]$, i.e., compute

$$y(t) = \int_0^t x(s)\, ds, \quad t \in [0, 1]. \tag{1.18}$$

In the inverse problem, we are given a continuously differentiable function
y on $[0, 1]$ with $y(0) = 0$ and want to determine $x = y'$. This means we
have to solve the integral equation $Kx = y$, where $K : C[0, 1] \to C[0, 1]$ is
defined by

$$(Kx)(t) := \int_0^t x(s)\, ds, \quad t \in [0, 1], \quad \text{for } x \in C[0, 1]. \tag{1.19}$$

Here we equip $C[0, 1]$ with the supremum norm $\|x\|_\infty := \max_{0 \le t \le 1} |x(t)|$. The
solution of $Kx = y$ is just the derivative $x = y'$, provided $y(0) = 0$ and y
is continuously differentiable! If x is the exact solution of $Kx = y$, and
if we perturb y in the norm $\|\cdot\|_\infty$, then the perturbed right-hand side
$\tilde{y}$ doesn't have to be differentiable, and even if it is the solution of the
perturbed problem is not necessarily close to the exact solution. We can,
for example, perturb y by $\delta \sin(t/\delta^2)$ for small δ. Then the error of the
data (with respect to $\|\cdot\|_\infty$) is δ and the error in the solution is $1/\delta$. The
problem $(K, C[0, 1], C[0, 1])$ is therefore ill-posed.

Now we choose a different space $Y := \{y \in C^1[0, 1] : y(0) = 0\}$ for the
right-hand side and equip Y with the stronger norm $\|x\|_{C^1} := \max_{0 \le t \le 1} |x'(t)|$.
If the right-hand side is perturbed with respect to this norm $\|\cdot\|_{C^1}$, then the
problem $(K, C[0, 1], Y)$ is well-posed because $K : C[0, 1] \to Y$ is boundedly invertible. This example again illustrates the fact that well-posedness
depends on the topology.

In the numerical treatment of integral equations, a discretization error cannot be avoided. For integral equations of the first kind, a "naive"
discretization usually leads to disastrous results as the following simple
example shows (see also [219]).

Example 1.16
The integral equation

$$\int_0^1 e^{ts} x(s)\, ds = y(t), \quad 0 \le t \le 1, \tag{1.20}$$

with $y(t) = (\exp(t+1) - 1)/(t+1)$, is uniquely solvable by $x(t) = \exp(t)$. We approximate the integral by the trapezoidal rule

$$\int_0^1 e^{ts} x(s)\, ds \approx h \left(\frac{1}{2} x(0) + \frac{1}{2} e^t x(1) + \sum_{j=1}^{n-1} e^{jht} x(jh) \right)$$

with $h := 1/n$. For $t = ih$, we obtain the linear system

$$h \left(\frac{1}{2} x_0 + \frac{1}{2} e^{ih} x_n + \sum_{j=1}^{n-1} e^{jih^2} x_j \right) = y(ih), \quad i = 0, \ldots, n. \quad (1.21)$$

Then x_i should be an approximation to $x(ih)$. The following table lists the error between the exact solution $x(t)$ and the approximate solution x_i for $t = 0$, 0.25, 0.5, 0.75, and 1. Here, i is chosen such that $ih = t$.

t	$n = 4$	$n = 8$	$n = 16$	$n = 32$
0	0.44	−3.08	1.08	−38.21
0.25	−0.67	−38.16	−25.17	50.91
0.5	0.95	−75.44	31.24	−116.45
0.75	−1.02	−22.15	20.03	103.45
1	1.09	−0.16	−4.23	−126.87

We see that the approximations have nothing to do with the true solution and become even worse for finer discretization schemes.

In the previous two examples, the problem was to solve integral equations of the first kind. Integral operators are *compact operators* in many natural topologies under very weak conditions on the kernels. The next theorem implies that linear equations of the form $Kx = y$ with compact operators K are *always* ill-posed.

Theorem 1.17
Let X, Y be normed spaces and $K : X \to Y$ be a linear compact operator with nullspace $\mathcal{N}(K) := \{ x \in X : Kx = 0 \}$. Let the dimension of the factor space $X/\mathcal{N}(K)$ be infinite. Then there exists a sequence $(x_n) \subset X$ such that $Kx_n \to 0$ but (x_n) does not converge. We can even choose (x_n) such that $\|x_n\| \to \infty$. In particular, if K is one-to-one, the inverse $K^{-1} : Y \supset K(X) \to X$ is unbounded.

Proof: We set $\mathcal{N} = \mathcal{N}(K)$ for abbreviation. The factor space $X/\mathcal{N}$ is a normed space with norm $\|[x]\| := \inf\{ \|x + z\| : z \in \mathcal{N} \}$ since the nullspace is closed. The induced operator $\tilde{K} : X/\mathcal{N} \to Y$, defined by $\tilde{K}([x]) :=$

Kx, $[x] \in X/\mathcal{N}$, is well-defined, compact, and one-to-one. The inverse $\tilde{K}^{-1} : Y \supset K(X) \rightarrow X/\mathcal{N}$ is unbounded since otherwise the identity $I = \tilde{K}^{-1}\tilde{K} : X/\mathcal{N} \rightarrow X/\mathcal{N}$ would be compact as a composition of a bounded and a compact operator. This would contradict the assumption that the dimension of $X/\mathcal{N}$ is infinite. Since $\tilde{K}^{-1}$ is unbounded, there exists a sequence $([z_n]) \subset X/\mathcal{N}$ with $Kz_n \rightarrow 0$ and $\|[z_n]\| = 1$. We choose $v_n \in \mathcal{N}$ such that $\|z_n + v_n\| \geq \frac{1}{2}$ and set $x_n := (z_n + v_n)/\sqrt{\|Kz_n\|}$. Then $Kx_n \rightarrow 0$ and $\|x_n\| \rightarrow \infty$.   □

1.3   The Worst-Case Error

We come back to Example 1.15 of the previous section: Determine $x \in C[0,1]$ such that $\int_0^t x(s)\,ds = y(t)$ for all $t \in [0,1]$. An obvious question is: How large could the error be in the worst case if the error in the right side y is at most δ? The answer is already given by Theorem 1.17: If the errors are measured in norms such that the integral operator is compact, then the solution error could be arbitrarily large. For the special Example 1.15, we have constructed explicit perturbations with this property.

However, the situation is different if additional information is available. Before we study the general case, we illustrate this observation for a model example.

Let y and $\tilde{y}$ be twice continuously differentiable and let a number $E > 0$ be available with

$$\|y''\|_\infty \leq E \quad \text{and} \quad \|\tilde{y}''\|_\infty \leq E. \tag{1.22}$$

Set $z := \tilde{y} - y$ and assume that $z'(0) = z(0) = 0$ and $z'(t) \geq 0$ for $t \in [0,1]$. Then we estimate the error $\tilde{x} - x$ in the solution of Example 1.15 by

$$|\tilde{x}(t) - x(t)|^2 = z'(t)^2 = \int_0^t \frac{d}{ds}\left[z'(s)^2\right]ds = 2\int_0^t z'(s)\,z''(s)\,ds$$

$$\leq 4E\int_0^t z'(s)\,ds = 4E\,z(t).$$

Therefore, under the above assumptions on $z = \tilde{y} - y$ we have shown that $\|\tilde{x} - x\|_\infty \leq 2\sqrt{E\,\delta}$ if $\|\tilde{y} - y\|_\infty \leq \delta$ and E is a bound as in (1.22). In this example, $2\sqrt{E\,\delta}$ is a bound on the worst-case error for an error δ in the data and the additional information $\|x'\|_\infty = \|y''\|_\infty \leq E$ on the solution. We define the following quite generally.

Definition 1.18
Let $K : X \to Y$ be a linear bounded operator between Banach spaces, $X_1 \subset X$ a subspace, and $\|\cdot\|_1$ a "stronger" norm on X_1, i.e., there exists $c > 0$ such that $\|x\| \leq c \|x\|_1$ for all $x \in X_1$. Then we define

$$\mathcal{F}\big(\delta, E, \|\cdot\|_1\big) := \sup \{\|x\| : x \in X_1, \; \|Kx\| \leq \delta, \; \|x\|_1 \leq E\}, \quad (1.23)$$

and call $\mathcal{F}\big(\delta, E, \|\cdot\|_1\big)$ the worst-case error *for the error δ in the data and a priori information $\|x\|_1 \leq E$.*

$\mathcal{F}\big(\delta, E, \|\cdot\|_1\big)$ depends on the operator K and the norms in X, Y, and X_1. It is desirable that this worst-case error not only converge to zero as δ tends to zero but that it be of order δ. This is certainly true (even without a priori information) for boundedly invertible operators, as is readily seen from the inequality $\|x\| \leq \|K^{-1}\| \, \|Kx\|$. For compact operators K, however, and norm $\|\cdot\|_1 = \|\cdot\|$, this worst-case error does not converge (see the following lemma), and one is forced to take a stronger norm $\|\cdot\|_1$.

Lemma 1.19
Let $K : X \to Y$ be linear and compact and assume that $X/\mathcal{N}(K)$ is infinite-dimensional. Then for every $E > 0$ there exists $c > 0$ and $\delta_0 > 0$ such that $\mathcal{F}\big(\delta, E, \|\cdot\|\big) \geq c$ for all $\delta \in (0, \delta_0)$.

Proof: Assume that there exists a sequence $\delta_n \to 0$ such that $\mathcal{F}\big(\delta_n, E, \|\cdot\|\big) \to 0$ as $n \to \infty$. Let $\tilde{K} : X/\mathcal{N}(K) \to Y$ be again the induced operator in the factor space. We show that $\tilde{K}^{-1}$ is bounded: Let $\tilde{K}\big([x_m]\big) = Kx_m \to 0$. Then there exists a subsequence $\big(x_{m_n}\big)$ with $\|Kx_{m_n}\| \leq \delta_n$ for all n. We set

$$z_n := \begin{cases} x_{m_n}, & \text{if } \|x_{m_n}\| \leq E, \\ E \, \|x_{m_n}\|^{-1} x_{m_n}, & \text{if } \|x_{m_n}\| > E. \end{cases}$$

Then $\|z_n\| \leq E$ and $\|Kz_n\| \leq \delta_n$ for all n. Since the worst-case error tends to zero, we also conclude that $\|z_n\| \to 0$. From this, we see that $z_n = x_{m_n}$ for sufficiently large n, i.e., $x_{m_n} \to 0$ as $n \to \infty$. This argument, applied to every subsequence of the original sequence (x_m), yields that x_m tends to zero for $m \to \infty$, i.e., $\tilde{K}^{-1}$ is bounded on $K(X)$. This, however, contradicts the assertion of Theorem 1.17. $\qquad\square$

In the following analysis, we will make use of the singular value decomposition of the operator K (see Appendix A, Definition A.49). Therefore, we assume from now on that X and Y are Hilbert spaces. In many applications X and Y are *Sobolev spaces*, i.e., spaces of measurable functions such that their (generalized) derivatives are square integrable:

$$H^p(a,b) := \left\{ x \in C^{p-1}[a,b] : x^{(p-1)}(t) = \alpha + \int_a^t \psi \, ds, \ \alpha \in \mathbb{R}, \ \psi \in L^2 \right\}$$

(1.24)

for $p \in \mathbb{N}$.

Example 1.20 *(Differentiation)*
As an example, we will study differentiation and set $X = Y = L^2(0,1)$,

$$Kx(t) = \int_0^t x(s) \, ds, \quad t \in (0,1), \ x \in L^2(0,1),$$

and

$$
\begin{aligned}
X_1 &:= \{x \in H^1(0,1) : x(1) = 0\}, & \text{(1.25a)} \\
X_2 &:= \{x \in H^2(0,1) : x(1) = 0, \ x'(0) = 0\}. & \text{(1.25b)}
\end{aligned}
$$

We define $\|x\|_1 := \|x'\|_{L^2}$ and $\|x\|_2 := \|x''\|_{L^2}$. Then the norms $\|\cdot\|_j$, $j = 1, 2$, are stronger than $\|\cdot\|_{L^2}$ (see Problem 1.2), and we can prove for every $E > 0$ and $\delta > 0$:

$$\mathcal{F}(\delta, E, \|\cdot\|_1) \leq \sqrt{\delta E} \quad \text{and} \quad \mathcal{F}(\delta, E, \|\cdot\|_2) \leq \delta^{2/3} E^{1/3}. \tag{1.26}$$

From this result, we observe that the possibility to reconstruct x is dependent on the smoothness of the solution. We will come back to this remark in a more general setting (Theorem 1.21). We will also see that these estimates are asymptotically sharp, i.e., the exponent of δ cannot be increased.

Proof of (1.26): First, assume that $x \in H^1(0,1)$ with $x(1) = 0$. Partial integration, which is easily seen to be allowed for H^1-functions and the Cauchy–Schwarz inequality, yields

$$
\begin{aligned}
\|x\|_{L^2}^2 &= \int_0^1 x(t)^2 dt \\
&= -\int_0^1 x'(t) \left[\int_0^t x(s) \, ds \right] dt + \left[x(t) \int_0^t x(s) \, ds \right]_{t=0}^{t=1} \\
&= -\int_0^1 x'(t) \, Kx(t) \, dt \ \leq \ \|Kx\|_{L^2} \|x'\|_{L^2}.
\end{aligned}
\tag{1.27}
$$

This yields the first estimate. Now let $x \in H^2(0,1)$ such that $x(1) = 0$ and $x'(0) = 0$. Using partial integration again, we estimate

$$
\begin{aligned}
\|x'\|_{L^2}^2 &= \int_0^1 x'(t)\, x'(t)\, dt \\
&= -\int_0^1 x(t)\, x''(t)\, dt + \left[x(t)\, x'(t) \right]_{t=0}^{t=1} \\
&= -\int_0^1 x(t)\, x''(t)\, dt \leq \|x\|_{L^2} \|x''\|_{L^2}.
\end{aligned}
$$

Now we substitute this into the right-hand side of (1.27):

$$
\|x\|_{L^2}^2 \leq \|Kx\|_{L^2} \|x'\|_{L^2} \leq \|Kx\|_{L^2} \sqrt{\|x\|_{L^2}} \sqrt{\|x''\|_{L^2}}.
$$

From this, the second estimate of (1.26) follows. $\qquad\square$

It is possible to prove these estimates for more general situations, provided smoothing properties of the operator K are known. The graph norms of suitable powers of K^*K play the role of $\|\cdot\|_1$. We refer to [144] for the derivation of the general theory and prove only a partial result.

Theorem 1.21
Let X and Y be Hilbert spaces, $K : X \to Y$ linear, compact, and one-to-one with dense range $K(X)$. Let $K^* : Y \to X$ be the adjoint operator.

(a) *Set $X_1 := K^*(Y)$ and $\|x\|_1 := \left\| (K^*)^{-1} x \right\|_Y$ for $x \in X_1$. Then $\mathcal{F}(\delta, E, \|\cdot\|_1) \leq \sqrt{\delta E}$. Furthermore, for every $E > 0$ there exists a sequence $\delta_n \to 0$ such that $\mathcal{F}(\delta_n, E, \|\cdot\|_1) = \sqrt{\delta_n E}$, i.e., this estimate is asymptotically sharp.*

(b) *Set $X_2 := K^*K(X)$ and $\|x\|_2 := \left\| (K^*K)^{-1} x \right\|_X$ for $x \in X_2$. Then $\mathcal{F}(\delta, E, \|\cdot\|_2) \leq \delta^{2/3} E^{1/3}$, and for every $E > 0$ there exists a sequence $\delta_n \to 0$ such that $\mathcal{F}(\delta_n, E, \|\cdot\|_2) = \delta_n^{2/3} E^{1/3}$.*

The norms $\|\cdot\|_1$ and $\|\cdot\|_2$ are well-defined because K^* and K^*K are one-to-one. In concrete examples, the assumptions $x \in K^*(Y)$ and $x \in K^*K(X)$ are *smoothness assumptions* on the exact solution x together with boundary conditions. In the preceding example, where $Kx(t) = \int_0^t x(s)\, ds$, the spaces $K^*(L^2(0,1))$ and $(K^*K)(L^2(0,1))$ coincide with the Sobolev spaces X_1 and X_2 defined in (1.25a) and (1.25b) (see Problem 1.3).

Proof of *Theorem 1.21:*   (a) Let $x = K^*z \in X_1$ with $\|Kx\|_Y \leq \delta$ and $\|x\|_1 \leq E$, i.e., $\|KK^*z\|_Y \leq \delta$ and $\|z\|_Y \leq E$. Then

$$\|x\|_X^2 = (K^*z, x)_X = (z, Kx)_Y \leq \|z\|_Y \|Kx\|_Y \leq E\,\delta.$$

This proves the first estimate. Now let (μ_n, x_n, y_n) be a singular system for K (see Appendix A, Theorem A.50). Set $\hat{x}_n = E K^* y_n = \mu_n E x_n$ and $\delta_n := \mu_n^2 E \to 0$. Then $\|\hat{x}_n\|_1 = E$, $\|K\hat{x}_n\| = \delta_n$, and $\|\hat{x}_n\| = \mu_n E = \sqrt{\delta_n E}$. This proves part (a). Part (b) is proven similarly.   □

   Next, we consider Example 1.9 again. We are given the parabolic initial boundary value problem

$$\frac{\partial u(x,t)}{\partial t} = \frac{\partial^2 u(x,t)}{\partial x^2}, \quad 0 < x < \pi, \ t > 0,$$

$$u(0,t) = u(\pi,t) = 0, \ t > 0, \quad u(x,0) = u_0(x), \ 0 < x < \pi.$$

In the inverse problem, we know the final temperature distribution $u(x,T)$, $0 \leq x \leq \pi$, and we want to determine the temperature $u(x,\tau)$ at time $\tau \in (0,T)$. As additional information we also assume the knowledge of $E > 0$ with $\|u(\cdot,0)\|_{L^2} \leq E$.

   The solution of the initial boundary value problem is given by the series

$$u(x,t) = \frac{2}{\pi} \sum_{n=1}^{\infty} e^{-n^2 t} \sin(nx) \int_0^{\pi} u_0(y) \sin(ny)\, dy, \quad 0 \leq x \leq \pi, \ t > 0.$$

We denote the unknown function by $v := u(\cdot, \tau)$, set $X = Y = L^2(0,\pi)$, and

$$X_1 := \left\{ v \in L^2(0,\pi) : v = \sum_{n=1}^{\infty} a_n e^{-n^2 \tau} \sin(n\cdot) \text{ with} \right.$$

$$\left. a_n = \frac{2}{\pi} \int_0^{\pi} u_0(y) \sin(ny)\, dy \text{ for some } u_0 \in L^2(0,\pi) \right\}$$

and $\|v\|_1 := \|u_0\|_{L^2}$ for $v \in X_1$. In this case, the operator $K : X \to Y$ is an integral operator with kernel

$$k(x,y) = \frac{2}{\pi} \sum_{n=1}^{\infty} e^{-n^2 (T-\tau)} \sin(nx) \sin(ny), \quad x, y \in [0,\pi],$$

(see Example 1.9). Then we have for any $\tau \in (0,T)$:

$$\mathcal{F}(\delta, E, \|\cdot\|_1) \leq E^{1-\tau/T} \delta^{\tau/T}. \tag{1.28}$$

This means that under the information $\|u(\cdot,0)\|_{L^2} \leq E$, the solution $u(\cdot,\tau)$ can be determined from the final temperature distribution $u(\cdot,T)$, the determination being better the closer τ is to T.

Proof of (1.28): Let $v \in X_1$ and

$$a_n := \frac{2}{\pi} \int_0^\pi u_0(y) \sin(ny)\,dy, \quad n \in \mathbb{N},$$

be the Fourier coefficients of u_0. From the definition of X_1 and

$$(Kv)(x) = \sum_{n=1}^\infty e^{-n^2 T} a_n \sin(nx),$$

we conclude that the Fourier coefficients of v are given by $\exp(-n^2\tau)\,a_n$ and those of Kv by $\exp(-n^2T)\,a_n$. Therefore, we have to maximize

$$\frac{\pi}{2} \sum_{n=1}^\infty |a_n|^2 e^{-2n^2\tau}$$

subject to the constraints

$$\frac{\pi}{2} \sum_{n=1}^\infty |a_n|^2 \leq E^2 \quad \text{and} \quad \frac{\pi}{2} \sum_{n=1}^\infty |a_n|^2 e^{-2n^2 T} \leq \delta^2.$$

From the Hölder inequality, we have (for $p, q > 1$ with $1/p + 1/q = 1$ to be specified in a moment):

$$\frac{\pi}{2} \sum_{n=1}^\infty |a_n|^2 e^{-2n^2\tau} = \frac{\pi}{2} \sum_{n=1}^\infty |a_n|^{2/q} \left(|a_n|^{2/p} e^{-2n^2\tau} \right)$$

$$\leq \left(\frac{\pi}{2} \sum_{n=1}^\infty |a_n|^2 \right)^{1/q} \left(\frac{\pi}{2} \sum_{n=1}^\infty |a_n|^2 e^{-2pn^2\tau} \right)^{1/p}.$$

We now choose $p = T/\tau$. Then $1/p = \tau/T$ and $1/q = 1 - \tau/T$. This yields the assertion. $\qquad\square$

The next chapter is devoted to the construction of regularization schemes that are *asymptotically optimal* in the sense that, under the information $x \in X_1$, $\|x\|_1 \leq E$, and $\|\tilde{y} - y\| \leq \delta$, an approximation $\tilde{x}$ and a constant $c > 0$ are constructed such that $\|\tilde{x} - x\|_\infty \leq c\,\mathcal{F}(\delta, E, \|\cdot\|_1)$.

As a first tutorial example, we consider the problem of numerical differentiation; see Examples 1.15 and 1.20.

Example 1.22

We fix $h \in (0, 1/2)$ and define the one-sided difference quotient by

$$
v(t) = \begin{cases} \frac{1}{h}\left[y(t+h) - y(t)\right], & 0 < t < 1/2, \\ \frac{1}{h}\left[y(t) - y(t-h)\right], & 1/2 < t < 1, \end{cases}
$$

for any $y \in L^2(0,1)$. First, we estimate $\|v - y'\|_{L^2}$ for smooth functions y, i.e., $y \in H^2(0,1)$. From Taylor's formula (see Problem 1.4), we have

$$
y(t \pm h) = y(t) \pm y'(t)\, h + \int_t^{t \pm h} (t \pm h - s)\, y''(s)\, ds,
$$

i.e.,

$$
v(t) - y'(t) = \frac{1}{h} \int_t^{t+h} (t + h - s)\, y''(s)\, ds
$$

$$
= \frac{1}{h} \int_0^h y''(t + h - \tau)\, \tau\, d\tau,
$$

for $t \in (0, 1/2)$ and analogously for $t \in (1/2, 1)$. Hence, we estimate

$$
h^2 \int_0^{1/2} \left| v(t) - y'(t) \right|^2 dt
$$

$$
= \int_0^h \int_0^h \tau s \left[\int_0^{1/2} y''(t + h - \tau)\, y''(t + h - s)\, dt \right] d\tau\, ds
$$

$$
\leq \int_0^h \int_0^h \tau s \sqrt{\int_0^{1/2} |y''(t + h - \tau)|^2 dt} \sqrt{\int_0^{1/2} |y''(t + h - s)|^2 dt}\, d\tau\, ds
$$

$$
\leq \|y''\|_{L^2}^2 \left[\int_0^h \tau\, d\tau \right]^2 = \frac{1}{4} h^4 \|y''\|_{L^2}^2,
$$

and analogously for $h^2 \int_{1/2}^1 |v(t) - y'(t)|^2\, dt$. Summing these estimates yields

$$
\|v - y'\|_{L^2} \leq \frac{1}{\sqrt{2}} E\, h,
$$

where E is some bound on $\|y''\|_{L_2}$.

Now we treat the situation with errors. Instead of $y(t)$ and $y(t \pm h)$, we measure $\tilde{y}(t)$ and $\tilde{y}(t \pm h)$, respectively. We assume that $\|\tilde{y} - y\|_{L^2} \leq \delta$. Instead of $v(t)$ we compute $\tilde{v}(t) = \pm\left[\tilde{y}(t \pm h) - \tilde{y}(t)\right]/h$ for $t \in (0, 1/2)$ or $t \in (1/2, 1)$ respectively. Since

$$|\tilde{v}(t) - v(t)| \leq \frac{|\tilde{y}(t \pm h) - y(\pm h)|}{h} + \frac{|\tilde{y}(t) - y(t)|}{h},$$

we conclude that $\|\tilde{v} - v\|_{L^2} \leq 2\sqrt{2}\,\delta/h$. Therefore, the total error due to the error on the right-hand side and the discretization error is

$$\|\tilde{v} - y'\|_{L^2} \leq \|\tilde{v} - v\|_{L^2} + \|v - y'\|_{L^2} \leq \frac{2\sqrt{2}\,\delta}{h} + \frac{1}{\sqrt{2}}\,E\,h. \quad (1.29)$$

By this estimate, it is desirable to choose the discretization parameter h as the minimum of the right-hand side of (1.29). Its minimum is obtained at $h = 2\sqrt{\delta/E}$. This results in the optimal error $\|\tilde{v} - y'\|_{L^2} \leq 2\sqrt{2E\,\delta}$.

Summarizing, we note that the discretization parameter h should be of order $\sqrt{\delta/E}$ if the derivative of a function is computed by the one-sided difference quotient. With this choice, the method is optimal under the information $\|x'\|_{L^2} \leq E$.

The two-sided difference quotient is optimal under the a priori information $\|x''\|_{L^2} \leq E$ and results in an algorithm of order $\delta^{2/3}$ (see Example 2.4 in the following chapter).

We have carried out the preceding analysis with respect to the L^2-norm rather that the maximum norm, mainly because we will present the general theory in Hilbert spaces. For this example, however, estimates with respect to $\|\cdot\|_\infty$ are simpler to derive (see the estimates preceding Definition 1.18 of the worst-case error).

The result of this example is of practical importance: For many algorithms using numerical derivatives (e.g., quasi-Newton methods in optimization), it is recommended that you choose the discretization parameter ε to be the square root of the floating-point precision of the computer since a one-sided difference quotient is used.

1.4   Problems

1.1 Show that equations (1.1) and (1.20) have at most one solution.
    Hint: Extend ρ in (1.1) by zero into $\mathbb{R}$ and apply the Fourier transform.

1.2 Let the Sobolev spaces X_1 and X_2 be defined by (1.25a) and (1.25b), respectively. Define the bilinear forms by

$$(x, y)_1 := \int_0^1 x'(t)\, y'(t)\, dt \quad \text{and} \quad (x, y)_2 := \int_0^1 x''(t)\, y''(t)\, dt$$

on X_1 and X_2, respectively. Prove that X_j are Hilbert spaces with respect to the inner products $(\cdot, \cdot)_j$, $j = 1, 2$, and that $\|x\|_{L^2} \leq \|x\|_j$ for all $x \in X_j$, $j = 1, 2$.

1.3 Let $K : L^2(0, 1) \rightarrow L^2(0, 1)$ be defined by (1.19). Show that the ranges $K^*\big(L^2(0, 1)\big)$ and $K^*K\big(L^2(0, 1)\big)$ coincide with the spaces X_1 and X_2 defined by (1.25a) and (1.25b), respectively.

1.4 Prove the following version of Taylor's formula by induction with respect to n and partial integration:

Let $y \in H^{n+1}(a, b)$ and $t, t + h \in [a, b]$. Then

$$y(t + h) = \sum_{k=0}^{n} \frac{y^{(k)}(t)}{k!}\, h^k + R_n(t; h),$$

where the error term is given by

$$R_n(t; h) = \frac{1}{n!} \int_t^{t+h} (t + h - s)^n\, y^{(n+1)}(s)\, ds.$$

2

Regularization Theory for Equations of the First Kind

We saw in the previous chapter that many inverse problems can be formulated as operator equations of the form

$$Kx = y,$$

where K is a linear compact operator between Hilbert spaces X and Y over the field $\mathbb{K} = \mathbb{R}$ or $\mathbb{C}$. We also saw that a successful reconstruction strategy requires additional a priori information about the solution.

This chapter is devoted to a systematic study of regularization strategies for solving $Kx = y$. In particular, we wish to investigate under which conditions they are *optimal*, i.e., of the same asymptotic order as the worst-case error. In Section 2.1, we will introduce the general concept of regularization. In Sections 2.2 and 2.3, we will study Tikhonov's method and the Landweber iteration as two of the most important regularization strategies. In these three sections, the regularization parameter $\alpha = \alpha(\delta)$ will be chosen a priori, i.e., before we start to compute the regularized solution. We will see that the optimal regularization parameter α depends on bounds of the exact solution; they are not known in advance. Therefore, it is advantageous to study strategies for the choice of α that depend on the numerical algorithm and are made during the algorithm (a posteriori). Different a posteriori choices will be studied in Sections 2.5–2.7.

All of them are motivated by the idea that it is certainly sufficient to compute an approximation $x^{\alpha,\delta}$ of the solution x such that the norm of the defect $Kx^{\alpha,\delta} - y^\delta$ is of the same order as the perturbation error δ of the right-hand side. The classical strategy, due to Morozov [154], determines α by solving a nonlinear scalar equation. To solve this equation, we still need a numerical algorithm such as the "regula falsi" or the Newton method. In Sections 2.6 and 2.7, we will investigate two well-known iterative algorithms for solving linear (or nonlinear) equations: Landweber's method (see [135]), which is the steepest descent method, and the conjugate gradient method. The choices of α are made implicitly by stopping the algorithm as soon as the defect $\left\|Kx^m - y^\delta\right\|$ is less than $\tau\delta$. Here, $\tau > 1$ is a given parameter.

Landweber's method and Morozov's discrepancy principle are easy to investigate theoretically since they can be formulated as *linear* regularization methods. The study of the conjugate gradient method is more difficult since the choice of α depends *nonlinearly* on the right-hand side y. Since the proofs in Section 2.7 are very technical, we postpone them to an appendix.

2.1   A General Regularization Theory

For simplicity, we assume throughout this chapter that the compact operator K is one-to-one. This is not a serious restriction since we can always replace the domain X by the orthogonal complement of the kernel of K. We make the assumption that there exists a solution $x \in X$ of the unperturbed equation $Kx = y$. In other words, we assume that $y \in K(X)$. The injectivity of K implies that this solution is unique.

In practice, the right-hand side $y \in Y$ is never known exactly but only up to an error of, say, $\delta > 0$. Therefore, we assume that we know $\delta > 0$ and $y^\delta \in Y$ with

$$\left\|y - y^\delta\right\| \leq \delta. \tag{2.1}$$

It is our aim to "solve" the perturbed equation

$$Kx^\delta = y^\delta. \tag{2.2}$$

In general, (2.2) is not solvable since we cannot assume that the measured data y^δ are in the range $K(X)$ of K. Therefore, the best we can hope is to determine an approximation $x^\delta \in X$ to the exact solution x that is "not much worse" than the worst-case error $\mathcal{F}(\delta, E, \|\cdot\|_1)$ of Definition 1.18.

An additional requirement is that the approximate solution x^δ should depend continuously on the data y^δ. In other words, it is our aim to construct a suitable bounded approximation $R : Y \to X$ of the (unbounded) inverse operator $K^{-1} : K(X) \to X$.

Definition 2.1
A regularization strategy is a family of linear and bounded operators

$$R_\alpha : Y \longrightarrow X, \quad \alpha > 0,$$

such that

$$\lim_{\alpha \to 0} R_\alpha K x \;=\; x \quad \text{for all } x \in X,$$

i.e., the operators $R_\alpha K$ converge pointwise to the identity.

From this definition and the compactness of K, we conclude the following.

Theorem 2.2
Let R_α be a regularization strategy for a compact operator $K : X \to Y$ where $\dim X = \infty$. Then we have

(1) *The operators R_α are not uniformly bounded, i.e., there exists a sequence (α_j) with $\|R_{\alpha_j}\| \to \infty$ for $j \to \infty$.*

(2) *The sequence $(R_\alpha K x)$ does not converge uniformly on bounded subsets of X, i.e., there is no convergence $R_\alpha K$ to the identity I in the operator norm.*

Proof: (1) Assume, on the contrary, that there exists $c > 0$ such that $\|R_\alpha\| \leq c$ for all $\alpha > 0$. From $R_\alpha y \to K^{-1}y$ $(\alpha \to 0)$ for all $y \in K(X)$ and $\|R_\alpha y\| \leq c\|y\|$ for $\alpha > 0$ we conclude that $\|K^{-1}y\| \leq c\|y\|$ for every $y \in K(X)$, i.e., K^{-1} is bounded. This implies that $I = K^{-1}K : X \to X$ is compact, a contradiction to $\dim X = \infty$.

(2) Assume that $R_\alpha K \to I$ in $\mathcal{L}(X, X)$. From the compactness of $R_\alpha K$ and Theorem A.32, we conclude that I is also compact, which again would imply that $\dim X < \infty$.   □

The notion of a regularization strategy is based on unperturbed data, i.e., the regularizer $R_\alpha y$ converges to x for the exact right-hand side $y = Kx$.

Now let $y \in K(X)$ be the exact right-hand side and $y^\delta \in Y$ be the measured data with $\|y - y^\delta\| \leq \delta$. We define

$$x^{\alpha,\delta} := R_\alpha y^\delta \tag{2.3}$$

as an approximation of the solution x of $Kx = y$. Then the error splits into two parts by the following obvious application of the triangle inequality:

$$
\begin{aligned}
\left\| x^{\alpha,\delta} - x \right\| &\leq \left\| R_\alpha y^\delta - R_\alpha y \right\| + \left\| R_\alpha y - x \right\| \\
&\leq \|R_\alpha\| \, \|y^\delta - y\| + \|R_\alpha K x - x\|
\end{aligned}
$$

and thus

$$\left\| x^{\alpha,\delta} - x \right\| \leq \delta \left\| R_\alpha \right\| + \left\| R_\alpha K x - x \right\|. \tag{2.4}$$

This is our fundamental estimate, which we will use often in the following.

We observe that the error between the exact and computed solutions consists of two parts: The first term on the right-hand side describes the error in the data multiplied by the "condition number" $\left\| R_\alpha \right\|$ of the regularized problem. By Theorem 2.2, this term tends to infinity as α tends to zero. The second term denotes the approximation error $\left\| (R_\alpha - K^{-1}) y \right\|$ at the exact right-hand side $y = Kx$. By the definition of a regularization strategy, this term tends to zero with α. The following figure illustrates the situation.

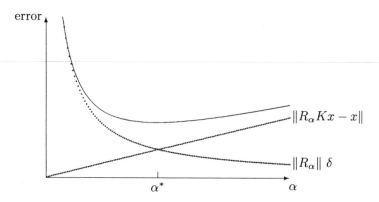

Figure 2.1: Behavior of the total error

We need a strategy to choose $\alpha = \alpha(\delta)$ dependent on δ in order to keep the total error as small as possible. This means that we would like to minimize

$$\delta \left\| R_\alpha \right\| + \left\| R_\alpha K x - x \right\|.$$

The procedure is the same in every concrete situation: One has to estimate the quantities $\left\| R_\alpha \right\|$ and $\left\| R_\alpha K x - x \right\|$ in terms of α and then minimize this upper bound with respect to α. Before we carry out these steps for two model examples, we introduce the following notation.

Definition 2.3

A regularization strategy $\alpha = \alpha(\delta)$ is called admissible *if*

$$\alpha(\delta) \longrightarrow 0 \quad and \quad \sup \left\{ \left\| R_{\alpha(\delta)} y^\delta - x \right\| : \left\| Kx - y^\delta \right\| \leq \delta \right\} \to 0, \quad \delta \to 0,$$

for every $x \in X$.

Example 2.4 *(Numerical differentiation by two-sided difference quotient)*
It is our aim to compute the derivative of a function by the two-sided
difference quotient (see Example 1.22 for the one-sided difference quotient).
Here $\alpha = h$ is the step size, and we define

$$R_h y(t) = \begin{cases} \frac{1}{h}\left[4\,y\left(t+\frac{h}{2}\right) - y(t+h) - 3\,y(t)\right], & 0 < t < \frac{h}{2}, \\ \frac{1}{h}\left[y\left(t+\frac{h}{2}\right) - y\left(t-\frac{h}{2}\right)\right], & \frac{h}{2} < t < 1 - \frac{h}{2}, \\ \frac{1}{h}\left[3\,y(t) + y(t-h) - 4\,y\left(t-\frac{h}{2}\right)\right], & 1 - \frac{h}{2} < t < 1, \end{cases}$$

for $y \in L^2(0,1)$. In order to prove that R_h defines a regularization strategy,
it suffices to show that $R_h K$ are uniformly bounded with respect to h in the
operator norm of $L^2(0,1)$ and that $\|R_h K x - x\|_{L^2}$ tends to zero for smooth
x (see Theorem A.27 of Appendix A). Later, we will show convergence for
$x \in H^2(0,1)$.

The fundamental theorem of calculus (or Taylor's formula from Problem 1.4 for $n = 0$) yields

$$R_h y(t) = \frac{1}{h} \int_{t-h/2}^{t+h/2} y'(s)\,ds = \frac{1}{h} \int_{-h/2}^{h/2} y'(s+t)\,ds, \quad \frac{h}{2} < t < 1 - \frac{h}{2},$$

and thus

$$\int_{h/2}^{1-h/2} |R_h y(t)|^2\,dt = \frac{1}{h^2} \int_{-h/2}^{h/2} \int_{-h/2}^{h/2} \int_{h/2}^{1-h/2} y'(s+t)\,y'(\tau+t)\,dt\,d\tau\,ds.$$

The Cauchy–Schwarz inequality yields

$$\int_{h/2}^{1-h/2} |R_h y(t)|^2\,dt \leq \|y'\|_{L^2}^2.$$

From $R_h y(t) = 4\left[y(t+h/2) - y(t)\right]/h - \left[y(t+h) - y(t)\right]/h$ for $0 < t < h/2$
and an analogous representation for $1 - h/2 < t < 1$, similar estimates yield
the existence of $c > 0$ with

$$\|R_h\,y\|_{L^2} \leq c\,\|y'\|_{L^2}$$

for all $y \in H^1(0,1)$. For $y = Kx$, $x \in L^2(0,1)$, the uniform boundedness of
$\left(R_h K\right)$ follows.

Now let $x \in H^2(0,1)$ and thus $y = Kx \in H^3(0,1)$. We apply Taylor's
formula (see Problem 1.4) in the form (first again for $h/2 < t < 1 - h/2$)

$$y(t \pm h/2) - y(t) \mp \frac{h}{2}\,y'(t) - \frac{h^2}{8}\,y''(t)$$

$$= \frac{1}{2} \int_0^{\pm h/2} s^2 \, y'''(t \pm h/2 - s) \, ds.$$

Subtracting the formulas for $+$ and $-$ yields

$$R_h y(t) - y'(t) = \frac{1}{2h} \int_0^{h/2} s^2 \left[y'''(t + h/2 - s) + y'''(t - h/2 + s) \right] ds,$$

and thus by changing the orders of integration und using the Cauchy–Schwarz inequality

$$\int_{h/2}^{1-h/2} |R_h y(t) - y'(t)|^2 \, dt \ \leq \ \frac{1}{h^2} \, \|y'''\|_{L^2}^2 \left(\int_0^{h/2} s^2 ds \right)^2$$

$$= \frac{1}{24^2} \, \|y'''\|_{L^2}^2 \, h^4.$$

Similar applications of Taylor's formula in the intervals $(0, h/2)$ and $(1 - h/2, 1)$ yield an estimate of the form

$$\|R_h K x - x\|_{L^2} = \|R_h y - y'\|_{L^2} \leq c_1 \, E \, h^2$$

for all $x \in H^2(0,1)$ with $\|x''\|_{L^2} \leq E$. Together with the uniform boundedness of $R_h K$, this implies that $R_h K x \to x$ for all $x \in L^2(0,1)$.

In order to apply the fundamental estimate (2.4), we must estimate the first term, i.e., the L^2-norm of R_h. It is easily checked that there exists $c_2 > 0$ with $\|R_h y\|_{L^2} \leq c_2 \|y\|_{L^2} /h$ for all $y \in L^2(0,1)$. Estimate (2.4) yields

$$\left\| R_h y^\delta - x \right\|_{L^2} \leq c_2 \frac{\delta}{h} + c_1 \, E \, h^2,$$

where E is a bound on $\|x''\|_{L^2} = \|y'''\|_{L^2}$. Minimization with respect to h of the expression on the right-hand side leads to

$$h(\delta) = c \sqrt[3]{\delta/E} \quad \text{and} \quad \left\| R_{h(\delta)} y^\delta - x \right\| \leq \tilde{c} \, E^{1/3} \, \delta^{2/3}$$

for some $c > 0$ and $\tilde{c} = c_2/c + c_1 c^2$.

We observe that this strategy is asymptotically optimal for the information $\|x''\|_{L^2} \leq E$ since it provides an approximation x^δ that is asymptotically not worse than the worst-case error (see Example 1.20).

The (one- or two-sided) difference quotient uses only local portions of the function y. An alternative approach is to first smooth the function y by mollification and then to differentiate the mollified function.

Example 2.5 *(Numerical differentiation by mollification)*
Again, we define the operator $Kx(t) = \int_0^t x(s)\,ds$, $t \in [0,1]$, but now as an operator from the (closed) subspace

$$L_0^2(0,1) := \left\{ z \in L^2(0,1) : \int_0^1 z(s)\,ds = 0 \right\}$$

of $L^2(0,1)$ into $L^2(0,1)$.

We define the Gaussian kernel ψ_α by

$$\psi_\alpha(t) = \frac{1}{\alpha\sqrt{\pi}} \exp\left(-t^2/\alpha^2\right), \quad t \in \mathbb{R},$$

where $\alpha > 0$ denotes a parameter. Then $\int_{-\infty}^{\infty} \psi_\alpha(t)\,dt = 1$, and the convolution

$$(\psi_\alpha * y)(t) := \int_{-\infty}^{\infty} \psi_\alpha(t-s)\,y(s)\,ds = \int_{-\infty}^{\infty} \psi_\alpha(s)\,y(t-s)\,ds, \quad t \in \mathbb{R},$$

exists and is an L^2–function for every $y \in L^2(\mathbb{R})$. Furthermore, by Young's inequality (see [32], p. 102), we have that

$$\|\psi_\alpha * y\|_{L^2} \leq \|\psi_\alpha\|_{L^1}\, \|y\|_{L^2} = \|y\|_{L^2} \quad \text{for all } y \in L^2(\mathbb{R}).$$

Therefore, the operators $y \mapsto \psi_\alpha * y$ are uniformly bounded in $L^2(\mathbb{R})$ with respect to α. We note that $\psi_\alpha * y$ is infinitely often differentiable on $\mathbb{R}$ for every $y \in L^2(\mathbb{R})$.

We will need the two convergence properties

$$\left\|\psi_\alpha * z - z\right\|_{L^2(\mathbb{R})} \to 0 \quad \text{as } \alpha \to 0 \quad \text{for every } z \in L^2(0,1) \qquad (2.5a)$$

and

$$\left\|\psi_\alpha * z - z\right\|_{L^2(\mathbb{R})} \leq \sqrt{2}\,\alpha\, \|z'\|_{L^2(0,1)} \qquad (2.5b)$$

for every $z \in H^1(0,1)$ with $z(0) = z(1) = 0$. Here and in the following we identify functions $z \in L^2(0,1)$ with functions $z \in L^2(\mathbb{R})$ where we think of them being extended by zero outside of $[0,1]$.

Proof of (2.5a), (2.5b): It is sufficient to prove (2.5b) because the space $\{z \in H^1(0,1) : z(0) = z(1) = 0\}$ is dense in $L^2(0,1)$, and the operators $z \mapsto \psi_\alpha * z$ are uniformly bounded from $L^2(0,1)$ into $L^2(\mathbb{R})$.

Let the Fourier transform be defined by

$$(\mathcal{F}z)(t) := \frac{1}{\sqrt{2\pi}} \int_{-\infty}^{\infty} z(s)\,e^{ist}ds, \quad t \in \mathbb{R},$$

for $z \in \mathcal{S}$, where the Schwarz space $\mathcal{S}$ is defined by

$$\mathcal{S} := \left\{ z \in C^\infty(\mathbb{R}) : \sup_{t \in \mathbb{R}} \left| t^p z^{(q)}(t) \right| < \infty \text{ for all } p, q \in \mathbb{N}_0 \right\}.$$

With this normalization, Plancherel's theorem and the convolution theorem take the form (see [32])

$$\|\mathcal{F}z\|_{L^2(\mathbb{R})} = \|z\|_{L^2(\mathbb{R})}, \quad \mathcal{F}(u * z)(t) = \sqrt{2\pi}\, (\mathcal{F}u)(t)\, (\mathcal{F}z)(t), \ t \in \mathbb{R},$$

for all $z, u \in \mathcal{S}$. Since $\mathcal{S}$ is dense in $L^2(\mathbb{R})$, these formulas hold also for $z \in L^2(\mathbb{R})$. Now we combine these properties and conclude that

$$\left\|\psi_\alpha * z - z\right\|_{L^2(\mathbb{R})} = \left\|\mathcal{F}(\psi_\alpha * z) - \mathcal{F}z\right\|_{L^2(\mathbb{R})} = \left\|\left[\sqrt{2\pi}\,\mathcal{F}(\psi_\alpha) - 1\right]\mathcal{F}z\right\|_{L^2(\mathbb{R})}$$

for every $z \in L^2(0,1)$. Partial integration yields that

$$\mathcal{F}(z')(t) = \frac{1}{\sqrt{2\pi}} \int_0^1 z'(s)\, e^{ist}\, ds = -\frac{it}{\sqrt{2\pi}} \int_0^1 z(s)\, e^{ist}\, ds = (-it)\, (\mathcal{F}z)(t)$$

for all $z \in H^1(0,1)$ with $z(0) = z(1) = 0$. We define the function φ_α by

$$\varphi_\alpha(t) := \frac{1}{it}\left[1 - \sqrt{2\pi}\,\mathcal{F}(\psi_\alpha)\right] = \frac{1}{it}\left[1 - e^{-\alpha^2 t^2/4}\right], \quad t \in \mathbb{R}.$$

Then we conclude that

$$\left\|\psi_\alpha * z - z\right\|_{L^2(\mathbb{R})} = \left\|\varphi_\alpha \mathcal{F}(z')\right\|_{L^2(\mathbb{R})} \le \left\|\varphi_\alpha\right\|_\infty \left\|\mathcal{F}(z')\right\|_{L^2(\mathbb{R})}$$

$$= \left\|\varphi_\alpha\right\|_\infty \left\|z'\right\|_{L^2(0,1)}.$$

From

$$\left|\varphi_\alpha(t)\right| = \frac{\alpha}{2} \frac{1}{|\alpha t/2|}\left[1 - e^{-(\alpha t/2)^2}\right]$$

and the elementary estimate $[1 - \exp(-\tau^2)]/\tau \le 2\sqrt{2}$ for all $\tau > 0$, the desired estimate (2.5b) follows.

After these preparations we define the regularization operators $R_\alpha : L^2(0,1) \to L_0^2(0,1)$ by

$$R_\alpha y(t) := \frac{d}{dt}\, (\psi_\alpha * y)(t) - \int_0^1 \frac{d}{ds}\, (\psi_\alpha * y)(s)\, ds$$

$$= (\psi'_\alpha * y)(t) - \int_0^1 (\psi'_\alpha * y)(s)\, ds$$

for $t \in (0,1)$ and $y \in L^2(0,1)$. First, we note that R_α is well-defined, i.e., maps $L^2(0,1)$ into $L_0^2(0,1)$ and is bounded. To prove that R_α is a regularization strategy, we proceed as in the previous example and show that

(i) $\|R_\alpha y\|_{L^2} \leq \frac{4}{\alpha\sqrt{\pi}} \|y\|_{L^2}$ for all $\alpha > 0$ and $y \in L^2(0,1)$,

(ii) $\|R_\alpha K x\|_{L^2} \leq 2\|x\|_{L^2}$ for all $\alpha > 0$ and $x \in L_0^2(0,1)$, i.e., the operators $R_\alpha K$ are uniformly bounded in $L_0^2(0,1)$, and

(iii) $\|R_\alpha K x - x\|_{L^2} \leq 2\sqrt{2}\,\alpha\,\|x'\|_{L^2}$ for all $\alpha > 0$ and $x \in H_0^1(0,1)$, where we have set

$$H_0^1(0,1) := \left\{ x \in H^1(0,1) : x(0) = x(1) = 0,\ \int_0^1 x(s)\,ds = 0 \right\}.$$

To prove part (i), we estimate with the Cauchy–Schwarz inequality

$$
\begin{aligned}
\|R_\alpha y\|_{L^2(0,1)} &\leq 2\|\psi_\alpha' * y\|_{L^2(0,1)} \leq 2\|\psi_\alpha' * y\|_{L^2(\mathbb{R})} \\
&\leq 2\|\psi_\alpha'\|_{L^1(\mathbb{R})} \|y\|_{L^2(0,1)} \leq \frac{4}{\alpha\sqrt{\pi}} \|y\|_{L^2(0,1)}
\end{aligned}
$$

for all $y \in L^2(0,1)$ since

$$\|\psi_\alpha'\|_{L^1(\mathbb{R})} = -2\int_0^\infty \psi_\alpha'(s)\,ds = 2\psi_\alpha(0) = \frac{2}{\alpha\sqrt{\pi}}.$$

This proves part (i).

Now let $y \in H^1(0,1)$ with $y(0) = y(1) = 0$. Then, by partial integration,

$$(\psi_\alpha' * y)(t) = \int_0^1 \psi_\alpha'(t-s)\,y(s)\,ds = \int_0^1 \psi_\alpha(t-s)\,y'(s)\,ds = (\psi_\alpha * y')(t).$$

Taking $y = Kx$, $x \in L_0^2(0,1)$ yields

$$R_\alpha K x(t) = (\psi_\alpha * x)(t) - \int_0^1 (\psi_\alpha * x)(s)\,ds.$$

Part (ii) now follows from Young's inequality.

Finally, we write

$$R_\alpha K x(t) - x(t) = (\psi_\alpha * x)(t) - x(t) - \int_0^1 [(\psi_\alpha * x)(s) - x(s)]\,ds$$

since $\int_0^1 x(s)\,ds = 0$. Therefore, by (2.5b),

$$\left\|R_\alpha Kx - x\right\|_{L^2(0,1)} \leq 2\left\|\psi_\alpha * x - x\right\|_{L^2(0,1)} \leq 2\sqrt{2}\,\alpha\,\|x'\|_{L^2}$$

for all $x \in H_0^1(0,1)$. This proves part (iii).

Now we conclude that $R_\alpha Kx$ converges to x for any $x \in L_0^2(0,1)$ by (ii), (iii), and the denseness of $H_0^1(0,1)$ in $L_0^2(0,1)$. Therefore, R_α defines a regularization strategy. From (i) and (iii) we rewrite the fundamental estimate (2.4) as

$$\left\|R_\alpha y^\delta - x\right\|_{L^2} \leq \frac{4\,\delta}{\alpha\sqrt{\pi}} + 2\sqrt{2}\,\alpha\,E$$

if $x \in H_0^1(0,1)$ with $\|x'\|_{L^2} \leq E$, $y = Kx$, and $y^\delta \in L^2(0,1)$ such that $\|y^\delta - y\|_{L^2} \leq \delta$. The choice $\alpha = c\sqrt{\delta/E}$ again leads to the optimal order $\mathcal{O}(\sqrt{\delta E})$.

For further applications of the mollification method, we refer to the monograph by Murio [158]. There exists an enormous number of publications on numerical differentiation. We mention only the papers [2, 49, 54, 127] and, for more general Volterra equations of the first kind, [19, 20, 55, 56, 140].

A convenient method to construct classes of admissible regularization strategies is given by filtering singular systems. Let $K : X \to Y$ be a linear compact operator, and let (μ_j, x_j, y_j) be a singular system for K (see Appendix A, Definition A.49, and Theorem A.50). As readily seen, the solution x of $Kx = y$ is given by Picard's theorem (see Theorem A.51 of Appendix A) as

$$x = \sum_{j=1}^{\infty} \frac{1}{\mu_j}\,(y, y_j)\,x_j \tag{2.6}$$

provided the series converges, i.e., $y \in K(X)$. This result illustrates again the influence of errors in y. We construct regularization strategies by damping the factors $1/\mu_j$.

Theorem 2.6
Let $K : X \to Y$ be compact with singular system (μ_j, x_j, y_j) and

$$q : (0, \infty) \times (0, \|K\|] \longrightarrow \mathbb{R}$$

be a function with the following properties:

(1) $|q(\alpha, \mu)| \leq 1$ for all $\alpha > 0$ and $0 < \mu \leq \|K\|$.

(2) For every $\alpha > 0$ there exists $c(\alpha)$ such that

$$|q(\alpha, \mu)| \leq c(\alpha)\,\mu \quad \text{for all } 0 < \mu \leq \|K\|.$$

(3a) $\lim\limits_{\alpha \to 0} q(\alpha, \mu) = 1$ *for every* $0 < \mu \leq \|K\|$.

Then the operator $R_\alpha : Y \to X$, $\alpha > 0$, *defined by*

$$R_\alpha y := \sum_{j=1}^{\infty} \frac{q(\alpha, \mu_j)}{\mu_j} (y, y_j) x_j, \quad y \in Y, \tag{2.7}$$

is a regularization strategy with $\|R_\alpha\| \leq c(\alpha)$. *A choice* $\alpha = \alpha(\delta)$ *is admissible if* $\alpha(\delta) \to 0$ *and* $\delta\, c(\alpha(\delta)) \to 0$ *as* $\delta \to 0$. *The function* q *is called a regularizing filter for* K.

Proof: The operators R_α are bounded since we have by assumption (2) that

$$
\begin{aligned}
\|R_\alpha y\|^2 &= \sum_{j=1}^{\infty} [q(\alpha, \mu_j)]^2 \frac{1}{\mu_j^2}\, |(y, y_j)|^2 \\
&\leq c(\alpha)^2 \sum_{j=1}^{\infty} |(y, y_j)|^2 \leq c(\alpha)^2 \|y\|^2,
\end{aligned}
$$

i.e., $\|R_\alpha\| \leq c(\alpha)$. From

$$R_\alpha Kx = \sum_{j=1}^{\infty} \frac{q(\alpha, \mu_j)}{\mu_j} (Kx, y_j) x_j, \quad x = \sum_{j=1}^{\infty} (x, x_j) x_j,$$

and $(Kx, y_j) = (x, K^* y_j) = \mu_j (x, x_j)$, we conclude that

$$\|R_\alpha Kx - x\|^2 = \sum_{j=1}^{\infty} [q(\alpha, \mu_j) - 1]^2\, |(x, x_j)|^2. \tag{2.8}$$

Here K^* denotes the adjoint of K (see Theorem A.23). This fundamental representation will be used quite often in the following. Now let $x \in X$ be arbitrary but fixed. For $\epsilon > 0$ there exists $N \in \mathbb{N}$ such that

$$\sum_{n=N+1}^{\infty} |(x, x_j)|^2 < \frac{\epsilon^2}{8}.$$

By (3a) there exists $\alpha_0 > 0$ such that

$$[q(\alpha, \mu_j) - 1]^2 < \frac{\epsilon^2}{2\|x\|^2} \quad \text{for all } j = 1, \ldots, N \text{ and } 0 < \alpha \leq \alpha_0.$$

With (1) we conclude that

$$\|R_\alpha Kx - x\|^2 = \sum_{j=1}^{N} [q(\alpha, \mu_j) - 1]^2\, |(x, x_j)|^2$$

$$+ \sum_{n=N+1}^{\infty} \left[q(\alpha,\mu_j) - 1\right]^2 |(x,x_j)|^2$$

$$< \frac{\epsilon^2}{2\,\|x\|^2} \sum_{j=1}^{N} |(x,x_j)|^2 + \frac{\epsilon^2}{2} \leq \epsilon^2$$

for all $0 < \alpha \leq \alpha_0$. Thus we have shown that

$$R_\alpha Kx \to x \quad (\alpha \to 0) \quad \text{for every } x \in X. \qquad \square$$

In this theorem, we showed convergence of $R_\alpha y$ to the solution x. As Examples 2.4 and 2.5 indicate, we are particularly interested in optimal strategies, i.e., those that converge of the same order as the worst-case error. We will see in the next theorem that a proper replacement of assumption (3a) leads to such optimal strategies. In parts (i) and (ii), we will assume that the solution x is in the range of K^* and K^*K, respectively. In concrete examples, these conditions correspond to smoothness assumptions and boundary conditions on the exact solution x (see Problem 1.3 for an example).

Theorem 2.7

Let assumptions (1) and (2) of the previous theorem hold.

(i) Let (3a) be replaced by the stronger assumption:

(3b) There exists $c_1 > 0$ with

$$|q(\alpha,\mu) - 1| \leq c_1 \frac{\sqrt{\alpha}}{\mu} \quad \text{for all } \alpha > 0 \text{ and } 0 < \mu \leq \|K\|.$$

If, furthermore, $x \in K^*(Y)$, then

$$\left\| R_\alpha Kx - x \right\| \leq c_1 \sqrt{\alpha} \, \|z\|, \qquad (2.9a)$$

where $x = K^*z$.

(ii) Let (3a) be replaced by the stronger assumption:

(3c) There exists $c_2 > 0$ with

$$|q(\alpha,\mu) - 1| \leq c_2 \frac{\alpha}{\mu^2} \quad \text{for all } \alpha > 0 \text{ and } 0 < \mu \leq \|K\|.$$

If, furthermore, $x \in K^*K(X)$, then

$$\left\| R_\alpha Kx - x \right\| \leq c_2 \alpha \, \|z\|, \qquad (2.9b)$$

where $x = K^*Kz$.

Proof: With $x = K^* z$ and $(x, x_j) = \mu_j(z, y_j)$, formula (2.8) takes the form

$$\|R_\alpha K x - x\|^2 = \sum_{j=1}^{\infty} [q(\alpha, \mu_j) - 1]^2 \mu_j^2 |(z, y_j)|^2 \leq c_1^2 \alpha \|z\|^2.$$

The case (ii) is proven analogously. $\square$

There are many examples of functions $q : (0, \infty) \times (0, \|K\|] \to \mathbb{R}$ that satisfy assumptions (1), (2), and (3a-c) of the preceding theorems. We will study two of the following three filter functions in the next sections in more detail.

Theorem 2.8

The following three functions q satisfy the assumptions (1), (2), and (3a–c) of Theorems 2.6 or 2.7, respectively:

(a) $q(\alpha, \mu) = \mu^2 / (\alpha + \mu^2)$. *This choice satisfies (2) with $c(\alpha) = 1/(2\sqrt{\alpha})$. Assumptions (3b) and (3c) hold with $c_1 = 1/2$ and $c_2 = 1$, respectively.*

(b) $q(\alpha, \mu) = 1 - (1 - a\mu^2)^{1/\alpha}$ *for some $0 < a < 1/\|K\|^2$. In this case (2) holds with $c(\alpha) = \sqrt{a/\alpha}$. (3b) and (3c) are satisfied with $c_1 = 1/\sqrt{2a}$ and $c_2 = 1/a$, respectively.*

(c) *Let q be defined by*

$$q(\alpha, \mu) = \begin{cases} 1, & \mu^2 \geq \alpha, \\ 0, & \mu^2 < \alpha. \end{cases}$$

In this case (2) holds with $c(\alpha) = 1/\sqrt{\alpha}$. (3b) and (3c) are satisfied with $c_1 = c_2 = 1$.

Therefore, all of the functions q defined in (a), (b), and (c) are regularizing filters that lead to optimal regularization strategies.

Proof: For all three cases, properties (1) and (3a) are obvious.

(a) Properties (2) and (3b) follow from the elementary estimate

$$\frac{\mu}{\alpha + \mu^2} \leq \frac{1}{2\sqrt{\alpha}} \quad \text{for all } \alpha, \mu > 0$$

since $1 - q(\alpha, \mu) = \alpha / (\alpha + \mu^2)$. Property (3c) is also obvious.

(b) Property (2) follows immediately from Bernoulli's inequality:

$$1 - (1 - a\mu^2)^{1/\alpha} \leq 1 - \left(1 - \frac{a\mu^2}{\alpha}\right) = \frac{a\mu^2}{\alpha},$$

thus $|q(\alpha, \mu)| \leq \sqrt{|q(\alpha, \mu)|} \leq \sqrt{a/\alpha}\,\mu$.

(3b) and (3c) follow from the elementary estimates

$$\mu\left(1 - a\mu^2\right)^\beta \leq \frac{1}{\sqrt{2a\beta}} \quad \text{and} \quad \mu^2\left(1 - a\mu^2\right)^\beta \leq \frac{1}{a\beta}$$

for all $\beta > 0$ and $0 \leq \mu \leq 1/\sqrt{a}$.

(c) For property (2) it is sufficient to consider the case $\mu^2 \geq \alpha$. In this case, $q(\alpha, \mu) = 1 \leq \mu/\sqrt{\alpha}$. For (3b) and (3c) we consider only the case $\mu^2 < \alpha$. Then $\mu\left(1 - q(\alpha, \mu)\right) = \mu \leq \sqrt{\alpha}$ and $\mu^2\left(1 - q(\alpha, \mu)\right) = \mu^2 \leq \alpha$.   □

We will see later that the regularization methods for the first two choices of q admit a characterization that avoids knowledge of the singular system. The choice (c) of q is called the *spectral cutoff*. The spectral cutoff solution $x^{\alpha, \delta} \in X$ is therefore defined by

$$x^{\alpha, \delta} = \sum_{\mu_j^2 \geq \alpha} \frac{1}{\mu_j}\left(y^\delta, y_j\right) x_j.$$

We combine the fundamental estimate (2.4) with the previous theorem and show the following result for the cutoff solution.

Theorem 2.9
Let $y^\delta \in Y$ be such that $\left\|y^\delta - y\right\| \leq \delta$, where $y = Kx$ denotes the exact right-hand side.

(a) Let $K : X \to Y$ be a compact and injective operator with singular system (μ_j, x_j, y_j). The operators

$$R_\alpha y := \sum_{\mu_j^2 \geq \alpha} \frac{1}{\mu_j}\left(y, y_j\right) x_j, \quad y \in Y, \tag{2.10}$$

define a regularization strategy with $\|R_\alpha\| \leq 1/\sqrt{\alpha}$. This strategy is admissible if $\alpha(\delta) \to 0$ $(\delta \to 0)$ and $\delta^2/\alpha(\delta) \to 0$ $(\delta \to 0)$.

(b) Let $x = K^*z \in K^*(Y)$ with $\|z\| \leq E$ and $c > 0$. For the choice $\alpha(\delta) = c\delta/E$, we have the estimate

$$\left\|x^{\alpha(\delta), \delta} - x\right\| \leq \left(\frac{1}{\sqrt{c}} + \sqrt{c}\right)\sqrt{\delta E}. \tag{2.11a}$$

(c) Let $x = K^*Kz \in K^*K(X)$ with $\|z\| \leq E$ and $c > 0$. The choice $\alpha(\delta) = c(\delta/E)^{2/3}$ leads to the estimate

$$\left\|x^{\alpha(\delta), \delta} - x\right\| \leq \left(\frac{1}{\sqrt{c}} + c\right) E^{1/3}\delta^{2/3}. \tag{2.11b}$$

Therefore, the spectral cutoff is optimal for the information $\left\|(K^*)^{-1}x\right\| \le$ *E or $\left\|(K^*K)^{-1}x\right\| \le E$, respectively (if K^* is one-to-one).*

Proof: Combining the fundamental estimate (2.4) with Theorems 2.7 and 2.8 yields the error estimate

$$\left\|x^{\alpha,\delta} - x\right\| \le \frac{\delta}{\sqrt{\alpha}} + \sqrt{\alpha}\,\|z\|$$

for part (b) and

$$\left\|x^{\alpha,\delta} - x\right\| \le \frac{\delta}{\sqrt{\alpha}} + \alpha\,\|z\|$$

for part (c). The choices $\alpha(\delta) = c\delta/E$ and $\alpha(\delta) = c(\delta/E)^{2/3}$ lead to the estimates (2.11a) and (2.11b), respectively.   □

The general regularization concept discussed in this section can be found in many books on inverse theory [15, 86, 144]. It was not the aim of this section to study the most general theory. During the last few years, this concept has been extended in several directions. For example, in [63] (see also [67]) the notions of strong and weak convergence and divergence are defined, and in [144] different notions of optimality of regularization schemes are discussed.

The idea of using filters has a long history [85, 217] and is very convenient for theoretical purposes. For given concrete integral operators, however, one often wants to avoid the computation of a singular system. In the next sections, we will give equivalent characterizations for the first two examples without using singular systems.

2.2 Tikhonov Regularization

A common method to deal with overdetermined finite linear systems of the form $Kx = y$ is to determine the best fit in the sense that one tries to minimize the defect $\|Kx - y\|$ with respect to $x \in X$ for some norm in Y. If X is infinite-dimensional and K is compact, this minimization problem is also ill-posed by the following lemma.

Lemma 2.10
Let X and Y be Hilbert spaces, $K : X \to Y$ be linear and bounded, and $y \in Y$. There exists $\hat{x} \in X$ with $\left\|K\hat{x} - y\right\| \le \left\|Kx - y\right\|$ for all $x \in X$ if and only if $\hat{x} \in X$ solves the normal equation $K^*K\hat{x} = K^*y$. Here, $K^* : Y \to X$ denotes the adjoint of K.

Proof: A simple application of the binomial theorem yields

$$
\begin{aligned}
\|Kx - y\|^2 - \|K\hat{x} - y\|^2 &= 2\,\mathrm{Re}\left(K\hat{x} - y, K(x - \hat{x})\right) + \|K(x - \hat{x})\|^2 \\
&= 2\,\mathrm{Re}\left(K^*(K\hat{x} - y), x - \hat{x}\right) + \|K(x - \hat{x})\|^2
\end{aligned}
$$

for all $x, \hat{x} \in X$. If $\hat{x}$ satisfies $K^*K\hat{x} = K^*y$, then $\|Kx - y\|^2 - \|K\hat{x} - y\|^2 \geq 0$, i.e., $\hat{x}$ minimizes $\|Kx - y\|$. If, on the other hand, $\hat{x}$ minimizes $\|Kx - y\|$, then we substitute $x = \hat{x} + tz$ for any $t > 0$ and $z \in X$ and arrive at

$$
0 \leq 2t\,\mathrm{Re}\left(K^*(K\hat{x} - y), z\right) + t^2\|Kz\|^2 .
$$

Division by $t > 0$ and $t \to 0$ yields $\mathrm{Re}\left(K^*(K\hat{x} - y), z\right) \geq 0$ for all $z \in X$, i.e., $K^*(K\hat{x} - y) = 0$, and $\hat{x}$ solves the normal equation. $\qquad\square$

As a consequence of this lemma we should penalize the defect (in the language of optimization theory) or replace the equation of the first kind $K^*K\hat{x} = K^*y$ with an equation of the second kind (in the language of integral equation theory). Both viewpoints lead to the following minimization problem.

Given the linear, bounded operator $K : X \to Y$ and $y \in Y$, determine $x^\alpha \in X$ that minimizes the *Tikhonov functional*

$$
J_\alpha(x) := \|Kx - y\|^2 + \alpha\|x\|^2 \quad \text{for } x \in X. \tag{2.12}
$$

We prove the following theorem.

Theorem 2.11
Let $K : X \to Y$ be a linear and bounded operator between Hilbert spaces and $\alpha > 0$. Then the Tikhonov functional J_α has a unique minimum $x^\alpha \in X$. This minimum x^α is the unique solution of the normal equation

$$
\alpha x^\alpha + K^*Kx^\alpha = K^*y. \tag{2.13}
$$

Proof: Let $(x_n) \subset X$ be a minimizing sequence, i.e., $J_\alpha(x_n) \to I := \inf_{x \in X} J_\alpha(x)$ as n tends to infinity. We show that (x_n) is a Cauchy sequence. Application of the binomial formula yields that

$$
\begin{aligned}
J_\alpha(x_n) + J_\alpha(x_m) &= 2J_\alpha\left(\frac{1}{2}(x_n + x_m)\right) \\
&\quad + \frac{1}{2}\|K(x_n - x_m)\|^2 + \frac{\alpha}{2}\|x_n - x_m\|^2 \\
&\geq 2I + \frac{\alpha}{2}\|x_n - x_m\|^2 .
\end{aligned}
$$

The left-hand side converges to $2I$ as n, m tend to infinity. This shows that (x_n) is a Cauchy sequence and thus convergent. Let $x^\alpha = \lim_{n \to \infty} x_n$, noting that $x^\alpha \in X$. From the continuity of J_α, we conclude that $J_\alpha(x_n) \to J_\alpha(x^\alpha)$, i.e., $J_\alpha(x^\alpha) = I$. This proves the existence of a minimum of J_α.

Now we use the following formula as in the proof of the previous lemma:

$$
\begin{aligned}
J_\alpha(x) - J_\alpha(x^\alpha) &= 2\,\mathrm{Re}\,(Kx^\alpha - y, K(x - x^\alpha)) + 2\alpha\,\mathrm{Re}\,(x^\alpha, x - x^\alpha) \\
&\quad + \|K(x - x^\alpha)\|^2 + \alpha\,\|x - x^\alpha\|^2 \\
&= 2\,\mathrm{Re}\,(K^*(Kx^\alpha - y) + \alpha x^\alpha,\, x - x^\alpha) \\
&\quad + \|K(x - x^\alpha)\|^2 + \alpha\,\|x - x^\alpha\|^2 \tag{2.14}
\end{aligned}
$$

for all $x \in X$. From this, the equivalence of the normal equation with the minimization problem for J_α is shown exactly as in the proof of Lemma 2.10. Finally, we show that $\alpha I + K^*K$ is one-to-one for every $\alpha > 0$. Let $\alpha x + K^*Kx = 0$. Multiplication by x yields $\alpha(x, x) + (Kx, Kx) = 0$, i.e., $x = 0$. □

The solution x^α of equation (2.13) can be written in the form $x^\alpha = R_\alpha y$ with

$$
R_\alpha := (\alpha I + K^*K)^{-1}K^* : Y \longrightarrow X. \tag{2.15}
$$

Choosing a singular system (μ_j, x_j, y_j) for the compact operator K, we see that $R_\alpha y$ has the representation

$$
R_\alpha y = \sum_{n=0}^{\infty} \frac{\mu_j}{\alpha + \mu_j^2}\,(y, y_j)\,x_j = \sum_{n=0}^{\infty} \frac{q(\alpha, \mu_j)}{\mu_j}\,(y, y_j)\,x_j, \quad y \in Y, \tag{2.16}
$$

with $q(\alpha, \mu) = \mu^2/(\alpha + \mu^2)$. This function q is exactly the filter function that was studied in Theorem 2.8, part (a). Therefore, applications of Theorems 2.6 and 2.7 yield the following.

Theorem 2.12

Let $K : X \to Y$ be a linear, compact operator and $\alpha > 0$.

(a) The operator $\alpha I + K^*K$ is boundedly invertible. The operators $R_\alpha : Y \to X$ from (2.15) form a regularization strategy with $\|R_\alpha\| \leq 1/(2\sqrt{\alpha})$. It is called the *Tikhonov regularization method*. $R_\alpha y^\delta$ is determined as the unique solution $x^{\alpha,\delta} \in X$ of the equation of the second kind

$$
\alpha\, x^{\alpha,\delta} + K^*Kx^{\alpha,\delta} = K^*y^\delta. \tag{2.17}
$$

Every choice $\alpha(\delta) \to 0$ $(\delta \to 0)$ with $\delta^2/\alpha(\delta) \to 0$ $(\delta \to 0)$ is admissible.

(b) Let $x = K^*z \in K^*(Y)$ with $\|z\| \leq E$. We choose $\alpha(\delta) = c\delta/E$ for some $c > 0$. Then the following estimate holds:

$$
\|x^{\alpha(\delta),\delta} - x\| \leq \frac{1}{2}(1/\sqrt{c} + \sqrt{c})\,\sqrt{\delta\,E}. \tag{2.18a}
$$

(c) Let $x = K^*Kz \in K^*K(X)$ with $\|z\| \leq E$. The choice $\alpha(\delta) = c(\delta/E)^{2/3}$ for some $c > 0$ leads to the error estimate

$$\left\|x^{\alpha(\delta),\delta} - x\right\| \leq \left(\frac{1}{2\sqrt{c}} + c\right) E^{1/3} \delta^{2/3}. \qquad (2.18b)$$

Therefore, Tikhonov's regularization method is optimal for the information $\|(K^*)^{-1}x\| \leq E$ or $\|(K^*K)^{-1}x\| \leq E$, respectively (provided K^* is one-to-one).

Proof: Combining the fundamental estimate (2.4) with Theorems 2.7 and 2.8 yields the error estimate

$$\left\|x^{\alpha,\delta} - x\right\| \leq \frac{\delta}{2\sqrt{\alpha}} + \frac{\sqrt{\alpha}}{2} \|z\|$$

for part (b) and

$$\left\|x^{\alpha,\delta} - x\right\| \leq \frac{\delta}{2\sqrt{\alpha}} + \alpha \|z\|$$

for part (c). The choices $\alpha(\delta) = c\delta/E$ and $\alpha(\delta) = c(\delta/E)^{2/3}$ lead to the estimates (2.18a) and (2.18b), respectively.   □

The eigenvalues of K tend to zero, and the eigenvalues of $\alpha I + K^*K$ are bounded away from zero by $\alpha > 0$.

From Theorem 2.12, we observe that α has to be chosen to depend on δ in such a way that it converges to zero as δ tends to zero but not as fast as δ^2. From parts (b) and (c), we conclude that the smoother the solution x is the slower α has to tend to zero. On the other hand, the convergence can be arbitrarily slow if no a priori assumption about the solution x (such as (b) or (c)) is available (see [198]).

It is surprising to note that the order of convergence of Tikhonov's regularization method cannot be improved. Indeed, we prove the following result.

Theorem 2.13
Let $K : X \to Y$ be linear, compact, and one-to-one such that the range $K(X)$ is infinite-dimensional. Furthermore, let $x \in X$, and assume that there exists a continuous function $\alpha : [0, \infty) \to [0, \infty)$ with $\alpha(0) = 0$ such that

$$\lim_{\delta \to 0} \left\|x^{\alpha(\delta),\delta} - x\right\| \delta^{-2/3} = 0$$

for every $y^\delta \in Y$ with $\|y^\delta - Kx\| \leq \delta$, where $x^{\alpha(\delta),\delta} \in X$ solves (2.17). Then $x = 0$.

Proof: Assume, on the contrary, that $x \neq 0$.

First, we show that $\alpha(\delta) \delta^{-2/3} \to 0$. Set $y = Kx$. From

$$\left(\alpha(\delta) I + K^* K\right)\left(x^{\alpha(\delta),\delta} - x\right) = K^*\left(y^\delta - y\right) - \alpha(\delta) x,$$

we estimate

$$|\alpha(\delta)| \, \|x\| \leq \|K\| \, \delta + \left(\alpha(\delta) + \|K\|^2\right) \left\|x^{\alpha(\delta),\delta} - x\right\|.$$

We multiply this equation by $\delta^{-2/3}$ and use the assumption that $x^{\alpha(\delta),\delta}$ tends to x faster than $\delta^{-2/3}$, i.e.,

$$\left\|x^{\alpha(\delta),\delta} - x\right\| \delta^{-2/3} \to 0.$$

This yields $\alpha(\delta) \delta^{-2/3} \to 0$.

In the second part, we construct a contradiction. Let $\left(\mu_j, x_j, y_j\right)$ be a singular system for K. Define

$$\delta_j := \mu_j^3 \quad \text{and} \quad y^{\delta_j} := y + \delta_j y_j, \quad j \in \mathbb{N}.$$

Then $\delta_j \to 0$ as $j \to \infty$ and, with $\alpha_j := \alpha(\delta_j)$,

$$
\begin{aligned}
x^{\alpha_j,\delta_j} - x &= \left(x^{\alpha_j,\delta_j} - x^{\alpha_j}\right) + \left(x^{\alpha_j} - x\right) \\
&= \left(\alpha_j I + K^* K\right)^{-1} K^* (\delta_j y_j) + \left(x^{\alpha_j} - x\right) \\
&= \frac{\delta_j \mu_j}{\alpha_j + \mu_j^2} x_j + \left(x^{\alpha_j} - x\right).
\end{aligned}
$$

Here x^{α_j} is the solution of Tikhonov's equation (2.17) for y^δ replaced by y. Since also $\left\|x^{\alpha_j} - x\right\| \delta_j^{-2/3} \to 0$, we conclude that

$$\frac{\delta_j^{1/3} \mu_j}{\alpha_j + \mu_j^2} \longrightarrow 0, \quad j \to \infty.$$

But, on the other hand,

$$\frac{\delta_j^{1/3} \mu_j}{\alpha_j + \mu_j^2} = \frac{\mu_j^2}{\alpha_j + \mu_j^2} = \left(1 + \alpha_j \delta_j^{-2/3}\right)^{-1} \longrightarrow 1, \quad j \to \infty.$$

This a contradiction. $\qquad\qquad\square$

This result shows that Tikhonov's regularization method is not optimal for stronger "smoothness" assumptions on the solution x, i.e., under the assumption $x \in \left(K^* K\right)^r (X)$ for some $r \in \mathbb{N}$, $r \geq 2$. This is in contrast to, e.g., Landweber's method or the conjugate gradient method, which will be discussed later.

The choice of α in Theorem 2.12 is made *a priori*, i.e., before starting the computation of x^α by solving the least squares problem. In Sections 2.6, 2.7, and 2.8, we will study *a posteriori* choices of α, i.e., choices of α made during the process of computing x^α.

It is possible to choose stronger norms in the penalty term of the Tikhonov functional. Instead of (2.12), one can minimize the functional

$$\left\| Kx - y^\delta \right\|^2 + \alpha \|x\|_1^2 \quad \text{on } X_1,$$

where $\|\cdot\|_1$ is a stronger norm (or only seminorm) on a subspace $X_1 \subset X$. This was originally done by Phillips [176] and Tikhonov [213, 214] (see also [74]) for linear integral equations of the first kind. They chose the seminorm $\|x\|_1 := \|x'\|_{L^2}$ or the H^1-norm $\|x\|_1 := \left(\|x\|_{L^2}^2 + \|x'\|_{L^2}^2 \right)^{1/2}$. By characterizing $\|\cdot\|_1$ through a singular system for K, one obtains similar convergence results as above in the stronger norm $\|\cdot\|_1$. For further aspects of regularization with differential operators or stronger norms, we refer to [50, 95, 142, 165] and the monographs [86, 87, 144]. The interpretation of regularization by smoothing norms in terms of reproducing kernel Hilbert spaces has been observed in [107].

2.3   Landweber Iteration

Landweber [135], Friedman [75], and Bialy [16] suggested rewriting the equation $Kx = y$ in the form $x = (I - a K^* K) x + a K^* y$ for some $a > 0$ and iterating this equation, i.e., computing

$$x^0 := 0 \quad \text{and} \quad x^m = (I - a K^* K) x^{m-1} + a K^* y \qquad (2.19)$$

for $m = 1, 2, \ldots$. This iteration scheme can be interpreted as the steepest descent algorithm applied to the quadratic functional $x \mapsto \|Kx - y\|^2$ as the following lemma shows.

Lemma 2.14

Let the sequence (x^m) be defined by (2.19) and define the functional $\psi :$ $X \to \mathbb{R}$ by $\psi(x) = \frac{1}{2} \|Kx - y\|^2$, $x \in X$. Then ψ is Fréchet differentiable in every $z \in X$ and

$$\psi'(z)x = \operatorname{Re}(Kz - y, Kx) = \operatorname{Re}(K^*(Kz - y), x), \quad x \in X. \quad (2.20)$$

The linear functional $\psi'(z)$ can be identified with $K^*(Kz - y) \in X$ in the Hilbert space X over the field $\mathbb{R}$. Therefore, $x^m = x^{m-1} - a K^*(Kx^{m-1} - y)$ is the steepest descent step with stepsize a.

Proof: The binomial formula yields

$$\psi(z + x) - \psi(z) - \text{Re}\,(Kz - y, Kx) = \frac{1}{2}\,\|Kx\|^2$$

and thus

$$\left|\psi(z + x) - \psi(z) - \text{Re}\,(Kz - y, Kx)\right| \leq \frac{1}{2}\,\|K\|^2\,\|x\|^2,$$

which proves that the mapping $x \mapsto \text{Re}\,(Kz - y, Kx)$ is the Fréchet derivative of ψ at z. □

Equation (2.19) is a linear recursion formula for x^m. By induction with respect to m, it is easily seen that x^m has the explicit form $x^m = R_m y$, where the operator $R_m : Y \to X$ is defined by

$$R_m := a \sum_{k=0}^{m-1} (I - aK^*K)^k K^* \quad \text{for } m = 1, 2, \ldots. \tag{2.21}$$

Choosing a singular system (μ_j, x_j, y_j) for the compact operator K, we see that $R_m y$ has the representation

$$R_m y = a \sum_{j=1}^{\infty} \mu_j \sum_{k=0}^{m-1} (1 - a\mu_j^2)^k\,(y, y_j)\,x_j$$

$$= \sum_{j=1}^{\infty} \frac{1}{\mu_j} \left[1 - (1 - a\mu_j^2)^m\right]\,(y, y_j)\,x_j$$

$$= \sum_{n=0}^{\infty} \frac{q(m, \mu_j)}{\mu_j}\,(y, y_j)\,x_j, \quad y \in Y, \tag{2.22}$$

with $q(m, \mu) = \left[1 - (1 - a\mu^2)^m\right]$. We studied this filter function q in Theorem 2.8, part (b), when we defined $\alpha = 1/m$. Therefore, applications of Theorems 2.6 and 2.7 yield the following result.

Theorem 2.15
(a) Again let $K : X \to Y$ be a compact operator and let $0 < a < 1/\|K\|^2$. Define the linear and bounded operators $R_m : Y \to X$ by (2.21). These operators R_m define a regularization strategy with discrete regularization parameter $\alpha = 1/m$, $m \in \mathbb{N}$, and $\|R_m\| \leq \sqrt{a\,m}$. The sequence $x^{m,\delta} = R_m y^\delta$ is computed by the iteration (2.19), i.e.,

$$x^{0,\delta} = 0 \quad \text{and} \quad x^{m,\delta} = (I - a\,K^*K)x^{m-1,\delta} + a\,K^*y^\delta \tag{2.23}$$

for $m = 1, 2, \ldots$. Every strategy $m(\delta) \to \infty$ ($\delta \to 0$) with $\delta^2\,m(\delta) \to 0$ ($\delta \to 0$) is admissible.

(b) Again let $x = K^*z \in K^*(Y)$ with $\|z\| \le E$ and $0 < c_1 < c_2$. For every choice $m(\delta)$ with $c_1 \frac{E}{\delta} \le m(\delta) \le c_2 \frac{E}{\delta}$, the following estimate holds:

$$\left\|x^{m(\delta),\delta} - x\right\| \le c_3 \sqrt{\delta}\, E \qquad (2.24a)$$

for some c_3 depending on c_1, c_2, and a. Therefore, the Landweber iteration is optimal for the information $\left\|(K^*)^{-1}x\right\| \le E$.

(c) Now let $x = K^*Kz \in K^*K(X)$ with $\|z\| \le E$ and $0 < c_1 < c_2$. For every choice $m(\delta)$ with $c_1(E/\delta)^{2/3} \le m(\delta) \le c_2(E/\delta)^{2/3}$, we have

$$\left\|x^{m(\delta),\delta} - x\right\| \le c_3\, E^{1/3}\, \delta^{2/3} \qquad (2.24b)$$

for some c_3 depending on c_1, c_2, and a. Therefore, the Landweber iteration is also optimal for the information $\left\|(K^*K)^{-1}x\right\| \le E$.

Proof: Combining the fundamental estimate (2.4) with the Theorems 2.7 and 2.8 yields the error estimate

$$\left\|x^{m,\delta} - x\right\| \le \delta\sqrt{a\,m} + \frac{1}{\sqrt{2a}}\,\|z\|$$

for part (b) and

$$\left\|x^{m,\delta} - x\right\| \le \delta\sqrt{a\,m} + \frac{1}{a}\,\|z\|$$

for part (c). Replacing m in the first term by the upper bound and in the second by the lower bound yields estimates (2.24a) and (2.24b), respectively. $\qquad\square$

The choice $x^0 = 0$ is made to simplify the analysis. In general, the explicit iteration x^m is given by

$$x^m = a\sum_{k=0}^{m-1}(I - aK^*K)^k K^*y + (I - aK^*K)^m x^0, \quad m = 1, 2, \ldots .$$

In this case, R_m is affine linear, i.e., of the form $R_m y = z^m + S_m y$, $y \in Y$, for some $z^m \in X$ and some linear operator $S_m : Y \to X$.

For this method, we observe again that high precision (ignoring the presence of errors) requires a large number m of iterations but stability forces us to keep m small enough.

It can be shown by the same arguments as earlier that Landweber's method is optimal also with respect to stronger norms. If $x \in (K^*K)^r(X)$ for some $r \in \mathbb{N}$, the following error estimate holds (see [144]):

$$\left\|x^{m(\delta),\delta} - x\right\| \le c\, E^{1/(2r+1)}\, \delta^{2r/(2r+1)},$$

where E is a bound on $(K^*K)^{-r}x$. Therefore, this situation is different from Tikhonov's regularization method (see Theorem 2.13).

We will come back to the Landweber iteration in the next chapter, where we will show that an optimal choice of $m(\delta)$ can be made a posteriori through a proper stopping rule.

In this section, we have studied only the particular cases $x \in K^*(Y)$ and $x \in K^*K(X)$, which correspond to two particular smoothness assumptions in concrete applications. It is possible to extend the theory to the case where $x \in (K^*K)^{\sigma/2}(X)$. Here, $(K^*K)^{\sigma/2}$ denotes the (fractional) power of the self-adjoint operator K^*K. We will come back to this generalization in Section 2.7 (see (2.42) and Problems 2.4 and 2.5).

Other possibilities for regularizing first kind equations $Kx = y$ with compact operators K, which we have not discussed, are methods using positivity or more general convexity constraints (see [21, 23, 190, 191]).

2.4   A Numerical Example

In this section, we demonstrate Tikhonov's regularization method for the following integral equation of the first kind:

$$\int_0^1 (1+ts)\, e^{ts}\, x(s)\, ds \;=\; e^t, \quad 0 \le t \le 1, \tag{2.25}$$

with unique solution $x(t) = 1$ (see Problem 2.1). The operator $K : L^2(0,1) \to L^2(0,1)$ is given by

$$(Kx)(t) \;=\; \int_0^1 (1+ts)\, e^{ts}\, x(s)\, ds$$

and is self-adjoint, i.e., $K^* = K$. We note that x does not belong to the range of K (see Problem 2.1). For the numerical evaluation of Kx, we use Simpson's rule. With $t_i = i/n$, $i = 0, \ldots, n$, n even, we replace $(Kx)(t_i)$ by

$$\sum_{j=0}^n w_j\,(1+t_it_j)\,e^{t_it_j}\,x(t_j) \quad \text{where} \quad w_j = \begin{cases} \frac{1}{3n}, & j = 0 \text{ or } n, \\ \frac{4}{3n}, & j = 1,3,\ldots,n-1, \\ \frac{2}{3n}, & j = 2,4,\ldots,n-2. \end{cases}$$

We note that the corresponding matrix A is not symmetric. This leads to the discretized Tikhonov equation $\alpha\, x^{\alpha,\delta} + A^2 x^{\alpha,\delta} = A\, y^\delta$. Here, $y^\delta =$

$(y_i^\delta) \in \mathbb{R}^{n+1}$ is a *perturbation* (uniformly distributed random vector) of the discrete right-hand $y_i = \exp(i/n)$ such that

$$|y - y^\delta|_2 := \sqrt{\frac{1}{n+1} \sum_{i=0}^{n} (y_i - y_i^\delta)^2} \leq \delta.$$

The average results of ten computations are given in the following tables, where we have listed the discrete norms $|1 - x^{\alpha,\delta}|_2$ of the errors between the exact solution $x(t) = 1$ and Tikhonov's approximation $x^{\alpha,\delta}$.

Table 2.1: Tikhonov regularization for $\delta = 0$

α	$n = 8$	$n = 16$
10^{-1}	$2.4 * 10^{-1}$	$2.3 * 10^{-1}$
10^{-2}	$7.2 * 10^{-2}$	$6.8 * 10^{-2}$
10^{-3}	$2.6 * 10^{-2}$	$2.4 * 10^{-2}$
10^{-4}	$1.3 * 10^{-2}$	$1.2 * 10^{-2}$
10^{-5}	$2.6 * 10^{-3}$	$2.3 * 10^{-3}$
10^{-6}	$9.3 * 10^{-4}$	$8.7 * 10^{-4}$
10^{-7}	$3.5 * 10^{-4}$	$4.4 * 10^{-4}$
10^{-8}	$1.3 * 10^{-3}$	$3.2 * 10^{-5}$
10^{-9}	$1.6 * 10^{-3}$	$9.3 * 10^{-5}$
10^{-10}	$3.9 * 10^{-3}$	$2.1 * 10^{-4}$

Table 2.2: Tikhonov regularization for $\delta > 0$

α	$\delta = 0.0001$	$\delta = 0.001$	$\delta = 0.01$	$\delta = 0.1$
10^{-1}	0.2317	0.2317	0.2310	0.2255
10^{-2}	0.0681	0.0677	0.0692	0.1194
10^{-3}	0.0238	0.0240	0.0268	0.1651
10^{-4}	0.0119	0.0127	0.1172	1.0218
10^{-5}	0.0031	0.0168	0.2553	3.0065
10^{-6}	0.0065	0.0909	0.6513	5.9854
10^{-7}	0.0470	0.2129	2.4573	30.595
10^{-8}	0.1018	0.8119	5.9775	
10^{-9}	0.1730	1.8985	16.587	
10^{-10}	1.0723	14.642		

In the first table, we have chosen $\delta = 0$, i.e., only the discretization error for Simpson's rule is responsible for the increase of the error for small α.

This difference between discretization parameters $n = 8$ and $n = 16$ is noticeable for $\alpha \leq 10^{-8}$. We refer to [219] for further examples.

In the second table, we always took $n = 16$ and observed that the total error first decreases with decreasing α up to an optimal value and then increases again. This is predicted by the theory, in particular by estimates (2.18a) and (2.18b).

In the following table, we list results corresponding to the iteration steps for Landweber's method with parameter $a = 0.5$ and again $n = 16$.

Table 2.3: Landweber iteration

m	$\delta = 0.0001$	$\delta = 0.001$	$\delta = 0.01$	$\delta = 0.1$
1	0.8097	0.8097	0.8088	0.8135
2	0.6274	0.6275	0.6278	0.6327
3	0.5331	0.5331	0.5333	0.5331
4	0.4312	0.4311	0.4322	0.4287
5	0.3898	0.3898	0.3912	0.3798
6	0.3354	0.3353	0.3360	0.3339
7	0.3193	0.3192	0.3202	0.3248
8	0.2905	0.2904	0.2912	0.2902
9	0.2838	0.2838	0.2845	0.2817
10	0.2675	0.2675	0.2677	0.2681
100	0.0473	0.0474	0.0476	0.0534
200	0.0248	0.0248	0.0253	0.0409
300	0.0242	0.0242	0.0249	0.0347
400	0.0241	0.0241	0.0246	0.0385
500	0.0239	0.0240	0.0243	0.0424

We observe that the error decreases quickly in the first few steps and then slows down. To compare Tikhonov's method and Landweber's iteration, we note that the error corresponding to iteration number m has to be compared with the error corresponding to $\alpha = 1/(2m)$ (see the estimates in the proofs of Theorems 2.15 and 2.12). Taking this into account, we observe that both methods are comparable where precision is concerned. We note, however, that the computation time of Landweber's method is considerably higher than for Tikhonov's method, in particular if the error δ is small. On the other hand, Landweber's method is stable with respect to perturbations of the right-hand side and gives very good results even for large errors δ.

We refer also to Section 3.5, where these regularization methods are compared with those to be discussed in the subsequent sections for Symm's integral equation.

2.5   The Discrepancy Principle of Morozov

The following three sections are devoted to a posteriori choices of the regularization parameter α. In this section, we will study a discrepancy principle based on the Tikhonov regularization method. Throughout this section, we assume again that $K : X \longrightarrow Y$ is a compact and injective operator between Hilbert spaces X and Y with dense range $K(X) \subset Y$. Again, we study the equation

$$Kx = y$$

for given $y \in Y$. The Tikhonov regularization of this equation was investigated in Section 2.2. It corresponds to the regularization operators

$$R_\alpha = (\alpha I + K^*K)^{-1}K^* \quad \text{for } \alpha > 0,$$

which approximate the unbounded inverse of K on $K(X)$. We have seen that $x^\alpha = R_\alpha y$ exists and is the unique minimum of the Tikhonov functional

$$J_\alpha(x) := \|Kx - y\|^2 + \alpha \|x\|^2, \quad x \in X, \quad \alpha > 0. \tag{2.26}$$

More facts about the dependence on α and y are proven in the following theorem.

Theorem 2.16
Let $y \in Y$, $\alpha > 0$, and x^α be the unique solution of the equation

$$\alpha x^\alpha + K^*Kx^\alpha = K^*y. \tag{2.27}$$

Then x^α depends continuously on y and α. The mapping $\alpha \mapsto \|x^\alpha\|$ is monotonously nonincreasing and

$$\lim_{\alpha \to \infty} x^\alpha = 0.$$

The mapping $\alpha \mapsto \|Kx^\alpha - y\|$ is monotonously nondecreasing and

$$\lim_{\alpha \to 0} Kx^\alpha = y.$$

If $K^*y \neq 0$, then strict monotonicity holds in both cases.

Proof: We proceed in four steps.

(i) Using the definition of J_α and the optimality of x^α, we conclude that

$$\alpha \|x^\alpha\|^2 \leq J_\alpha(x^\alpha) \leq J_\alpha(0) = \|y\|^2,$$

i.e., $\|x^\alpha\| \leq \|y\| / \sqrt{\alpha}$. This proves that $x^\alpha \to 0$ as $\alpha \to \infty$.

(ii) We choose $\alpha > 0$ and $\beta > 0$ and subtract the equations for x^α and x^β:

$$\alpha\left(x^\alpha - x^\beta\right) + K^*K\left(x^\alpha - x^\beta\right) + (\alpha - \beta)x^\beta = 0. \qquad (2.28)$$

Multiplication by $\left(x^\alpha - x^\beta\right)$ yields

$$\alpha\left\|x^\alpha - x^\beta\right\|^2 + \left\|K\left(x^\alpha - x^\beta\right)\right\|^2 = (\beta - \alpha)\left(x^\beta, x^\alpha - x^\beta\right). \qquad (2.29)$$

From this equation, we first conclude that

$$\alpha\left\|x^\alpha - x^\beta\right\|^2 \leq |\beta - \alpha|\left|\left(x^\beta, x^\alpha - x^\beta\right)\right| \leq |\beta - \alpha|\left\|x^\beta\right\|\left\|x^\alpha - x^\beta\right\|,$$

i.e.,

$$\alpha\left\|x^\alpha - x^\beta\right\| \leq |\beta - \alpha|\left\|x^\beta\right\| \leq |\beta - \alpha|\frac{\|y\|}{\sqrt{\beta}}.$$

This proves the continuity of the mapping $\alpha \mapsto x^\alpha$.

(iii) Now let $\beta > \alpha > 0$. From (2.29) we conclude that $\left(x^\beta, x^\alpha - x^\beta\right) \geq 0$. Thus $\left\|x^\beta\right\|^2 \leq \left(x^\beta, x^\alpha\right) \leq \left\|x^\beta\right\|\left\|x^\alpha\right\|$, i.e., $\left\|x^\beta\right\| \leq \left\|x^\alpha\right\|$, which proves monotonicity of $\alpha \mapsto \left\|x^\alpha\right\|$.

(iv) We multiply the normal equation for x^β by $\left(x^\alpha - x^\beta\right)$. This yields

$$\beta\left(x^\beta, x^\alpha - x^\beta\right) + \left(Kx^\beta - y, K(x^\alpha - x^\beta)\right) = 0.$$

Now let $\alpha > \beta$. From (2.29), we see that $\left(x^\beta, x^\alpha - x^\beta\right) \leq 0$, i.e.,

$$0 \leq \left(Kx^\beta - y, K(x^\alpha - x^\beta)\right) = \left(Kx^\beta - y, Kx^\alpha - y\right) - \left\|Kx^\beta - y\right\|^2.$$

The Cauchy–Schwarz inequality yields $\left\|Kx^\beta - y\right\| \leq \left\|Kx^\alpha - y\right\|$.

(v) Finally, let $\varepsilon > 0$. Since the range of K is dense in Y, there exists $x \in X$ with $\|Kx - y\|^2 \leq \varepsilon^2/2$. Choose α_0 such that $\alpha_0\|x\|^2 \leq \varepsilon^2/2$. Then

$$\left\|Kx^\alpha - y\right\|^2 \leq J_\alpha(x^\alpha) \leq J_\alpha(x) \leq \varepsilon^2,$$

i.e., $\left\|Kx^\alpha - y\right\| \leq \varepsilon$ for all $\alpha \leq \alpha_0$. $\qquad\square$

Now we consider the determination of $\alpha(\delta)$ from the discrepancy principle, see [154, 155, 156].

We compute $\alpha = \alpha(\delta) > 0$ such that the corresponding Tikhonov solution $x^{\alpha,\delta}$, i.e., the solution of the equation

$$\alpha x^{\alpha,\delta} + K^*Kx^{\alpha,\delta} = K^*y^\delta,$$

that is, the minimum of

$$J_{\alpha,\delta}(x) := \left\|Kx - y^\delta\right\|^2 + \alpha\|x\|^2,$$

satisfies the equation

$$\left\| Kx^{\alpha,\delta} - y^\delta \right\| = \delta. \tag{2.30}$$

Note that this choice of α by the discrepancy principle guarantees that, on the one side, the error of the defect is δ and, on the other side, α is not too small.

Equation (2.30) is uniquely solvable, provided $\left\| y^\delta - y \right\| \leq \delta < \left\| y^\delta \right\|$, since by the previous theorem

$$\lim_{\alpha \to \infty} \left\| Kx^{\alpha,\delta} - y^\delta \right\| = \left\| y^\delta \right\| > \delta$$

and

$$\lim_{\alpha \to 0} \left\| Kx^{\alpha,\delta} - y^\delta \right\| = 0 < \delta.$$

Furthermore, $\alpha \mapsto \left\| Kx^{\alpha,\delta} - y^\delta \right\|$ is continuous and strictly increasing.

Theorem 2.17

Let $K : X \to Y$ be linear, compact and one-to-one with dense range in Y. Let $Kx = y$ with $x \in X$, $y \in Y$, $y^\delta \in Y$ such that $\left\| y - y^\delta \right\| \leq \delta < \left\| y^\delta \right\|$. Let the Tikhonov solution $x^{\alpha(\delta)}$ satisfy $\left\| Kx^{\alpha(\delta),\delta} - y^\delta \right\| = \delta$ for all $\delta \in (0, \delta_0)$. Then

(a) $x^{\alpha(\delta),\delta} \to x$ for $\delta \to 0$, i.e., the discrepancy principle is admissible.

(b) Let $x = K^* z \in K^*(Y)$ with $\|z\| \leq E$. Then

$$\left\| x^{\alpha(\delta),\delta} - x \right\| \leq 2\sqrt{\delta\,E}.$$

Therefore, the discrepancy principle is an optimal regularization strategy under the information $\left\| (K^*)^{-1} x \right\| \leq E$.

Proof: $x^\delta := x^{\alpha(\delta),\delta}$ minimizes the Tikhonov functional

$$J^{(\delta)}(x) := J_{\alpha(\delta),\delta}(x) = \alpha(\delta)\|x\|^2 + \left\| Kx - y^\delta \right\|^2.$$

Therefore, we conclude that

$$
\begin{aligned}
\alpha(\delta)\left\| x^\delta \right\|^2 + \delta^2 &= J^{(\delta)}(x^\delta) \leq J^{(\delta)}(x) \\
&= \alpha(\delta)\|x\|^2 + \left\| y - y^\delta \right\|^2 \\
&\leq \alpha(\delta)\|x\|^2 + \delta^2,
\end{aligned}
$$

and hence $\left\| x^\delta \right\| \leq \|x\|$ for all $\delta > 0$. This yields the following important estimate:

$$
\begin{aligned}
\left\| x^\delta - x \right\|^2 &= \left\| x^\delta \right\|^2 - 2\,\mathrm{Re}\,(x^\delta, x) + \|x\|^2 \\
&\leq 2\left[\|x\|^2 - \mathrm{Re}\,(x^\delta, x) \right] = 2\,\mathrm{Re}\,(x - x^\delta, x).
\end{aligned}
$$

First, we prove part (b): Let $x = K^*z$, $z \in Y$. Then

$$
\begin{aligned}
\left\|x^\delta - x\right\|^2 &\leq 2 \operatorname{Re}\left(x - x^\delta, K^*z\right) = 2 \operatorname{Re}\left(y - Kx^\delta, z\right) \\
&\leq 2 \operatorname{Re}\left(y - y^\delta, z\right) + 2 \operatorname{Re}\left(y^\delta - Kx^\delta, z\right) \\
&\leq 2\delta\|z\| + 2\delta\|z\| = 4\delta\|z\| \leq 4\delta E.
\end{aligned}
$$

(a) Now let $x \in X$ and $\varepsilon > 0$ be arbitrary. The range $K^*(Y)$ is dense in X since K is one-to-one. Therefore, there exists $\hat{x} = K^*z \in K^*(Y)$ such that $\|\hat{x} - x\| \leq \varepsilon/3$. Then we conclude by similar arguments as above that

$$
\begin{aligned}
\left\|x^\delta - x\right\|^2 &\leq 2 \operatorname{Re}\left(x - x^\delta, x - \hat{x}\right) + 2 \operatorname{Re}\left(x - x^\delta, K^*z\right) \\
&\leq 2\left\|x - x^\delta\right\| \frac{\varepsilon}{3} + 2 \operatorname{Re}\left(y - Kx^\delta, z\right) \\
&\leq 2\left\|x - x^\delta\right\| \frac{\varepsilon}{3} + 4\delta\|z\|.
\end{aligned}
$$

This can be rewritten as $\left(\left\|x - x^\delta\right\| - \varepsilon/3\right)^2 \leq \varepsilon^2/9 + 4\delta\|z\|$.

Now we choose $\delta > 0$ such that the right-hand side is less than $4\varepsilon^2/9$. Taking the square root, we conclude that $\left\|x - x^\delta\right\| \leq \varepsilon$ for this δ. $\square$

The condition $\left\|y^\delta\right\| > \delta$ certainly makes sense since otherwise the right-hand side would be less than the error level δ, and $x^\delta = 0$ would be an acceptable approximation to x.

The determination of $\alpha(\delta)$ is thus equivalent to the problem of finding the zero of the monotone function $\varphi(\alpha) := \left\|Kx^{\alpha,\delta} - y^\delta\right\|^2 - \delta^2$ (for fixed $\delta > 0$). It is not necessary to satisfy the equation $\left\|Kx^{\alpha,\delta} - y^\delta\right\| = \delta$ exactly. An inclusion of the form

$$
c_1\delta \leq \left\|Kx^{\alpha,\delta} - y^\delta\right\| \leq c_2\delta
$$

is sufficient to prove the assertions of the previous theorem.

The computation of $\alpha(\delta)$ can be carried out with Newton's method. The derivative of the mapping $\alpha \mapsto x^{\alpha,\delta}$ is given by the solution of the equation $\left(\alpha I + K^*K\right)\frac{d}{d\alpha}x^{\alpha,\delta} = -x^{\alpha,\delta}$, as is easily seen by differentiating (2.27) with respect to α.

In the following theorem, we prove that the order of convergence $\mathcal{O}(\sqrt{\delta})$ is best possible for the discrepancy principle. Therefore, by the results of Example 1.20, it cannot be optimal under the information $\left\|(K^*K)^{-1}x\right\| \leq E$.

Theorem 2.18

Let K be one-to-one and compact, and let $\alpha(\delta)$ be chosen by the discrepancy principle. Assume that for every $x \in K^*K(X)$, $y = Kx \neq 0$, and all sequences $\delta_n \to 0$ and $y^{\delta_n} \in Y$ with $\|y - y^{\delta_n}\| \leq \delta_n$ and $\|y^{\delta_n}\| > \delta_n$ for all n, the corresponding Tikhonov solutions $x^n = x^{\alpha(\delta_n),\delta_n}$ converge to x faster than $\sqrt{\delta_n}$, i.e.,

$$\frac{1}{\sqrt{\delta_n}} \|x^n - x\| \longrightarrow 0 \quad \text{as } n \to \infty.$$

Then the range $K(X)$ has to be finite-dimensional.

Proof: We show first that the choice of $\alpha(\delta)$ by the discrepancy principle implies the boundedness of $\alpha(\delta)/\delta$. Abbreviating $x^\delta := x^{\alpha(\delta),\delta}$, we write

$$\begin{aligned}
\|y^\delta\| - \delta &= \|y^\delta\| - \|y^\delta - Kx^\delta\| \leq \|Kx^\delta\| \\
&= \frac{1}{\alpha(\delta)} \|KK^*(y^\delta - Kx^\delta)\| \leq \frac{\delta}{\alpha(\delta)} \|K\|^2,
\end{aligned}$$

where we applied K to (2.27). This yields $\alpha(\delta) \leq \delta \|K\|^2 / (\|y^\delta\| - \delta)$.

From $\|y^\delta\| \geq \|y\| - \|y - y^\delta\| \geq \|y\| - \delta$, we conclude also that $\|y^\delta\| - \delta$ is bounded away from zero for sufficiently small δ. Thus we have shown that there exists $c > 0$ with $\alpha(\delta) \leq c\delta$ for all sufficiently small δ.

Now we assume that $\dim K(X) = \infty$ and construct a contradiction. Let (μ_j, x_j, y_j) be a singular system of K and define

$$y := y_1 \quad \text{and} \quad y^{\delta_n} := y_1 + \delta_n y_n \quad \text{with} \quad \delta_n := \mu_n^2.$$

Then $\delta_n \to 0$ as $n \to \infty$, $y \in K(K^*K)^k(X)$ for every $k \in \mathbb{N}$ and $\|y^{\delta_n} - y\| = \delta_n < \sqrt{1 + \delta_n^2} = \|y^{\delta_n}\|$. Therefore, the assumptions for the discrepancy principle are satisfied. The solutions of $Kx = y$ and $\alpha(\delta_n)x^n + K^*Kx^n = K^*y^{\delta_n}$ are given by

$$x = \frac{1}{\mu_1} x_1 \quad \text{and} \quad x^n = \frac{\mu_1}{\alpha(\delta_n) + \mu_1^2} x_1 + \frac{\mu_n \delta_n}{\alpha(\delta_n) + \mu_n^2} x_n,$$

respectively. $\alpha(\delta_n)$ has to be chosen such that $\|Kx^n - y^{\delta_n}\| = \delta_n$. We compute

$$x^n - x = -\frac{\alpha(\delta_n)}{\mu_1 (\alpha(\delta_n) + \mu_1^2)} x_1 + \frac{\mu_n \delta_n}{\alpha(\delta_n) + \mu_n^2} x_n$$

and hence for $n \geq 2$

$$\|x^n - x\| \geq \frac{\mu_n \delta_n}{\alpha(\delta_n) + \mu_n^2} = \sqrt{\delta_n} \frac{1}{1 + \alpha(\delta_n)/\delta_n} \geq \sqrt{\delta_n} \frac{1}{1 + c}.$$

This contradicts $\|x^n - x\| = o(\sqrt{\delta_n})$. $\qquad\square$

We remark that the estimate $\alpha(\delta) \leq \delta \|K\|^2 / (\|y^\delta\| - \delta)$ derived in the previous proof suggests to use $\delta \|K\|^2 / (\|y^\delta\| - \delta)$ as a starting value for Newton's method to determine $\alpha(\delta)$!

There has been an enormous effort to modify the original discrepancy principle while still retaining optimal orders of convergence. We refer to [65, 71, 78, 169] and, recently, [193].

2.6 Landweber's Iteration Method with Stopping Rule

It is very natural to use the following stopping criteria, which can be implemented in every iterative algorithm for the solution of $Kx = y$.

Let $r > 1$ be a fixed number. Stop the algorithm at the first occurrence of $m \in \mathbb{N}_0$ with $\|Kx^{m,\delta} - y^\delta\| \leq r\,\delta$. The following theorem shows that this choice of m is possible for Landweber's method and leads to an admissible and even optimal regularization strategy.

Theorem 2.19
Let $K : X \to Y$ be linear, compact, and one-to-one with dense range. Let $r > 1$ and $y^\delta \in Y$ be perturbations with $\|y - y^\delta\| \leq \delta$ and $\|y^\delta\| \geq r\delta$ for all $\delta \in (0, \delta_0)$. Let the sequence $x^{m,\delta}$, $m = 0, 1, 2, \ldots$, be determined by Landweber's method, i.e.,

$$x^{m+1,\delta} = x^{m,\delta} + a\,K^* \left(y - Kx^{m,\delta} \right), \quad m = 0, 1, 2, \ldots , \qquad (2.31)$$

for some $0 < a < 1/\|K\|^2$. Then the following assertions hold:

(1) $\lim\limits_{m \to \infty} \|Kx^{m,\delta} - y^\delta\| = 0$ *for every $\delta > 0$, i.e., the following stopping rule is well-defined: Let $m = m(\delta) \in \mathbb{N}_0$ be the smallest integer with $\|Kx^{m,\delta} - y^\delta\| \leq r\delta$.*

(2) $\delta^2 m(\delta) \to 0$ *for $\delta \to 0$, i.e., this choice of $m(\delta)$ is admissible. Therefore, by the assertions of Theorem 2.15, the sequence $x^{m(\delta),\delta}$ converges to x.*

(3) If $x = K^* z \in K^*(Y)$ or $x = K^* K z \in K^* K(X)$ for some z with $\|z\| \leq E$, then we have the following orders of convergence:

$$\left\| x^{m(\delta),\delta} - x \right\| \leq c\sqrt{E\,\delta} \qquad or \qquad (2.32a)$$

$$\left\| x^{m(\delta),\delta} - x \right\| \leq c E^{1/3} \delta^{2/3}, \qquad (2.32b)$$

respectively, for some $c > 0$. This means that this choice of $m(\delta)$ is optimal.

Proof: In (2.22), we showed the representation

$$R_m y = \sum_{j=1}^{\infty} \frac{1 - (1 - a\mu_j^2)^m}{\mu_j} (y, y_j) x_j$$

for every $y \in Y$ and thus

$$\|KR_m y - y\|^2 = \sum_{j=1}^{\infty} (1 - a\mu_j^2)^{2m} |(y, y_j)|^2.$$

From $|1 - a\mu_j^2| < 1$, we conclude that $\|KR_m - I\| \leq 1$. Application to y^δ instead of y yields

$$\|Kx^{m,\delta} - y^\delta\|^2 = \sum_{j=1}^{\infty} (1 - a\mu_j^2)^{2m} |(y^\delta, y_j)|^2.$$

(1) Let $\varepsilon > 0$ be given. Choose $J \in \mathbb{N}$ with

$$\sum_{j=J+1}^{\infty} |(y^\delta, y_j)|^2 < \frac{\varepsilon^2}{2}.$$

Since $|1 - a\mu_j^2|^{2m} \to 0$ as $m \to \infty$ uniformly for $j = 1, \dots, J$, we can find $m_0 \in \mathbb{N}$ with

$$\sum_{j=1}^{J} (1 - a\mu_j^2)^{2m} |(y^\delta, y_j)|^2$$

$$\leq \max_{j=1,\dots,J} (1 - a\mu_j^2)^{2m} \sum_{j=1}^{J} |(y^\delta, y_j)|^2 \leq \frac{\varepsilon^2}{2} \quad \text{for } m \geq m_0.$$

This implies that $\|Kx^{m,\delta} - y^\delta\|^2 \leq \varepsilon^2$ for $m \geq m_0$, i.e., that the method is admissible.

It is sufficient to prove assertion (2) only for the case $m(\delta) \to \infty$. We set $m := m(\delta)$ for abbreviation. By the choice of $m(\delta)$, we have for $y = Kx$

$$\|KR_{m-1}y - y\| \geq \|KR_{m-1}y^\delta - y^\delta\| - \|(KR_{m-1} - I)(y - y^\delta)\|$$
$$\geq r\delta - \|KR_{m-1} - I\| \delta \geq (r - 1)\delta,$$

and hence

$$m(r-1)^2 \delta^2 \leq m \sum_{j=1}^{\infty} (1 - a\mu_j^2)^{2m-2} |(y, y_j)|^2$$

$$= \sum_{j=1}^{\infty} m (1 - a\mu_j^2)^{2m-2} \mu_j^2 |(x, x_j)|^2. \qquad (2.33)$$

We show that the series converges to zero as $\delta \to 0$. (The dependence on δ is hidden in m.) First we note that $m\mu^2(1 - a\mu^2)^{2m-2} \le 1/(2a)$ for all $m \ge 2$ and all $\mu \ge 0$. Now we again split the series into a finite sum and a remaining series and estimate in the "long tail" the expression $m(1 - a\mu_j^2)^{2m-2}\mu_j^2$ by $1/(2a)$ and note that $m(1 - a\mu_j^2)^{2m-2}$ tends to zero as $m \to \infty$ uniformly in $j \in \{1, \ldots, J\}$. This proves convergence and thus part (2).

For part (3) we remind the reader of the fundamental estimate (2.4), which we need in the following form (see Theorem 2.15, part (a)):

$$\left\|x^{m,\delta} - x\right\| \le \delta\sqrt{am} + \left\|R_m K x - x\right\|. \tag{2.34}$$

First, let $x = K^* z$ and $\|z\| \le E$. Writing $m = m(\delta)$ again we conclude from (2.33) that

$$(r - 1)^2\delta^2 m^2 \le \sum_{j=1}^{\infty} m^2\left(1 - a\mu_j^2\right)^{2m-2}\mu_j^4\left|(z, y_j)\right|^2.$$

The estimate $m^2\mu^4\left(1 - a\mu^2\right)^{2m-2} \le 1/a^2$ for all $m \ge 2$ and $0 \le \mu \le 1/\sqrt{a}$ yields

$$(r - 1)^2\delta^2 m^2 \le \frac{1}{a^2}\|z\|^2,$$

i.e., we have shown the upper bound

$$m(\delta) \le \frac{1}{a(r - 1)}\frac{E}{\delta}.$$

Now we estimate the second term on the right-hand side of (2.34). From the Cauchy–Schwarz inequality, we conclude that

$$\|(I - R_m K)x\|^2 = \sum_{j=1}^{\infty}\mu_j^2\left(1 - a\mu_j^2\right)^{2m}|(z, y_j)|^2$$

$$= \sum_{j=1}^{\infty}\left[\mu_j^2\left(1 - a\mu_j^2\right)^m|(z, y_j)|\right]\left[\left(1 - a\mu_j^2\right)^m|(z, y_j)|\right]$$

$$\le \sqrt{\sum_{j=1}^{\infty}\mu_j^4\left(1 - a\mu_j^2\right)^{2m}|(z, y_j)|^2}\sqrt{\sum_{j=1}^{\infty}\underbrace{\left(1 - a\mu_j^2\right)^{2m}}_{\le 1}|(z, y_j)|^2}$$

$$\le \|KR_m y - y\|\,\|z\| \le E\left[\|(I - KR_m)(y - y^\delta)\| + \|(I - KR_m)y^\delta\|\right]$$

$$\le E(1 + r)\,\delta.$$

Therefore, we conclude from (2.34) that

$$\left\|x^{m(\delta),\delta} - x\right\| \le \delta\sqrt{am(\delta)} + \left\|R_{m(\delta)} K x - x\right\| \le c\sqrt{E}\,\delta.$$

Now let $x = K^* K z$ with $\|z\| \leq E$. By the same arguments as earlier, we conclude that

$$(r-1)^2 \delta^2 \leq \sum_{j=1}^{\infty} \left(1 - a\mu_j^2\right)^{2m-2} \mu_j^6 \left|(z, y_j)\right|^2.$$

Now we use the estimate $m^3 \mu^6 \left(1 - a\mu^2\right)^{2m-2} \leq 27/(8a^3)$ for all $m \geq 2$, $0 \leq \mu \leq 1/\sqrt{a}$, and we arrive at

$$(r-1)^2 \delta^2 \leq \frac{27}{8 a^3 m^3} \|z\|^2 \; ,$$

i.e.,

$$m(\delta) \leq c E^{2/3} \delta^{-2/3},$$

for some $c > 0$. To prove the second estimate (2.32b), we use *Hölder's inequality*

$$\sum_{j=1}^{\infty} |a_j b_j| \leq \left[\sum_{j=1}^{\infty} |a_j|^p\right]^{1/p} \left[\sum_{j=1}^{\infty} |b_j|^q\right]^{1/q} \; , \tag{2.35}$$

where $p, q > 1$ with $1/p + 1/q = 1$. With $p = 3/2$ and $q = 3$, we conclude that

$$\|(I - R_m K)x\|^2 = \sum_{j=1}^{\infty} \mu_j^4 \left(1 - a\mu_j^2\right)^{2m} |(z, x_j)|^2$$

$$= \sum_{j=1}^{\infty} \left[\mu_j^4 \left(1 - a\mu_j^2\right)^{4m/3} |(z, x_j)|^{4/3}\right] \left[\left(1 - a\mu_j^2\right)^{2m/3} |(z, x_j)|^{2/3}\right]$$

$$\leq \left[\sum_{j=1}^{\infty} \mu_j^6 \left(1 - a\mu_j^2\right)^{2m} |(z, x_j)|^2\right]^{2/3} \left[\sum_{j=1}^{\infty} \underbrace{\left(1 - a\mu_j^2\right)^{2m}}_{\leq 1} |(z, x_j)|^2\right]^{1/3}$$

$$\leq \|K R_m y - y\|^{4/3} \|z\|^{2/3} \; ,$$

i.e.,

$$\|(I - R_m K)x\| \leq E^{1/3} (1 + r)^{2/3} \delta^{2/3}.$$

Therefore, (2.34) yields

$$\left\|x^{m(\delta), \delta} - x\right\| \leq \delta \sqrt{a\, m(\delta)} + \left\|R_{m(\delta)} K x - x\right\| \leq c E^{1/3} \delta^{2/3}. \qquad \square$$

It is also possible to formulate a similar stopping criterion for Morozov's discrepancy principle. Choose an arbitrary monotonic decreasing sequence $(\alpha_m) \subset \mathbb{R}$ with $\lim_{m \to \infty} \alpha_m = 0$. Determine $m = m(\delta)$ as the smallest integer m with $\left\| Kx^{\alpha_m,\delta} - y^\delta \right\| \leq r\delta$. For details, we refer the reader to [68] or [144].

One can construct more general classes of methods through the spectral representation of the solution x.

Comparing the regularizer x^δ of Landweber's method with the true solution x, we observe that the function $\varphi(\mu) = 1/\mu$, $\mu > 0$, is approximated by the polynomial $\mathbb{P}_m(\mu) = \left[1 - (1 - a\mu^2)^m \right]/\mu$. It is certainly possible to choose better polynomial approximations of the function $\mu \mapsto 1/\mu$. Orthogonal polynomials are particularly useful. This leads to the $\nu-$methods; see [17, 94], or [96].

A common feature of these methods that is very crucial in the analysis is the fact that all of the polynomials $\mathbb{P}_m$ are independent of y and y^δ. For the important conjugate gradient algorithm discussed in the next section, this is not the case, and that makes an error analysis much more difficult to obtain.

2.7 The Conjugate Gradient Method

In this section, we will study the regularizing properties of the conjugate gradient method. Since the proofs of the theorems are rather technical, we will only state the results and transfer the proofs to an appendix.

First, we recall the conjugate gradient method for least squares problems for overdetermined systems of linear equations of the form $Kx = y$. Here, $K \in \mathbb{R}^{m \times n}$ and $y \in \mathbb{R}^m$ with $m \geq n$ are given. Since it is hopeless to satisfy all equations simultaneously, one minimizes the defect $f(x) := \left\| Kx - y \right\|^2$, $x \in \mathbb{R}^n$, where $\|\cdot\|$ denotes the Euclidean norm in $\mathbb{R}^m$. Standard algorithms for solving least squares problems are the QR-algorithm or the conjugate gradient method; see [51, 82, 106]. Since we assume that the latter is known for systems of equations, we formulate it now for the operator equation $Kx = y$, where $K : X \rightarrow Y$ is a bounded, linear, and injective operator between Hilbert spaces X and Y with adjoint $K^* : Y \rightarrow X$.

Define again the function

$$f(x) \; := \; \left\| Kx - y \right\|^2 \; = \; (Kx - y, Kx - y), \quad x \in X.$$

We abbreviate $\nabla f(x) := 2\, K^*(Kx - y) \in X$ and note that $\nabla f(x)$ is indeed the Riesz representation of the Fréchet derivative $f'(x; \cdot)$ of f at x (see Lemma 2.14). We call two elements $p, q \in X$ K-conjugate if $\left(Kp, Kq \right) = 0$. If K is one-to-one, this bilinear form has the properties of an inner product on X.

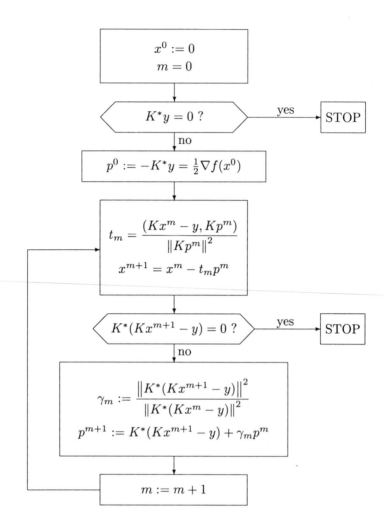

Figure 2.2: The conjugate gradient method

Theorem 2.20 *(Fletcher–Reeves)*
Let $K : X \to Y$ be a bounded, linear and injective operator between Hilbert spaces X and Y. Then the conjugate gradient method is well-defined and either stops or produces sequences (x^k), $(p^k) \subset X$ with the properties

$$\left(\nabla f(x^k), \nabla f(x^j)\right) = 0 \quad \text{for all } j \neq k \tag{2.36a}$$

and

$$\left(Kp^k, Kp^j\right) = 0 \quad \text{for all } j \neq k, \tag{2.36b}$$

i.e., the gradients are orthogonal and the directions p^k are K-conjugate. Furthermore,

$$\left(\nabla f(x^j), K^*Kp^k\right) \;=\; 0 \quad \text{for all } j < k. \tag{2.36c}$$

The following theorem gives an interesting and different interpretation of the elements x^m.

Theorem 2.21

Let (x^m) and (p^m) be the sequences of the conjugate gradient method. Define the space $V_m := \text{span}\{p^0, \ldots, p^m\}$. Then we have the following equivalent characterizations of V_m:

$$
\begin{aligned}
V_m \;&=\; \text{span}\left\{\nabla f(x^0), \ldots, \nabla f(x^m)\right\} & (2.37a)\\
&=\; \text{span}\left\{p^0, K^*Kp^0, \ldots, (K^*K)^m p^0\right\} & (2.37b)
\end{aligned}
$$

for $m = 0, 1, \ldots$. The space V_m is called a Krylov space. Furthermore, x^m is the minimum of f on V_{m-1} for every $m \geq 1$.

By this result, we can write x^m in the form

$$x^m \;=\; -\mathbb{P}_{m-1}(K^*K)p^0 \;=\; \mathbb{P}_{m-1}(K^*K)K^*y \tag{2.38}$$

with a well-defined polynomial $\mathbb{P}_{m-1} \in \mathcal{P}_{m-1}$ of degree $m-1$. Analogously, we write the defect in the form

$$
\begin{aligned}
y - Kx^m \;&=\; y - K\mathbb{P}_{m-1}(K^*K)K^*y \;=\; y - KK^*\mathbb{P}_{m-1}(KK^*)y\\
&=\; \mathbb{Q}_m(KK^*)y
\end{aligned}
$$

with the polynomial $\mathbb{Q}_m(t) := 1 - t\,\mathbb{P}_{m-1}(t)$ of degree m.

Let $\left(\mu_j, x_j, y_j\right)$ be a singular system for K. If it happens that

$$y \;=\; \sum_{j=1}^{N} \alpha_j y_j \;\in\; W_N := \text{span}\{y_1, \ldots, y_N\}$$

for some $N \in \mathbb{N}$, then all iterates $x^m \in A_N := \text{span}\{x_1, \ldots, x_N\}$ since

$$x^m \;=\; \mathbb{P}_{m-1}(K^*K)K^*y \;=\; \sum_{j=1}^{N} \alpha_j\,\mathbb{P}_{m-1}(\mu_j^2)\,\mu_j\,x_j.$$

In this exceptional case, the algorithm terminates after at most N iterations since the dimension of A_N is at most N and the gradients $\nabla f(x^i)$ are orthogonal to each other. This is the reason why the conjugate gradient method applied to matrix equations stops after finitely many iterations. For operator equations in infinite-dimensional Hilbert spaces, this method produces sequences of, in general, infinitely many elements.

The following characterizations of $\mathbb{Q}_m$ will be useful.

Lemma 2.22

(a) *The polynomial* $\mathbb{Q}_m$ *minimizes the functional*

$$H(\mathbb{Q}) := \|\mathbb{Q}(KK^*)y\|^2 \quad on \quad \{\mathbb{Q} \in \mathcal{P}_m : \mathbb{Q}(0) = 1\}$$

and satisfies

$$H(\mathbb{Q}_m) = \|Kx^m - y\|^2.$$

(b) *For* $k \neq \ell$, *the following orthogonality relation holds:*

$$\langle \mathbb{Q}_k, \mathbb{Q}_\ell \rangle := \sum_{j=1}^{\infty} \mu_j^2 \, \mathbb{Q}_k(\mu_j^2) \, \mathbb{Q}_\ell(\mu_j^2) \, |(y, y_j)|^2 = 0. \qquad (2.39)$$

If $y \notin \operatorname{span}\{y_1, \ldots, y_N\}$ *for any* $N \in \mathbb{N}$, *then* $\langle \cdot, \cdot \rangle$ *defines an inner product on the space* $\mathcal{P}$ *of all polynomials.*

Without a priori information, the sequence (x^m) does not converge to the solution x of $Kx = y$. The images, however, do converge to y. This is the subject of the next theorem.

Theorem 2.23

Let K *and* K^* *be one-to-one, and assume that the conjugate gradient method does not stop after finitely many steps. Then*

$$Kx^m \longrightarrow y \quad as \quad m \to \infty$$

for every $y \in Y$.

We give a proof of this theorem since it is a simple conclusion of the previous lemma.

Proof: Let $\mathbb{Q} \in \mathcal{P}_m$ be an arbitrary polynomial with $\mathbb{Q}(0) = 1$. Then, by the previous lemma,

$$\|Kx^m - y\|^2 = H(\mathbb{Q}_m) \leq H(\mathbb{Q}) = \sum_{j=1}^{\infty} \mathbb{Q}(\mu_j^2)^2 \, |(y, y_j)|^2. \qquad (2.40)$$

Now let $\varepsilon > 0$ be arbitrary. Choose $J \in \mathbb{N}$ such that

$$\sum_{j=J+1}^{\infty} |(y, y_j)|^2 < \frac{\varepsilon^2}{2}$$

and choose a function $R \in C[0, \mu_1^2]$ with $R(0) = 1$, $\|R\|_\infty \leq 1$ and $R(\mu_j^2) = 0$ for $j = 1, 2, \ldots, J$. By the theorem of Weierstrass, there exist polynomials $\tilde{\mathbb{Q}}_m \in \mathcal{P}_m$ with $\|R - \tilde{\mathbb{Q}}_m\|_\infty \to 0$ as $m \to \infty$. We set $\hat{\mathbb{Q}}_m = \tilde{\mathbb{Q}}_m / \tilde{\mathbb{Q}}_m(0)$, which is possible for sufficiently large m since $R(0) = 1$. Then $\hat{\mathbb{Q}}_m$ converges to R as $m \to \infty$ and $\hat{\mathbb{Q}}_m(0) = 1$. Substituting this into (2.40) yields

$$
\begin{aligned}
\|Kx^m - y\|^2 \;\;\leq\;\; & H(\hat{\mathbb{Q}}_m) \\
\leq\;\; & \sum_{j=1}^{J} \big|\hat{\mathbb{Q}}_m(\mu_j^2) - \underbrace{R(\mu_j^2)}_{=0}\big|^2 |(y, y_j)|^2 \;+\; \|\hat{\mathbb{Q}}_m\|_\infty^2 \sum_{j>J} |(y, y_j)|^2 \\
\leq\;\; & \|\hat{\mathbb{Q}}_m - R\|_\infty^2 \|y\|^2 \;+\; \frac{\varepsilon^2}{2} \|\hat{\mathbb{Q}}_m\|_\infty^2.
\end{aligned}
$$

This expression is less than ε^2 for sufficiently large m. $\qquad\square$

Now we return to the regularization of the operator equation $Kx = y$. The operator $\mathbb{P}_{m-1}(K^*K)K^* : Y \to X$ corresponds to the regularization operator R_α of the general theory. But this operator certainly depends on the right-hand side y. The mapping $y \mapsto \mathbb{P}_{m-1}(K^*K)K^*y$ is therefore nonlinear.

So far, we have formulated and studied the conjugate gradient method for unperturbed right-hand sides. Now we will consider the situation where we know only an approximation y^δ of y such that $\|y^\delta - y\| \leq \delta$. We apply the algorithm to y^δ instead of y. This yields a sequence $x^{m,\delta}$ and polynomials $\mathbb{P}_m^\delta$ and $\mathbb{Q}_m^\delta$. There is no a priori strategy $m = m(\delta)$ such that $x^{m(\delta),\delta}$ converges to x as δ tends to zero; see [57]. An a posteriori choice as in the previous section, however, again leads to an optimal strategy. We stop the algorithm with the smallest $m = m(\delta)$ such that the defect $\|Kx^{m,\delta} - y^\delta\| \leq \tau\delta$, where $\tau > 1$ is some given parameter. From now on, we make the assumption that y^δ is never a finite linear combination of the y_j. Then, by Theorem 2.23, the defect tends to zero, and this stopping rule is well-defined. We want to show that the choice $m = m(\delta)$ leads to an optimal algorithm. The following analysis, which we learned from [97], is more elementary than, e.g., in [17, 143], or [144]. We carry out the complete analysis but, again, postpone the proofs to an appendix since they are rather technical.

We recall that by our stopping rule

$$
\|Kx^{m(\delta),\delta} - y^\delta\| \;\leq\; \tau\delta \;<\; \|Kx^{m(\delta)-1,\delta} - y^\delta\|. \tag{2.41}
$$

At this stage, we wish to formulate the smoothness assumptions $x \in K^*(Y)$ and $x \in K^*K(X)$, respectively, in a more general way since the analysis will not become more difficult. We introduce the subspaces X^σ of X for $\sigma \in \mathbb{R}$, $\sigma \geq 0$, by

$$
X^\sigma \;:=\; (K^*K)^{\sigma/2}(X) \;:=\; \left\{ x \in X : \sum_{j=1}^{\infty} \mu_j^{-2\sigma} |(x, x_j)|^2 < \infty \right\} \tag{2.42}
$$

with norm

$$\|x\|_\sigma := \sqrt{\sum_{j=1}^{\infty} \mu_j^{-2\sigma} |(x, x_j)|^2}.$$

In general, these spaces depend on K through (μ_j, x_j, y_j).

It is easily seen that $(X^\sigma, \|\cdot\|_\sigma)$ are Hilbert spaces. For $\sigma < 0$ the spaces X^σ are defined as completions of X with respect to the norm $\|\cdot\|_\sigma$ (see Appendix A, Theorem A.10). Then $(X^\sigma, \|\cdot\|_\sigma)$ forms a scale of Hilbert spaces with the following properties: $\sigma_1 < \sigma_2$ implies that $X^{\sigma_2} \subset X^{\sigma_1}$, and the inclusion is compact. Furthermore, $\|x\|_{-1} = \|Kx\|$ for $x \in X^{-1}$ (see Problem 2.3).

The assumptions $x \in X^\sigma$, $\|x\|_\sigma \leq E$ generalize the former assumptions $x = K^*z \in K^*(Y)$, $\|z\| \leq E$ (set $\sigma = 1$) and $x = K^*Kz \in K^*K(X)$, $\|z\| \leq E$ (set $\sigma = 2$). The following theorem establishes the optimality of the conjugate gradient method with this stopping rule.

Theorem 2.24

Assume that y and y^δ do not belong to the linear span of finitely many y_j. Let the sequence $x^{m(\delta),\delta}$ be constructed by the conjugate gradient method with stopping rule (2.41) for fixed parameter $\tau > 1$. Let $x \in X^\sigma$ for some $\sigma > 0$ and $\|x\|_\sigma \leq E$. Then there exists $c > 0$ with

$$\left\| x - x^{m(\delta),\delta} \right\| \leq c\, \delta^{\sigma/(\sigma+1)}\, E^{1/(\sigma+1)}. \tag{2.43}$$

For $\sigma = 1$ and $\sigma = 2$, respectively, this estimate is of the same order as for Landweber's method. It is also optimal for any $\sigma > 0$ under the a priori information $\left\| (K^*K)^{\sigma/2} \right\| \leq E$ (see Problem 2.4).

There is a much simpler implementation of the conjugate gradient method for self-adjoint positive definite operators $K : X \to X$. For such K there exists a unique self-adjoint positive operator $A : X \to X$ with $A^2 = K$. Let $Kx = y$ and set $z := Ax$, i.e., $Az = y$. We apply the conjugate gradient method to the equation $Ax = z$ without knowing z. In the process of the algorithm, only the elements $A^*z = y$, $\|Ap^m\|^2 = (Kp^m, p^m)$, and $A^*(Ax^m - z) = Kx^m - y$ have to be computed. The square root A and the quantity z do not have to be known explicitly, and the method is much simpler to implement.

Actually, the conjugate gradient method presented here is only one member of a large class of conjugate gradient methods. For a detailed study of these methods in connection with ill-posed problems, we refer to [80, 99, 167, 168, 170] and, in particular, the recent work [98].

2.8 Problems

2.1 Let $K : L^2(0,1) \to L^2(0,1)$ be the integral operator

$$Kx(t) := \int_0^1 (1+ts)\, e^{ts}\, x(s)\, ds, \quad 0 < t < 1.$$

Show by induction that

$$\frac{d^n}{dt^n} Kx(t) = \int_0^1 (n+1+ts)\, s^n\, e^{ts}\, x(s)\, ds, \quad 0 < t < 1,\ n = 0, 1, \dots\ .$$

Prove that K is one-to-one and that the constant functions do not belong to the range of K.

2.2 Apply Tikhonov's method of Section 2.2 to the integral equation

$$\int_0^t x(s)\, ds = y(t), \quad 0 \le t \le 1.$$

Prove that for $y \in H^1(0,1)$ with $y(0) = 0$ Tikhonov's solution x^α is given by the solution of the boundary value problem

$$-\alpha\, \ddot{x}(t) + x(t) = \dot{y}(t),\ 0 < t < 1, \quad x(1) = 0,\ \dot{x}(0) = 0.$$

2.3 Let $K : X \to Y$ be compact and one-to-one. Let the spaces X^σ be defined by (2.42). Prove that X^{σ_2} is compactly embedded in X^{σ_1} for $\sigma_2 > \sigma_1$.

2.4 Let $\sigma > 0$ and $\mathcal{F}(\delta, E, \|\cdot\|_\sigma)$ be the worst-case error for K for the information $\|x\|_\sigma \le E$ (see Definition 1.18). Prove that

$$\mathcal{F}(\delta, E, \|\cdot\|_\sigma) \le \delta^{\sigma/(1+\sigma)}\, E^{1/(1+\sigma)}.$$

Hint: Use a singular system for the representations of $\|x\|$, $\|x\|_\sigma$, and $\|Kx\|$ and apply Hölder's inequality (2.35).

2.5 Let $q : (0,\infty) \times (0, \|K\|] \to \mathbb{R}$ be a filter function with the properties (1) and (2) of Theorem 2.6 where (3a) is replaced by:

(3d) There exists $c > 0$ and $\sigma > 0$ with

$$|q(\alpha, \mu) - 1| \le c \left(\frac{\sqrt{\alpha}}{\mu}\right)^\sigma \quad \text{for all } \alpha > 0 \text{ and } 0 < \mu \le \|K\|.$$

Let $R_\alpha : Y \to X$ denote the corresponding operator and assume, furthermore, that $x \in X^\sigma$ (see (2.42). Prove the following error estimate:

$$\|R_\alpha K x - x\| \leq c \alpha^{\sigma/2} \|x\|_\sigma.$$

Show that R_α is asymptotically optimal for $x \in X^\sigma$ (see Problem 2.4) if, in addition, there exists $\tilde c > 0$ with $|q(\alpha, \mu)| \leq \tilde c \mu / \sqrt{\alpha}$ for all α, μ.

2.6 The *iterated Tikhonov regularization* (see [66, 120]) $x^{m,\alpha,\delta}$ is defined by

$$x^{0,\alpha,\delta} = 0, \quad (\alpha I + K^* K) x^{m+1,\alpha,\delta} = K^* y^\delta + \alpha x^{m,\alpha,\delta}$$

for $m = 0, 1, 2, \ldots$.

(a) Show that $q^m(\alpha, \mu) := 1 - \left(1 - \mu^2/(\alpha + \mu^2)\right)^m$, $m = 1, 2, \ldots$, is the corresponding filter function.

(b) Prove that this filter function leads to a regularizing operator R_α^m with $\|R_\alpha^m\| \leq m/(2\sqrt{\alpha})$ and satisfies (3d) from Problem 2.5.

2.7 Fix y^δ with $\|y - y^\delta\| \leq \delta$ and let $x^{\alpha,\delta}$ be the Tikhonov solution corresponding to $\alpha > 0$. The curve

$$\alpha \mapsto \begin{pmatrix} f(\alpha) \\ g(\alpha) \end{pmatrix} := \begin{pmatrix} \|K x^{\alpha,\delta} - y^\delta\|^2 \\ \|x^{\alpha,\delta}\|^2 \end{pmatrix}, \quad \alpha > 0,$$

in $\mathbb{R}^2$ is called an *L-curve* since it has often the shape of the letter L; see [69, 100, 101].

Show by using a singular system that $f'(\alpha) = -\alpha g'(\alpha)$. Furthermore, compute the curvature

$$C(\alpha) := \frac{|f'(\alpha) g''(\alpha) - g'(\alpha) f''(\alpha)|}{\left(f'(\alpha)^2 + g'(\alpha)^2\right)^{3/2}}$$

and show that the curvature increases monotonously for $0 < \alpha \leq 1/\|K\|^2$.

3
Regularization by Discretization

In this chapter, we will study a different approach to regularizing operator equations of the form $Kx = y$, where x and y are elements of certain function spaces. This approach is motivated by the fact that for the numerical treatment of such equations one has to discretize the continuous problem and reduce it to a finite system of (linear or nonlinear) equations. We will see in this chapter that the discretization schemes themselves are regularization strategies in the sense of Chapter 2.

In Section 3.1, we will study the general concept of projection methods and give a necessary and sufficient condition for convergence. Although we have in mind the treatment of integral equations of the first kind, we will treat the general case where K is a linear, bounded, not necessarily compact operator between (real or complex) Banach or Hilbert spaces. Section 3.2 is devoted to Galerkin methods. As special cases, we will study least squares and dual least squares methods in Subsections 3.2.1 and 3.2.2. In Subsection 3.2.3, we will investigate the Bubnov–Galerkin method for the case where the operator satisfies a Gårding's inequality. In Section 3.3, we will illustrate the Galerkin methods for Symm's integral equation of the first kind. This equation arises in potential theory and serves as a model equation for more complicated situations. Section 3.4 is devoted to collocation methods. We will restrict ourselves to the moment method in Subsection 3.4.1 and to collocation by piecewise constant functions in Subsection 3.4.2, where the analysis is carried out only for Symm's integral equation. In Section 3.5, we will present numerical results for various regularization techniques (Tikhonov, Landweber, conjugate gradient, projection, and collocation methods) tested for Dirichlet boundary value problems for the Laplacian in an ellipse. Finally, we will study the Backus–Gilbert method in Section 3.6. Although not very popular among mathematicians, this method is extensively used by scientists in geophysics and other applied sciences. The general ideas of Sections 3.1 and 3.2 can also be found in, for example, [15, 130, 144].

3.1 Projection Methods

First, we recall the definition of a projection operator.

Definition 3.1
Let X be a normed space over the field $\mathbb{K}$ where $\mathbb{K} = \mathbb{R}$ or $\mathbb{K} = \mathbb{C}$. Let $U \subset X$ be a closed subspace. A linear bounded operator $P : X \to X$ is called a projection operator *on U if*

- *$Px \in U$ for all $x \in X$ and*

- *$Px = x$ for all $x \in U$.*

We now summarize some obvious properties of projection operators.

Theorem 3.2
Every nontrivial projection operator satisfies $P^2 = P$ and $\|P\| \geq 1$.

Proof: $P^2 x = P(Px) = Px$ follows from $Px \in U$. Furthermore, $\|P\| = \|P^2\| \leq \|P\|^2$ and $P \neq 0$. This implies $\|P\| \geq 1$. $\qquad\square$

In the following two examples, we introduce the most important projection operators.

Example 3.3
(a) (*Orthogonal projection*) Let X be a pre-Hilbert space over $\mathbb{K} = \mathbb{R}$ or $\mathbb{K} = \mathbb{C}$ and $U \subset X$ be a complete subspace. Let $Px \in U$ be the best approximation to x in U, i.e., Px satisfies

$$\|Px - x\| \leq \|u - x\| \quad \text{for all } u \in U. \tag{3.1}$$

By the projection theorem (Theorem A.13 of Appendix A), $P : X \to U$ is linear and $Px \in U$ is characterized by the abstract "normal equation" $(x - Px, u) = 0$ for all $u \in U$, i.e., $x - Px \in U^\perp$. In this example, by the binomial theorem we have

$$
\begin{aligned}
\|x\|^2 &= \|Px + (x - Px)\|^2 \\
&= \|Px\|^2 + \|x - Px\|^2 + 2 \underbrace{\operatorname{Re}(x - Px, Px)}_{=0} \geq \|Px\|^2,
\end{aligned}
$$

i.e., $\|P\| = 1$. Important examples of subspaces U are spaces of splines or finite elements.

(b) (*Interpolation operator*) Let $X = C[a, b]$ be the space of real valued continuous functions on $[a, b]$ supplied with the supremum norm. Then X is a normed space over $\mathbb{R}$. Let $U = \operatorname{span}\{u_1, \ldots, u_n\}$ be an n-dimensional subspace and $t_1, \ldots, t_n \in [a, b]$ such that the interpolation problem in U is uniquely solvable, i.e., $\det[u_j(t_k)] \neq 0$. We define $Px \in U$ by the interpolant of $x \in C[a, b]$ in U, i.e., $u = Px \in U$ satisfies $u(t_j) = x(t_j)$ for all $j = 1, \ldots, n$. Then $P : X \to U$ is a projection operator.

Examples for U are spaces of algebraic or trigonometric polynomials. As a drawback of these choices, we note that from the results of Faber (see [163]) the interpolating polynomials of continuous functions x do not, in general, converge to x as the degree of the polynomials tends to infinity. For smooth periodic functions, however, trigonometric interpolation at equidistant points converges with optimal order of convergence. We will use this fact in Subsection 3.2.2. Here, as an example, we recall the interpolation by linear splines. For simplicity, we formulate only the case where the endpoints are included in the set of interpolation points.

Let $a = t_1 < \cdots < t_n = b$ be given points, and let $U \subset C[a, b]$ be defined by

$$
\begin{aligned}
U &= \mathcal{S}_1(t_1, \ldots, t_n) \\
&:= \left\{ x \in C[a, b] : x|_{[t_j, t_{j+1}]} \in \mathcal{P}_1, \ j = 1, \ldots, n - 1 \right\}, \quad (3.2)
\end{aligned}
$$

where $\mathcal{P}_1$ denotes the space of polynomials of degree at most one. Then the interpolation operator $Q_n : C[a, b] \to \mathcal{S}_1(t_1, \ldots, t_n)$ is given by

$$
Q_n x = \sum_{j=1}^n x(t_j) \, \hat{y}_j \quad \text{for } x \in C[a, b],
$$

where the basis functions $\hat{y}_j \in \mathcal{S}_1(t_1, \ldots, t_n)$ are defined by

$$
\hat{y}_j(t) = \begin{cases}
\dfrac{t - t_{j-1}}{t_j - t_{j-1}}, & t \in [t_{j-1}, t_j] \quad (\text{if } j \geq 2), \\[2ex]
\dfrac{t_{j+1} - t}{t_{j+1} - t_j}, & t \in [t_j, t_{j+1}] \quad (\text{if } j \leq n - 1), \\[2ex]
0, & t \notin [t_{j-1}, t_{j+1}],
\end{cases} \quad (3.3)
$$

for $j = 1, \ldots, n$. In this example $\|Q_n\|_\infty = 1$ (see Problem 3.1).

For general interpolation operators, $\|Q_n\|$ exceeds one and $\|Q_n\|$ is not bounded with respect to n. This follows from the theorems of Faber and Banach–Steinhaus (see Theorem A.26).

Now we define the class of projection methods.

Definition 3.4

Let X and Y be Banach spaces and $K : X \to Y$ be bounded and one-to-one. Furthermore, let $X_n \subset X$ and $Y_n \subset Y$ be finite-dimensional subspaces of dimension n and $Q_n : Y \to Y_n$ be a projection operator. For given $y \in Y$ the projection method for solving the equation $Kx = y$ is to solve the equation

$$Q_n K x_n = Q_n y \quad \text{for } x_n \in X_n. \tag{3.4}$$

Let $\{\hat{x}_1, \ldots, \hat{x}_n\}$ and $\{\hat{y}_1, \ldots, \hat{y}_n\}$ be bases of X_n and Y_n, respectively. Then we can represent $Q_n y$ and every $Q_n K \hat{x}_j$, $j = 1, \ldots, n$, in the forms

$$Q_n y = \sum_{i=1}^{n} \beta_i \hat{y}_i \quad \text{and} \quad Q_n K \hat{x}_j = \sum_{i=1}^{n} A_{ij} \hat{y}_i, \quad j = 1, \ldots, n, \tag{3.5}$$

with β_i, $A_{ij} \in \mathbb{K}$. The linear combination $x_n = \sum_{j=1}^{n} \alpha_j \hat{x}_j$ solves (3.4) if and only if $\alpha = (\alpha_1, \ldots, \alpha_n)^\top \in \mathbb{K}^n$ solves the finite system of linear equations

$$\sum_{i=1}^{n} A_{ij} \alpha_j = \beta_i, \quad i = 1, \ldots, n, \quad \text{i.e.,} \quad A\alpha = \beta. \tag{3.6}$$

The orthogonal projection and the interpolation operator from Example 3.3 lead to the following important classes of projection methods, which will be studied in more detail in the next sections.

Example 3.5

Let $K : X \to Y$ be bounded and one-to-one.

(a) (*Galerkin method*) Let X and Y be pre-Hilbert spaces and $X_n \subset X$ and $Y_n \subset Y$ be finite-dimensional subspaces with $\dim X_n = \dim Y_n = n$. Let $Q_n : Y \to Y_n$ be the *orthogonal* projection. Then the projected equation $Q_n K x_n = Q_n y$ is equivalent to

$$(K x_n, z_n) = (y, z_n) \quad \text{for all } z_n \in Y_n. \tag{3.7a}$$

Again let $X_n = \text{span}\{\hat{x}_1, \ldots, \hat{x}_n\}$ and $Y_n = \text{span}\{\hat{y}_1, \ldots, \hat{y}_n\}$. Looking for a solution of (3.7a) in the form $x_n = \sum_{j=1}^{n} \alpha_j \hat{x}_j$ leads to the system

$$\sum_{j=1}^{n} \alpha_j (K \hat{x}_j, \hat{y}_i) = (y, \hat{y}_i) \quad \text{for } i = 1, \ldots, n, \tag{3.7b}$$

or $A\alpha = \beta$, where $A_{ij} := (K\hat{x}_j, \hat{y}_i)$ and $\beta_i = (y, \hat{y}_i)$.

(b) (*Collocation method*) Let X be a Banach space, $Y = C[a,b]$, and $K : X \to C[a,b]$ be a bounded operator. Let $a = t_1 < \cdots < t_n = b$ be given points (*collocation points*) and $Y_n = S_1(t_1, \ldots, t_n)$ be the corresponding space (3.2) of linear splines with interpolation operator $Q_n y = \sum_{j=1}^{n} y(t_j)\hat{y}_j$. Let $y \in C[a,b]$ and some n-dimensional subspace $X_n \subset X$ be given. Then $Q_n K x_n = Q_n y$ is equivalent to

$$(Kx_n)(t_i) = y(t_i) \quad \text{for all } i = 1, \ldots, n. \tag{3.8a}$$

If we denote by $\{\hat{x}_1, \ldots, \hat{x}_n\}$ a basis of X_n, then looking for a solution of (3.8a) in the form $x_n = \sum_{j=1}^{n} \alpha_j \hat{x}_j$ leads to the finite linear system

$$\sum_{j=1}^{n} \alpha_j K\hat{x}_j(t_i) = y(t_i), \quad i = 1, \ldots, n, \tag{3.8b}$$

or $A\alpha = \beta$, where $A_{ij} = K\hat{x}_j(t_i)$ and $\beta_i = y(t_i)$.

We are particularly interested in the study of integral equations of the form

$$\int_a^b k(t,s)\, x(s)\, ds = y(t), \quad t \in [a,b], \tag{3.9}$$

in $L^2(a,b)$ or $C[a,b]$ for some continuous or weakly singular function k. (3.7b) and (3.8b) now take the form

$$A\alpha = \beta, \tag{3.10}$$

where $x = \sum_{j=1}^{n} \alpha_j \hat{x}_j$ and

$$A_{ij} = \int_a^b \int_a^b k(t,s)\, \hat{x}_j(s)\, \hat{y}_i(t)\, ds\, dt \tag{3.11a}$$

$$\beta_i = \int_a^b y(t)\, \hat{y}_i(t)\, dt \tag{3.11b}$$

for the Galerkin method, and

$$A_{ij} = \int_a^b k(t_i, s)\, \hat{x}_j(s)\, ds \tag{3.11c}$$

$$\beta_i = y(t_i) \tag{3.11d}$$

for the collocation method.

Comparing the systems of equations in (3.10), we observe that the computation of the matrix elements (3.11c) is less expensive than for those of (3.11a) due to the double integration for every matrix element in (3.11a). For this reason, collocation methods are generally easier to implement than Galerkin methods. On the other hand, Galerkin methods have convergence properties of high order in weak norms (superconvergence) that are of practical importance in many cases, such as boundary element methods for the solution of boundary value problems.

For the remaining part of this section, we make the following assumption.

Assumption 3.6: *Let $K : X \to Y$ be a linear, bounded, and injective operator between Banach spaces, $X_n \subset X$ and $Y_n \subset Y$ be finite-dimensional subspaces of dimension n, and $Q_n : Y \to Y_n$ be a projection operator. We assume that $\bigcup_{n \in \mathbb{N}} X_n$ is dense in X and that $Q_n K|_{X_n} : X_n \to Y_n$ is one-to-one and thus invertible. Let $x \in X$ be the solution of*

$$K x = y. \tag{3.12}$$

By $x_n \in X_n$, we denote the unique solutions of the equations

$$Q_n K x_n = Q_n y \tag{3.13}$$

for $n \in \mathbb{N}$.

We can represent the solutions $x_n \in X_n$ of (3.13) in the form $x_n = R_n y$, where $R_n : Y \to X_n \subset X$ is defined by

$$R_n := \left(Q_n K|_{X_n} \right)^{-1} Q_n : Y \longrightarrow X_n \subset X. \tag{3.14}$$

The projection method is called *convergent* if the approximate solutions $x_n \in X_n$ of (3.13) converge to the exact solution $x \in X$ of (3.12) for every $y \in K(X)$, i.e., if

$$R_n K x = \left(Q_n K|_{X_n} \right)^{-1} Q_n K x \longrightarrow x, \quad n \to \infty, \tag{3.15}$$

for every $x \in X$.

We observe that this definition of convergence coincides with Definition 2.1 of a regularization strategy for the equation $K x = y$ with regularization parameter $\alpha = 1/n$. Therefore, the projection method converges if and only if R_n is a regularization strategy for the equation $K x = y$.

Obviously, we can only expect convergence if we require that $\bigcup_{n \in \mathbb{N}} X_n$ is dense in X and $Q_n y \to y$ for all $y \in K(X)$. But, in general, this is not sufficient for convergence if K is compact. We have to assume the following boundedness condition.

Theorem 3.7

Let Assumption 3.6 be satisfied. The solution $x_n = R_n y \in X_n$ of (3.13) converges to x for every $y = Kx$ if and only if there exists $c > 0$ such that

$$\|R_n K\| \leq c \quad \text{for all } n \in \mathbb{N}. \tag{3.16}$$

If (3.16) is satisfied the following error estimate holds:

$$\|x_n - x\| \leq (1 + c) \min_{z_n \in X_n} \|z_n - x\| \tag{3.17}$$

with the same constant c as in (3.16).

Proof: Let the projection method be convergent. Then $R_n K x \to x$ for every $x \in X$. The assertion follows directly from the principle of uniform boundedness (Theorem A.26 of Appendix A).

Now let $\|R_n K\|$ be bounded. The operator $R_n K$ is a projection operator onto X_n since for $z_n \in X_n$ we have $R_n K z_n = (Q_n K|_{X_n})^{-1} Q_n K z_n = z_n$. Thus we conclude that

$$x_n - x = (R_n K - I)x = (R_n K - I)(x - z_n) \quad \text{for all } z_n \in X_n.$$

This yields

$$\|x_n - x\| \leq (c + 1) \|x - z_n\| \quad \text{for all } z_n \in X_n$$

and proves (3.17). Convergence $x_n \to x$ follows since $\bigcup_{n \in \mathbb{N}} X_n$ is dense in X. $\qquad \square$

So far, we have considered the case where the right-hand side is known exactly. Now we study the case where the right-hand side is known only approximately, i.e., we assume the knowledge of $y^\delta \in Y$ with $\|y^\delta - y\| \leq \delta$. We understand the operator R_n from (3.14) as a regularization operator in the sense of the previous chapter. We have to distinguish between two kinds of errors on the right-hand side. The first kind measures the error in the norm of Y and corresponds to the kind of perturbation discussed in Chapter 2. We will call this the *continuous perturbation* of the right-hand side. In this case, the norm of Y plays an essential role. A simple application of the triangle inequality yields with $x_n^\delta := R_n y^\delta$:

$$\begin{aligned}
\|x_n^\delta - x\| &\leq \|x_n^\delta - R_n y\| + \|R_n y - x\| \\
&\leq \|R_n\| \|y^\delta - y\| + \|R_n K x - x\|.
\end{aligned} \tag{3.18}$$

This estimate corresponds to the fundamental estimate from (2.4). The first term reflects the illposedness of the equation: The (continuous) error

δ of the right-hand side is multiplied by the norm of R_n. The second term describes the discretization error with exact data.

In practice one solves the discrete system (3.6) where the vector β is replaced by a perturbed vector $\beta^\delta \in \mathbb{K}^n$ with

$$\left|\beta^\delta - \beta\right|^2 \;=\; \sum_{j=1}^n \left|\beta_j^\delta - \beta_j\right|^2 \;\leq\; \delta^2.$$

We will call this the *discrete perturbation* of the right-hand side. Instead of (3.6) one solves $A\alpha^\delta = \beta^\delta$ and defines $x_n^\delta \in X_n$ by

$$x_n^\delta \;=\; \sum_{j=1}^n \alpha_j^\delta \, \hat{x}_j.$$

In this case, the choices of basis functions $\hat{x}_j \in X_n$ and $\hat{y}_j \in Y_n$ are essential rather than the norm of Y. We will see, however, that the condition number of A reflects the ill-conditioning of the equation $Kx = y$. For a general discussion of the condition number of discretized integral equations of the first kind, we refer to [199].

The last theorem of this general section is a perturbation result: It is sufficient to study the question of convergence for the "principal part" of an operator K. In particular, if the projection method converges for an operator K, then convergence and the error estimates hold also for $K + C$, where C is compact relative to K (i.e., $K^{-1}C$ is compact).

Theorem 3.8
Let Assumption 3.6 hold. Let $C : X \to Y$ be a linear operator with $C(X) \subset K(X)$ such that $K + C$ is one-to-one and $K^{-1}C$ is compact in X. Assume, furthermore, that the projection method converges for K, i.e., that $R_n K x \to x$ for every $x \in X$, where again

$$R_n \;=\; \left[Q_n K|_{X_n}\right]^{-1} Q_n.$$

Then it converges also for $K + C$, i.e.,

$$\left[Q_n(K + C)|_{X_n}\right]^{-1} Q_n(K + C)x \;\longrightarrow\; x \quad \text{for all } x \in X.$$

Let $x \in X$ be the solution of $(K + C)x = y$ and $x_n^\delta \in X_n$ be the solution of the corresponding projected equation $Q_n(K + C)x_n^\delta = y_n^\delta$ for some $y_n^\delta \in Y_n$. Then there exists $c > 0$ with

$$\left\| x - x_n^\delta \right\| \;\leq\; c \left[\left\| K^{-1}Cx - R_n Cx \right\| \;+\; \left\| K^{-1}y - R_n y_n^\delta \right\| \right] \qquad (3.19)$$

for all sufficiently large n and $\delta > 0$.

Proof: We have to compare the equations for x_n^δ and x, i.e., the equations $Q_n(K+C)x_n^\delta = y_n^\delta$ and $(K+C)x = y$, which we rewrite in the forms

$$x_n^\delta + R_n C x_n^\delta = R_n y_n^\delta, \tag{3.20a}$$

$$x + K^{-1}Cx = K^{-1}y. \tag{3.20b}$$

The operators $R_n C = [R_n K]K^{-1}C$ converge to $K^{-1}C$ in the operator norm since $R_n K x \to x$ for every $x \in X$ and $K^{-1}C$ is compact in X (see Theorem A.32 of Appendix A). Furthermore, $I + K^{-1}C = K^{-1}(K+C)$ is an isomorphism in X. We apply the general Theorem A.35 of Appendix A to equations (3.20a) and (3.20b). This yields the assertion. □

The first term on the right-hand side of (3.19) is just the error of the projection method for the equation $Kx = Cx$ without perturbation of the right-hand side. By Theorem 3.7, this is estimated by

$$\left\|K^{-1}Cx - R_n Cx\right\| \leq (1+c)\min_{z_n \in X_n}\left\|K^{-1}Cx - z_n\right\|.$$

The second term on the right-hand side of (3.19) is the error for the equation $Kx = y$. This theorem includes both the continuous and the discrete perturbations of the right-hand side. For the continuous case we set $y_n^\delta = Q_n y^\delta$, while in the discrete case we set $y_n^\delta = \sum_{i=1}^n \beta_i^\delta \hat{y}_i$.

3.2   Galerkin Methods

In this section, we assume that X and Y are (real or complex) Hilbert spaces; $K : X \to Y$ is linear, bounded, and one-to-one; $X_n \subset X$ and $Y_n \subset Y$ are finite-dimensional subspaces with $\dim X_n = \dim Y_n = n$; and $Q_n : Y \to Y_n$ is the orthogonal projection operator onto Y_n. Then equation $Q_n K x_n = Q_n y$ reduces to the Galerkin equations (see Example 3.5)

$$(K x_n, z_n) = (y, z_n) \quad \text{for all } z_n \in Y_n. \tag{3.21}$$

If we choose bases $\{\hat{x}_1, \ldots, \hat{x}_n\}$ and $\{\hat{y}_1, \ldots, \hat{y}_n\}$ of X_n and Y_n, respectively, then this leads to a finite system for the coefficients of $x_n = \sum_{j=1}^n \alpha_j \hat{x}_j$ (see (3.7b)):

$$\sum_{i=1}^n A_{ij}\,\alpha_j = \beta_i, \quad i = 1, \ldots, n, \tag{3.22}$$

where

$$A_{ij} = (K\hat{x}_j, \hat{y}_i) \quad \text{and} \quad \beta_i = (y, \hat{y}_i). \tag{3.23}$$

We observe that A_{ij} and β_i coincide with the definitions (3.5) only if the set $\{\hat{y}_j : j = 1, \ldots, n\}$ forms an orthonormal basis of Y_n.

The Galerkin method is also known as the *Petrov–Galerkin method* (see [175]) since Petrov was the first to consider the general situation of (3.21). The special case $X = Y$ and $X_n = Y_n$ was studied by Bubnov in 1913 and later by Galerkin in 1915 (see [76]). For this reason, this special case is also known as the *Bubnov–Galerkin method*. In the case when the operator K is self-adjoint and positive definite, we will see that the Bubnow-Galerkin method coincides with the *Rayleigh–Ritz method*, see [184] and [179].

In the following theorem, we prove error estimates for the Galerkin method of the form (3.18). They differ only in the first term, which corresponds to the perturbation of the right-hand side. The second term bounds the error for the exact right-hand side and tends to zero, provided assumption (3.16) of Theorem 3.7 is satisfied.

Theorem 3.9

Let $Kx = y$ and assume that the Galerkin equations (3.21) are uniquely solvable for every right-hand side.

(a) Let $y^\delta \in Y$ with $\|y - y^\delta\| \leq \delta$ be given and $x_n^\delta \in X_n$ be the solution of

$$\left(Kx_n^\delta, z_n\right) = \left(y^\delta, z_n\right) \quad \text{for all } z_n \in Y_n. \tag{3.24}$$

Then the following error estimate holds:

$$\left\|x_n^\delta - x\right\| \leq \delta \left\|R_n\right\| + \left\|R_n Kx - x\right\|. \tag{3.25}$$

(b) Let A and β be given by (3.23) and $\beta^\delta \in \mathbb{K}^n$ with $|\beta - \beta^\delta| \leq \delta$, where $|\cdot|$ denotes the Euclidean norm in $\mathbb{K}^n$. Let $\alpha^\delta \in \mathbb{K}^n$ be the solution of $A\alpha^\delta = \beta^\delta$. Define $x_n^\delta = \sum_{j=1}^n \alpha_j^\delta \hat{x}_j \in X_n$. Then the following error estimates hold:

$$\left\|x_n^\delta - x\right\| \leq \frac{a_n}{\lambda_n} \delta + \left\|R_n Kx - x\right\|, \tag{3.26a}$$

$$\left\|x_n^\delta - x\right\| \leq b_n \left\|R_n\right\| \delta + \left\|R_n Kx - x\right\|, \tag{3.26b}$$

where

$$a_n = \max\left\{\left\|\sum_{j=1}^n \rho_j \hat{x}_j\right\| : \sum_{j=1}^n |\rho_j|^2 = 1\right\}, \tag{3.27a}$$

$$b_n = \max\left\{\sqrt{\sum_{j=1}^n |\rho_j|^2} : \left\|\sum_{j=1}^n \rho_j \hat{y}_j\right\| = 1\right\}, \tag{3.27b}$$

and $\lambda_n > 0$ denotes the smallest singular value of the matrix A.

In the first term of the right-hand side of (3.26a) and (3.26b), we observe the dependence on the choice of basis functions $\{\hat{x}_j : j = 1, \ldots, n\}$ and $\{\hat{y}_j : j = 1, \ldots, n\}$, respectively. We note that $a_n = 1$ or $b_n = 1$ if the sets $\{\hat{x}_j : j = 1, \ldots, n\}$ or $\{\hat{y}_j : j = 1, \ldots, n\}$, respectively, form an orthonormal system.

Proof: Part (a) is a direct consequence of (3.18).

(b) From the triangle inequality $\|x_n^\delta - x\| \leq \|x_n^\delta - R_n y\| + \|R_n y - x\|$, we observe that it is sufficient to estimate the first term.

We note that $R_n y = \sum_{j=1}^n \alpha_j \hat{x}_j$, where α satisfies the linear system $A\alpha = \beta$. Writing $x_n^\delta - R_n y = \sum_{j=1}^n (\alpha_j^\delta - \alpha_j)\hat{x}_j$, we estimate

$$\begin{aligned}
\|x_n^\delta - R_n y\| &\leq a_n |\alpha^\delta - \alpha| = a_n |A^{-1}(\beta^\delta - \beta)| \\
&\leq a_n |A^{-1}|_2 |\beta^\delta - \beta| \leq \frac{a_n}{\lambda_n} \delta,
\end{aligned}$$

where $|A^{-1}|_2$ denotes the spectral norm of A^{-1}, i.e., the smallest singular value of A. This yields (3.26a).

Now we choose $y_n^\delta \in Y_n$ such that $(y_n^\delta, \hat{y}_i) = \beta_i^\delta$ for $i = 1, \ldots, n$. Then $R_n y_n^\delta = x_n^\delta$ and thus

$$\begin{aligned}
\|x_n^\delta - R_n y\| &\leq \|R_n\| \|y_n^\delta - Q_n y\| = \|R_n\| \sup_{z_n \in Y_n} \frac{(y_n^\delta - Q_n y, z_n)}{\|z_n\|} \\
&= \|R_n\| \sup_{\rho_j} \frac{\sum_{j=1}^n \rho_j (y_n^\delta - Q_n y, \hat{y}_j)}{\left\|\sum_{j=1}^n \rho_j \hat{y}_j\right\|} \\
&= \|R_n\| \sup_{\rho_j} \frac{\sum_{j=1}^n \rho_j (\beta_j^\delta - \beta_j)}{\left\|\sum_{j=1}^n \rho_j \hat{y}_j\right\|} \\
&\leq \|R_n\| |\beta^\delta - \beta| \sup_{\rho_j} \frac{\sqrt{\sum_{j=1}^n \rho_j^2}}{\left\|\sum_{j=1}^n \rho_j \hat{y}_j\right\|} \\
&\leq \|R_n\| b_n \delta. \qquad\qquad\qquad\qquad\qquad\qquad \square
\end{aligned}$$

We point out again that the Galerkin method is convergent only if the boundedness assumption (3.16) of Theorem 3.7 is satisfied.

In the following three subsections we will derive error estimates for three special choices for the finite-dimensional subspaces X_n and Y_n. The cases where X_n and Y_n are coupled by $Y_n = K(X_n)$ or $X_n = K^*(Y_n)$ will lead to the *least squares method* or the *dual least squares method*, respectively. Here, $K^* : Y \to X$ denotes the adjoint of K. In Subsection 3.2.3, we will study the Bubnov–Galerkin method for the case where K satisfies Gårding's inequality. In all of the subsections, we formulate the Galerkin equations for the perturbed cases first without using particular bases and then with respect to given bases in X_n and Y_n.

3.2.1 The Least Squares Method

An obvious method to solve an equation of the kind $Kx = y$ is the following: Given a finite-dimensional subspace $X_n \subset X$, determine $x_n \in X_n$ such that

$$\|Kx_n - y\| \leq \|Kz_n - y\| \quad \text{for all } z_n \in X_n. \tag{3.28}$$

Existence and uniqueness of $x_n \in X_n$ follow easily since X_n is finite-dimensional and K is one-to-one. The solution $x_n \in X_n$ of this least squares problem is characterized by

$$\bigl(Kx_n, Kz_n\bigr) = \bigl(y, Kz_n\bigr) \quad \text{for all } z_n \in X_n. \tag{3.29a}$$

We observe that this method is a special case of the Galerkin method when we set $Y_n := K(X_n)$.

Choosing a basis $\{\hat{x}_j : j = 1, \dots, n\}$ of X_n leads to the finite system

$$\sum_{j=1}^{n} \alpha_j \bigl(K\hat{x}_j, K\hat{x}_i\bigr) = \beta_i = \bigl(y, K\hat{x}_i\bigr) \quad \text{for } i = 1, \dots, n, \tag{3.29b}$$

or $A\alpha = \beta$. The corresponding matrix $A \in \mathbb{K}^{n \times n}$ with $A_{ij} = \bigl(K\hat{x}_j, K\hat{x}_i\bigr)$ is symmetric (if $\mathbb{K} = \mathbb{R}$) or hermitean (if $\mathbb{K} = \mathbb{C}$) and positive definite since K is also one-to-one.

Again, we study the case where the right-hand side is perturbed by an error. For continuous perturbations, let $x_n^\delta \in X_n$ be the solution of

$$\bigl(Kx_n^\delta, Kz_n\bigr) = \bigl(y^\delta, Kz_n\bigr) \quad \text{for all } z_n \in X_n, \tag{3.30a}$$

where $y^\delta \in Y$ is the perturbed right-hand side with $\|y^\delta - y\| \leq \delta$.

For the discrete perturbation, we assume that $\beta \in \mathbb{K}^n$ is replaced by $\beta^\delta \in \mathbb{K}^n$ with $|\beta^\delta - \beta| \leq \delta$, where $|\cdot|$ denotes the Euclidean norm in $\mathbb{K}^n$. This leads to the following finite system of equations for the coefficients of $x_n^\delta = \sum_{j=1}^{n} \alpha_j^\delta \hat{x}_j$:

$$\sum_{j=1}^{n} \alpha_j^\delta \bigl(K\hat{x}_j, K\hat{x}_i\bigr) = \beta_i^\delta \quad \text{for } i = 1, \dots, n. \tag{3.30b}$$

This system is uniquely solvable since the matrix A is positive definite. For least squares methods, the boundedness condition (3.16) is not satisfied without additional assumptions. We refer to [201] or [130], Problem 17.2, for an example. However, we can prove the following theorem.

Theorem 3.10

Let $K : X \to Y$ be a linear, bounded, and injective operator between Hilbert spaces and $X_n \subset X$ be finite-dimensional subspaces such that $\bigcup_{n \in \mathbb{N}} X_n$ is dense in X. Let $x \in X$ be the solution of $Kx = y$ and $x_n^\delta \in X_n$ be the least squares solution from (3.30a) or (3.30b). Define

$$\sigma_n := \max\{\|z_n\| : z_n \in X_n, \ \|Kz_n\| = 1\} \qquad (3.31)$$

and let there exist $c > 0$, independent of n, such that

$$\min_{z_n \in X_n} \left\{ \|x - z_n\| + \sigma_n \|K(x - z_n)\| \right\} \leq c\|x\| \quad \text{for all } x \in X. \qquad (3.32)$$

Then the least squares method is convergent and $\|R_n\| \leq \sigma_n$. In this case, we have the error estimate

$$\left\| x - x_n^\delta \right\| \leq r_n \sigma_n \delta + \tilde{c} \min\{\|x - z_n\| : z_n \in X_n\} \qquad (3.33)$$

for some $\tilde{c} > 0$. Here, $r_n = 1$ if $x_n^\delta \in X_n$ solves (3.30a), i.e., δ measures the continuous perturbation $\|y^\delta - y\|$. If δ measures the discrete error $|\beta^\delta - \beta|$ in the Euclidean norm and $x_n^\delta = \sum_{j=1}^n \alpha_j^\delta \hat{x}_j \in X_n$, where α^δ solves (3.30b), then r_n is given by

$$r_n = \max\left\{ \sqrt{\sum_{j=1}^n |\rho_j|^2} : \left\| K\left(\sum_{j=1}^n \rho_j \hat{x}_j \right) \right\| = 1 \right\}. \qquad (3.34)$$

Proof: We will prove that $\|R_n K\|$ is bounded uniformly in n. Let $x \in X$ and $x_n := R_n Kx$. Then x_n satisfies $\left(Kx_n, Kz_n\right) = \left(Kx, Kz_n\right)$ for all $z_n \in X_n$. This yields

$$
\begin{aligned}
\|K(x_n - z_n)\|^2 &= \left(K(x_n - z_n), K(x_n - z_n)\right) \\
&= \left(K(x - z_n), K(x_n - z_n)\right) \\
&\leq \|K(x - z_n)\| \, \|K(x_n - z_n)\|
\end{aligned}
$$

and thus $\|K(x_n - z_n)\| \leq \|K(x - z_n)\|$ for all $z_n \in X_n$. Using this and the definition of σ_n, we conclude that

$$\|x_n - z_n\| \leq \sigma_n \|K(x_n - z_n)\| \leq \sigma_n \|K(x - z_n)\|,$$

and thus

$$
\begin{aligned}
\|x_n\| &\leq \|x_n - z_n\| + \|z_n - x\| + \|x\| \\
&\leq \|x\| + \left[\|z_n - x\| + \sigma_n \|K(x - z_n)\| \right].
\end{aligned}
$$

This holds for all $z_n \in X_n$. Taking the minimum, we have by assumption (3.32) that $\|x_n\| \leq (1 + c) \|x\|$. Thus the boundedness condition (3.16) is satisfied. Application of Theorem 3.7 proves convergence.

Analogously we prove the estimate for $\|R_n\|$. Let $y \in Y$ and set $x_n := R_n y$. Then from (3.29a) we have that

$$\|Kx_n\|^2 = (Kx_n, Kx_n) = (y, Kx_n) \leq \|y\| \, \|Kx_n\|$$

and thus

$$\|x_n\| \leq \sigma_n \|Kx_n\| \leq \sigma_n \|y\| \, .$$

This proves the estimate $\|R_n\| \leq \sigma_n$.

The error estimates (3.33) follow directly from Theorem 3.9 and the estimates (3.25) or (3.26b) for $\hat{y}_j = K\hat{x}_j$.   □

For further numerical aspects of least squares methods, we refer to [58, 61, 83, 117, 150, 151, 162, 161].

3.2.2 The Dual Least Squares Method

As a next example for a Galerkin method, we study the dual least squares method. We will see that the boundedness condition (3.16) is always satisfied.

Given any finite-dimensional subspace $Y_n \subset Y$, determine $u_n \in Y_n$ such that

$$(KK^* u_n, z_n) = (y, z_n) \quad \text{for all } z_n \in Y_n, \tag{3.35}$$

where $K^* : Y \to X$ denotes the adjoint of K. Then $x_n := K^* u_n$ is called the *dual least squares solution*. It is a special case of the Galerkin method when we set $X_n := K^*(Y_n)$. Writing equation (3.35) for $y = Kx$ in the form

$$(K^* u_n, K^* z_n) = (x, K^* z_n) \quad \text{for all } z_n \in Y_n,$$

we observe that the dual least squares method is just the least squares method for the equation $K^* u = x$. This explains the name.

We assume again that the right-hand side is perturbed. Let $y^\delta \in Y$ with $\|y^\delta - y\| \leq \delta$. Instead of equation (3.35), one determines $x_n^\delta = K^* u_n^\delta \in X_n$ with

$$(K^* u_n^\delta, K^* z_n) = (y^\delta, z_n) \quad \text{for all } z_n \in Y_n. \tag{3.36}$$

For discrete perturbations, we choose a basis $\{\hat{y}_j : j = 1, \ldots, n\}$ of Y_n and assume that the right-hand sides $\beta_i = (y, \hat{y}_i)$, $i = 1, \ldots, n$, of the Galerkin equations are perturbed by a vector $\beta^\delta \in \mathbb{K}^n$ with $|\beta^\delta - \beta| \le \delta$ where $|\cdot|$ denotes the Euclidean norm in $\mathbb{K}^n$. Instead of (3.35), we determine

$$x_n^\delta = K^* u_n^\delta = \sum_{j=1}^n \alpha_j^\delta K^* \hat{y}_j,$$

where $\alpha^\delta \in \mathbb{K}^n$ solves

$$\sum_{j=1}^n \alpha_j^\delta \left(K^* \hat{y}_j, K^* \hat{y}_i \right) = \beta_i^\delta \quad \text{for } i = 1, \ldots, n. \tag{3.37}$$

First we show that equations (3.36) and (3.37) are uniquely solvable. $K^* : Y \to X$ is one-to-one since the range $K(X)$ is dense in Y. Thus the dimensions of Y_n and X_n coincide and K^* is an isomorphism from Y_n onto X_n. It is sufficient to prove uniqueness of a solution to (3.36). Let $u_n \in Y_n$ with $\left(K^* u_n, K^* z_n \right) = 0$ for all $z_n \in Y_n$. For $z_n = u_n$ we conclude that $0 = \left(K^* u_n, K^* u_n \right) = \| K^* u_n \|^2$, i.e., $K^* u_n = 0$ or $u_n = 0$.

Convergence and error estimates are proven in the following theorem.

Theorem 3.11
Let X and Y be Hilbert spaces and $K : X \to Y$ be linear, bounded, and one-to-one such that the range $K(X)$ is dense in Y. Let $Y_n \subset Y$ be finite-dimensional subspaces such that $\bigcup_{n \in \mathbb{N}} Y_n$ is dense in Y. Let $x \in X$ be the solution of $Kx = y$. Then the Galerkin equations (3.36) and (3.37) are uniquely solvable for every right-hand side and every $n \in \mathbb{N}$. The dual least squares method is convergent and

$$\| R_n \| \le \sigma_n := \max \{ \| z_n \| : z_n \in Y_n, \ \| K^* z_n \| = 1 \}. \tag{3.38}$$

Furthermore, we have the error estimates

$$\left\| x - x_n^\delta \right\| \le r_n \sigma_n \delta + c \min \{ \| x - z_n \| : z_n \in K^*(Y_n) \} \tag{3.39}$$

for some $c > 0$. Here, $r_n = 1$ if $x_n^\delta \in X_n$ solves (3.36), i.e., δ measures the norm $\| y^\delta - y \|$ in Y. If δ measures the discrete error $|\beta^\delta - \beta|$ and $x_n^\delta = \sum_{j=1}^n \alpha_j^\delta K^* \hat{y}_j \in X_n$, where α^δ solves (3.37), then r_n is given by

$$r_n = \max \left\{ \sqrt{\sum_{j=1}^n |\rho_j|^2} : \left\| \sum_{j=1}^n \rho_j \hat{y}_j \right\| = 1 \right\}. \tag{3.40}$$

We note that $r_n = 1$ if $\{ \hat{y}_j : j = 1, \ldots, n \}$ forms an orthonormal system in Y.

Proof: We have seen already that (3.36) and (3.37) are uniquely solvable for every right-hand side and every $n \in \mathbb{N}$.

Now we prove the estimate $\|R_n K\| \leq 1$, which is condition (3.16) with $c = 1$. Let $x \in X$ and set $x_n := R_n K x \in X_n$. Then $x_n = K^* u_n$, and $u_n \in Y_n$ satisfies

$$(K^* u_n, K^* z_n) = (Kx, z_n) \quad \text{for all } z_n \in Y_n.$$

For $z_n = u_n$ this implies

$$\|x_n\|^2 = \|K^* u_n\|^2 = (Kx, u_n) = (x, K^* u_n) \leq \|x\| \, \|x_n\|,$$

which proves the desired estimate. If we replace Kx by y in the preceding arguments, we have

$$\|x_n\|^2 \leq \|y\| \, \|u_n\| \leq \sigma_n \|y\| \, \|K^* u_n\| = \sigma_n \|y\| \, \|x_n\|,$$

which proves (3.38).

Finally, we show that $\bigcup_{n \in \mathbb{N}} X_n$ is dense in X. Let $x \in X$ and $\varepsilon > 0$. Since $K^*(Y)$ is dense in X, there exists $y \in Y$ with $\|x - K^* y\| < \varepsilon/2$. Since $\bigcup_{n \in \mathbb{N}} Y_n$ is dense in Y, there exists $y_n \in Y_n$ with $\|y - y_n\| < \varepsilon/(2 \|K\|)$. The triangle inequality yields that for $x_n := K^* y_n \in X_n$

$$\|x - x_n\| \leq \|x - K^* y\| + \|K^*(y - y_n)\| \leq \varepsilon.$$

Application of Theorem 3.9 and (3.25) and (3.26b) proves (3.39).   □

3.2.3   The Bubnov–Galerkin Method for Coercive Operators

In this subsection, we assume that $K : X \to X$ is a linear and bounded operator and X_n, $n \in \mathbb{N}$, are finite-dimensional subspaces. The Galerkin method reduces to the problem of determining $x_n \in X_n$ such that

$$(Kx_n, z_n) = (y, z_n) \quad \text{for all } z_n \in X_n. \tag{3.41}$$

This special case is called the *Bubnov–Galerkin method*. Again, we consider two kinds of perturbations of the right-hand side. If $y^\delta \in X$ with $\|y^\delta - y\| \leq \delta$ is a perturbed right-hand side, then instead of (3.41) we study the equation

$$(Kx_n^\delta, z_n) = (y^\delta, z_n) \quad \text{for all } z_n \in X_n. \tag{3.42}$$

The other possibility is to choose a basis $\{\hat{x}_j : j = 1, \ldots, n\}$ of X_n and assume that the right-hand sides $\beta_i = (y, \hat{x}_i)$, $i = 1, \ldots, n$, of the Galerkin equations are perturbed by a vector $\beta^\delta \in \mathbb{K}^n$ with $\left|\beta^\delta - \beta\right| \leq \delta$, where $|\cdot|$ denotes again the Euclidean norm in $\mathbb{K}^n$. In this case, instead of (3.41), we have to solve

$$\sum_{j=1}^{n} \alpha_j^\delta \left(K\hat{x}_j, \hat{x}_i\right) = \beta_i^\delta \quad \text{for } i = 1, \ldots, n, \tag{3.43}$$

for $\alpha^\delta \in \mathbb{K}^n$ and set $x_n^\delta = \sum_{j=1}^n \alpha_j^\delta \hat{x}_j$.

Before we prove a convergence result for this method, we will briefly describe the Rayleigh–Ritz method and show that it is a special case of the Bubnov–Galerkin method.

Let $K : X \to X$ also be self-adjoint and positive definite, i.e., $(Kx, y) = (x, Ky)$ and $(Kx, x) > 0$ for all $x, y \in X$ with $x \neq 0$. We define the functional

$$\psi(z) := \left(Kz, z\right) - 2\operatorname{Re}(y, z) \quad \text{for } z \in X. \tag{3.44}$$

From the equation

$$\psi(z) - \psi(x) = 2\operatorname{Re}\left(Kx - y, z - x\right) + \left(K(z - x), z - x\right) \tag{3.45}$$

and the positivity of K, we easily conclude (see Problem 3.2) that $x \in X$ is the unique minimum of ψ if and only if x solves $Kx = y$. The Rayleigh–Ritz method is to minimize ψ over the finite-dimensional subspace X_n. From (3.45), we see that if $x_n \in X_n$ minimizes ψ on X_n, then, for $z_n = x_n \pm \varepsilon u_n$ with $u_n \in X_n$ and $\varepsilon > 0$, we have that

$$0 \leq \psi(z_n) - \psi(x_n) = \pm\varepsilon\, 2\operatorname{Re}\left(Kx_n - y, u_n\right) + \varepsilon^2\left(Ku_n, u_n\right)$$

for all $u_n \in X_n$. Dividing by $\varepsilon > 0$ and letting $\varepsilon \to 0$ yields that $x_n \in X_n$ satisfies the Galerkin equation (3.41). If, on the other hand, $x_n \in X_n$ solves (3.41), then from (3.45)

$$\psi(z_n) - \psi(x_n) = \left(K(z_n - x_n), z_n - x_n\right) \geq 0$$

for all $z_n \in X_n$. Therefore, the Rayleigh–Ritz method is identical to the Bubnov–Galerkin method.

Now we will generalize the Rayleigh–Ritz method and study the Bubnov–Galerkin method for the important class of coercive operators. Before we can formulate the results, we briefly recall some definitions.

Definition 3.12

Let V be a reflexive Banach space with dual space V^*. We denote the norms *in V and V^* by $\|\cdot\|_V$ and $\|\cdot\|_{V*}$, respectively. A linear bounded operator $K : V^* \to V$ is called* coercive *if there exists $\gamma > 0$ with*

$$\mathrm{Re}\,\langle x, Kx \rangle \ \geq\ \gamma\,\|x\|_{V*}^2 \quad \text{for all } x \in V^*, \tag{3.46}$$

where $\langle \cdot, \cdot \rangle$ denotes the dual pairing in (V^*, V). The operator K satifies Gårding's inequality *if there exists a linear compact operator $C : V^* \to V$ such that $K + C$ is coercive, i.e.,*

$$\mathrm{Re}\,\langle x, Kx \rangle \ \geq\ \gamma\,\|x\|_{V*}^2 \ -\ \mathrm{Re}\,\langle x, Cx \rangle \quad \text{for all } x \in V^*.$$

By the same arguments as in the proof of the Lax–Milgram theorem (see [103]), it can be shown that every coercive operator is an isomorphism from V^* onto V. Coercive operators play an important role in the study of partial differential equations and integral equations by variational methods. In the usual definition, the roles of V and V^* are interchanged. For integral operators that are "smoothing," our definition seems more appropriate. However, both definitions are equivalent in the sense that the inverse operator $K^{-1} : V \to V^*$ is coercive in the usual sense with γ replaced by $\gamma/\|K\|^2$.

Definition 3.13

A Gelfand triple *(V, X, V^*) consists of a reflexive Banach space V, a Hilbert space X, and the dual V^* of V such that*

(a) V is a dense subspace of X, and

(b) the imbedding $J : V \to X$ is bounded.

We write $V \subset X \subset V^*$ since we can identify X with a dense subspace of V^*. This identification is given by the dual operator $J^* : X \to V^*$ of J, where we identify the dual of the Hilbert space X by itself. From $(x, y) = \langle J^* x, y \rangle$, for all $x \in X$ and $y \in V$ we see that with this identification the dual pairing $\langle \cdot, \cdot \rangle$ in (V^*, V) is an extension of the inner product $(\cdot, \cdot)$ in X, i.e, we write

$$\langle x, y \rangle \ =\ (x, y) \quad \text{for all } x \in X \text{ and } y \in V.$$

Furthermore, we have the estimates

$$|\langle x, y \rangle| \ \leq\ \|x\|_{V*}\,\|y\|_V \quad \text{for all } x \in V^*,\ y \in V,$$

and thus

$$|(x, y)| \ \leq\ \|x\|_{V*}\,\|y\|_V \quad \text{for all } x \in X,\ y \in V.$$

J^* is one-to-one and has dense range (see Problem 3.3). As before, we denote the norm in X by $\|\cdot\|$.

Now we can prove the main theorem about convergence of the Bubnov–Galerkin method for coercive operators.

Theorem 3.14

Let (V, X, V^*) be a Gelfand triple, and $X_n \subset V$ be finite-dimensional subspaces such that $\bigcup_{n \in \mathbb{N}} X_n$ is dense in X. Let $K : V^* \to V$ be coercive with constant $\gamma > 0$. Let $x \in X$ be the solution of $Kx = y$. Then we have the following:

(a) There exist unique solutions of the Galerkin equations (3.41)–(3.43), and the Bubnov–Galerkin method converges in V^* with

$$\|x - x_n\|_{V^*} \leq c \min\{\|x - z_n\|_{V^*} : z_n \in X_n\} \tag{3.47}$$

for some $c > 0$.

(b) Define $\rho_n > 0$ by

$$\rho_n = \max\{\|u\| : u \in X_n, \, \|u\|_{V^*} = 1\} \tag{3.48}$$

and the orthogonal projection operator P_n from X onto X_n. The Bubnov–Galerkin method converges in X if there exists $c > 0$ with

$$\|u - P_n u\|_{V^*} \leq \frac{c}{\rho_n} \|u\| \quad \text{for all } u \in X. \tag{3.49}$$

In this case, we have the estimates

$$\|R_n\| \leq \frac{1}{\gamma} \rho_n^2 \tag{3.50}$$

and

$$\|x - x_n^\delta\| \leq c \left[r_n \rho_n^2 + \min\{\|x - z_n\| : z_n \in X_n\} \right] \tag{3.51}$$

for some $c > 0$. Here, $r_n = 1$ if $x_n^\delta \in X_n$ solves (3.42), i.e., δ measures the norm $\|y^\delta - y\|$ in X. If δ measures the discrete error $|\beta^\delta - \beta|$ in the Euclidean norm and $x_n^\delta = \sum_{j=1}^{n} \alpha_j^\delta \hat{x}_j \in X_n$, where α^δ solves (3.43), then r_n is given by

$$r_n = \max\left\{ \sqrt{\sum_{j=1}^{n} |\rho_j|^2} : \left\| \sum_{j=1}^{n} \rho_j \hat{x}_j \right\| = 1 \right\}. \tag{3.52}$$

Again, we note that $r_n = 1$ if $\{\hat{x}_j : j = 1, \ldots, n\}$ forms an orthonormal system in X.

Proof: (a) We will apply Theorem 3.7 to the equations $Kx = y$, $x \in V^*$, and $P_n K x_n = P_n y$, $x_n \in X_n$, where we consider K as an operator from V^* into V. We observe that the orthogonal projection operator P_n is also bounded from V into X_n. This follows from the observation that on the finite-dimensional space X_n the norms $\|\cdot\|$ and $\|\cdot\|_V$ are equivalent and thus

$$\|P_n u\|_V \leq c\|P_n u\| \leq c\|u\| \leq \tilde{c}\|u\|_V \quad \text{for } u \in V.$$

The constants c, and thus $\tilde{c}$, depend on n. Since V is dense in X and X is dense in V^*, we conclude that also $\bigcup_{n\in\mathbb{N}} X_n$ is dense in V^*. To apply Theorem 3.7, we have to show that (3.41) is uniquely solvable in V^* and that $R_n K : V^* \to X_n \subset V^*$ is uniformly bounded with respect to n.

Since (3.41) is a finite-dimensional quadratic system, it is sufficient to prove uniqueness. Let $x_n \in X_n$ satisfy (3.41) for $y = 0$. Since K is coercive, we have that

$$\gamma\|x_n\|_{V^*}^2 \leq \operatorname{Re}\langle x_n, K x_n\rangle = \operatorname{Re}(K x, x_n) = 0;$$

thus $x_n = 0$.

Now let $x \in V^*$ and set $x_n = R_n K x$. Then $x_n \in X_n$ satisfies

$$(K x_n, z_n) = (K x, z_n) \quad \text{for all } z_n \in X_n. \tag{3.53}$$

Again, we conclude that

$$\gamma\|x_n\|_{V^*}^2 \leq \operatorname{Re}\langle x_n, K x_n\rangle = \operatorname{Re}(K x, x_n) \leq \|K x\|_V \|x_n\|_{V^*}$$

and thus

$$\|x_n\|_{V^*} \leq \frac{1}{\gamma}\|K x\|_V \leq \frac{1}{\gamma}\|K\|_{\mathcal{L}(V^*,V)} \|x\|_{V^*}.$$

Since this holds for all $x \in V^*$, we conclude that

$$\|R_n K\|_{\mathcal{L}(V^*,V^*)} \leq \frac{1}{\gamma}\|K\|_{\mathcal{L}(V^*,V)}.$$

Then the assumptions of Theorem 3.7 are satisfied for $K : V^* \to V$.

(b) Now let $x \in X$ and $x_n = R_n K x$. Using the estimates (3.47) and (3.49), we conclude that

$$
\begin{aligned}
\|x - x_n\| &\leq \|x - P_n x\| + \|P_n x - x_n\| \\
&\leq \|x - P_n x\| + \rho_n \|P_n x - x_n\|_{V^*} \\
&\leq \|x - P_n x\| + \rho_n \|P_n x - x\|_{V^*} + \rho_n \|x - x_n\|_{V^*} \\
&\leq \|x - P_n x\| + \rho_n \|P_n x - x\|_{V^*} + c\rho_n \min_{z_n \in X_n} \|x - z_n\|_{V^*} \\
&\leq \|x - P_n x\| + (c+1)\rho_n \|P_n x - x\|_{V^*} \\
&\leq 2\|x\| + c_1\|x\| = (2 + c_1)\|x\|,
\end{aligned}
$$

and thus $\|x_n\| \leq \|x_n - x\| + \|x\| \leq (3 + c_1) \|x\|$. Application of Theorem 3.7 yields convergence in X.

Finally, we prove the estimate of R_n in $\mathcal{L}(X, X)$. Let $y \in X$ and $x_n = R_n y$. We estimate

$$\gamma \|x_n\|_{V^*}^2 \; \leq \; \mathrm{Re} \langle x_n, K x_n \rangle \; = \; \mathrm{Re}\,(y, x_n) \; \leq \; \|y\|\, \|x_n\| \; \leq \; \rho_n \|y\|\, \|x_n\|_{V^*}$$

and thus

$$\|x_n\| \; \leq \; \rho_n \|x_n\|_{V^*} \; \leq \; \frac{1}{\gamma} \rho_n^2 \, \|y\|\,,$$

which proves the estimate (3.50). Application of Theorem 3.9 and the estimates (3.25) and (3.26b) proves (3.51). □

From our general perturbation theorem (Theorem 3.8), we observe that the assumption of K being coercive can be weakened. It is sufficient to assume that K is one-to-one and satisfies Gårding's inequality. We formulate the result in the next theorem.

Theorem 3.15
The assertions of Theorem 3.14 also hold if $K : V^* \to V$ is one-to-one and satisfies Gårding's inequality with some compact operator $C : V^* \to V$.

For further reading, we refer to [164] and the monographs [15, 130, 144].

3.3 Application to Symm's Integral Equation of the First Kind

In this section, we will apply the Galerkin methods to an integral equation of the first kind, which occurs in potential theory. We study the Dirichlet problem for the *Laplace equation*

$$\Delta u \; = \; 0 \quad \text{in } \Omega, \qquad u = f \quad \text{on } \partial\Omega, \tag{3.54}$$

where $\Omega \subset \mathbb{R}^2$ is some bounded, simply connected region with analytic boundary $\partial\Omega$ and $f \in C(\partial\Omega)$ is some given function. The simple layer potential

$$u(x) \; = \; -\frac{1}{\pi} \int\limits_{\partial\Omega} \varphi(y) \ln |x - y| \; ds(y), \qquad x \in \Omega, \tag{3.55}$$

solves the boundary value problem (3.54) if and only if the density $\varphi \in C(\partial\Omega)$ solves *Symm's equation*

$$-\frac{1}{\pi} \int\limits_{\partial\Omega} \varphi(y) \ln |x - y| \; ds(y) \; = \; f(x) \quad \text{for } x \in \partial\Omega; \tag{3.56}$$

see [37]. It is well known (see [111]) that in general the corresponding integral operator is not one-to-one. One has to make assumptions on the transfinite diameter of Ω; see [226]. We give a more elementary assumption in the following theorem.

Theorem 3.16
Suppose there exists $z_0 \in \Omega$ with $|x - z_0| \neq 1$ for all $x \in \partial\Omega$. Then the only solution $\varphi \in C(\partial\Omega)$ of Symm's equation (3.56) for $f = 0$ is $\varphi = 0$, i.e., the integral operator is one-to-one.

Proof: We give a more elementary proof than in [112], but we still need a few results from potential theory.

From the continuity of $x \mapsto |x - z_0|$, we conclude that either $|x - z_0| < 1$ for all $x \in \partial\Omega$ or $|x - z_0| > 1$ for all $x \in \partial\Omega$. Assume first that $|x - z_0| < 1$ for all $x \in \partial\Omega$ and choose a disc $A \subset \Omega$ with $|x - z| < 1$ for all $x \in \partial\Omega$ and $z \in A$. Let $\varphi \in C(\partial\Omega)$ satisfy (3.56) for $f = 0$ and define u by

$$u(x) = -\frac{1}{\pi} \int_{\partial\Omega} \varphi(y) \ln|x - y| \, ds(y) \quad \text{for } x \in \mathbb{R}^2.$$

From potential theory (see [37]), we conclude that u is continuous in $\mathbb{R}^2$, harmonic in $\mathbb{R}^2 \setminus \partial\Omega$, and vanishes on $\partial\Omega$. The maximum principle for harmonic functions implies that u vanishes in Ω. We will show that u also vanishes in the exterior Ω^e of Ω. The main part is to prove that

$$\hat{\varphi} := \int_{\partial\Omega} \varphi(y) \, ds(y) = 0.$$

Without loss of generality, we can assume that $\hat{\varphi} \geq 0$. We study the function v defined by

$$\begin{aligned}
v(x) &:= u(x) + \frac{\hat{\varphi}}{\pi} \ln|x - z| \\
&= \frac{1}{\pi} \int_{\partial\Omega} \varphi(y) \ln\frac{|x - z|}{|x - y|} \, ds(y), \quad x \in \Omega^e,
\end{aligned}$$

for some $z \in A$. From the choice of A, we have that

$$v(x) = \frac{\hat{\varphi}}{\pi} \ln|x - z| \leq 0 \quad \text{for } x \in \partial\Omega.$$

Furthermore, $v(x) \to 0$ as $|x|$ tends to infinity. The maximum principle applied to v in Ω^e yields that $v(x) \leq 0$ for all $x \in \Omega^e$. Now we study the asymptotic behavior of v. Elementary calculations show that

$$\frac{|x - z|}{|x - y|} = 1 + \frac{1}{|x|} \hat{x} \cdot (y - z) + \mathcal{O}(1/|x|^2)$$

for $|x| \to \infty$ uniformly in $y \in \partial\Omega$, $z \in A$, and $\hat{x} := x/|x|$. This implies that

$$v(x) = \frac{1}{\pi |x|} \hat{x} \cdot \int_{\partial\Omega} \varphi(y) (y - z) \, ds(y) + \mathcal{O}(1/|x|^2)$$

and thus

$$\hat{x} \cdot \int_{\partial\Omega} \varphi(y) (y - z) \, ds(y) \leq 0 \quad \text{for all } |\hat{x}| = 1.$$

This implies that

$$\int_{\partial\Omega} \varphi(y) \, y \, ds(y) = z \int_{\partial\Omega} \varphi(y) \, ds(y).$$

Since this holds for all $z \in A$, we conclude that $\int_{\partial\Omega} \varphi(y) \, ds(y) = 0$.

Now we see from the definition of v (for any fixed $z \in A$) that

$$u(x) = v(x) \to 0 \quad \text{as } |x| \to \infty.$$

The maximum principle again yields $u = 0$ in Ω^e.

Finally, the jump conditions of the normal derivative of the simple layer potential operator (see [37]) yield

$$2\varphi(x) = \lim_{\varepsilon \to 0+} \left[\nabla u(x - \varepsilon\nu(x)) - \nabla u(x + \varepsilon\nu(x)) \right] \cdot \nu(x) = 0$$

for $x \in \partial\Omega$, where $\nu(x)$ denotes the unit normal vector at $x \in \partial\Omega$ directed into the exterior of Ω.

This ends the proof for the case that $\max_{x \in \partial\Omega} |x - z_0| < 1$. The case $\min_{x \in \partial\Omega} |x - z_0| > 1$ is settled by the same arguments. $\square$

Now we assume that the boundary $\partial\Omega$ has a parametrization of the form

$$x = \gamma(s), \quad s \in [0, 2\pi],$$

for some 2π-periodic analytic function $\gamma : [0, 2\pi] \to \mathbb{R}^2$, which satisfies $|\dot{\gamma}(s)| > 0$ for all $s \in [0, 2\pi]$. Then Symm's equation (3.56) takes the form

$$-\frac{1}{\pi} \int_0^{2\pi} \psi(s) \ln|\gamma(t) - \gamma(s)| \, ds = f(\gamma(t)) \quad \text{for } t \in [0, 2\pi] \quad (3.57)$$

for the transformed density $\psi(s) := \varphi(\gamma(s)) |\dot{\gamma}(s)|$, $s \in [0, 2\pi]$.

For the special case where Ω is the disc with center 0 and radius $a > 0$, we have $\gamma_a(s) = a \left(\cos s, \sin s \right)$ and thus

$$\ln|\gamma_a(t) - \gamma_a(s)| = \ln a + \frac{1}{2} \ln \left(4 \sin^2 \frac{t - s}{2} \right). \quad (3.58)$$

For general boundaries, we can split the kernel in the form

$$-\frac{1}{\pi}\ln|\gamma(t)-\gamma(s)| = -\frac{1}{2\pi}\ln\left(4\sin^2\frac{t-s}{2}\right) + k(t,s), \quad t\neq s, \quad (3.59)$$

for some function k that is analytic for $t\neq s$. From the mean value theorem, we conclude that

$$\lim_{s\to t}k(t,s) = -\frac{1}{\pi}\ln|\dot\gamma(t)|.$$

This implies that k has an analytic continuation onto $[0,2\pi]\times[0,2\pi]$. With this, splitting the integral equation (3.57) takes the form

$$-\frac{1}{2\pi}\int_0^{2\pi}\psi(s)\ln\left(4\sin^2\frac{t-s}{2}\right)ds + \int_0^{2\pi}\psi(s)\,k(t,s)\,ds = f(\gamma(t)) \quad (3.60)$$

for $t\in[0,2\pi]$. We want to apply the results of the previous section on Galerkin methods to this integral equation.

As the Hilbert space X, we choose $X = L^2(0,2\pi)$. The operators K, K_0, and C are defined by

$$K\psi(t) = -\frac{1}{\pi}\int_0^{2\pi}\psi(s)\ln|\gamma(t)-\gamma(s)|\,ds, \quad (3.61a)$$

$$K_0\psi(t) = -\frac{1}{2\pi}\int_0^{2\pi}\psi(s)\left[\ln\left(4\sin^2\frac{t-s}{2}\right)-1\right]ds, \quad (3.61b)$$

$$C\psi = K\psi - K_0\psi \quad (3.61c)$$

for $t\in[0,2\pi]$ and $\psi\in L^2(0,1)$. First, we observe that K, K_0, and C are well defined and compact operators in $L^2(0,1)$ since the kernels are weakly singular (see Theorem A.33 of Appendix A). They are also self-adjoint in $L^2(0,1)$.

We define the finite-dimensional subspaces X_n and Y_n as the spaces of truncated Fourier series:

$$X_n = Y_n = \left\{\sum_{j=-n}^{n}\alpha_j\,e^{ijt} : \alpha_j\in\mathbb{C}\right\}. \quad (3.62)$$

To investigate the mapping properties, we need the following technical result (see [131]).

Lemma 3.17

$$\frac{1}{2\pi}\int_0^{2\pi}e^{ins}\ln\left(4\sin^2\frac{s}{2}\right)ds = \begin{cases} -1/|n|, & n\in\mathbb{Z},\ n\neq 0, \\ 0, & n=0. \end{cases} \quad (3.63)$$

Proof: It suffices to study the case $n \in \mathbb{N}_0$. First let $n \in \mathbb{N}$. Integrating the geometric sum

$$1 + 2 \sum_{j=1}^{n-1} e^{ijs} + e^{ins} = i\left(1 - e^{ins}\right) \cot \frac{s}{2}, \quad 0 < s < 2\pi,$$

yields

$$\int_0^{2\pi} \left(e^{ins} - 1\right) \cot \frac{s}{2}\, ds = 2\pi\, i.$$

Integration of the identity

$$\frac{d}{ds}\left[\left(e^{ins} - 1\right) \ln\left(4\sin^2 \frac{s}{2}\right)\right] = in\, e^{ins} \ln\left(4\sin^2 \frac{s}{2}\right) + \left(e^{ins} - 1\right) \cot \frac{s}{2}$$

yields

$$\int_0^{2\pi} e^{ins} \ln\left(4\sin^2 \frac{s}{2}\right) ds = -\frac{1}{in} \int_0^{2\pi} \left(e^{ins} - 1\right) \cot \frac{s}{2}\, ds = -\frac{2\pi}{n},$$

which proves the assertion for $n \in \mathbb{N}$.

It remains to study the case where $n = 0$. Define

$$I := \int_0^{2\pi} \ln\left(4\sin^2 \frac{s}{2}\right) ds.$$

Then we conclude that

$$2I = \int_0^{2\pi} \ln\left(4\sin^2 \frac{s}{2}\right) ds + \int_0^{2\pi} \ln\left(4\cos^2 \frac{s}{2}\right) ds$$

$$= \int_0^{2\pi} \ln\left(16 \sin^2 \frac{s}{2} \cos^2 \frac{s}{2}\right) ds$$

$$= \int_0^{2\pi} \ln\left(4\sin^2 s\right) ds = \frac{1}{2}\int_0^{4\pi} \ln\left(4\sin^2 \frac{s}{2}\right) ds = I$$

and thus $I = 0$. □

This lemma shows that the functions

$$\hat{\psi}_n(t) := e^{int}, \quad t \in [0, 2\pi], \quad n \in \mathbb{Z}, \tag{3.64}$$

are eigenfunctions of K_0:

$$K_0\hat{\psi}_n = \frac{1}{|n|}\hat{\psi}_n \quad \text{for } n \neq 0 \quad \text{and} \tag{3.65a}$$

$$K_0\hat{\psi}_0 = \hat{\psi}_0. \tag{3.65b}$$

Now can prove the mapping properties of the operators.

Theorem 3.18
*Suppose there exists $z_0 \in \Omega$ with $|x - z_0| \neq 1$ for all $x \in \partial\Omega$. Let the opera-
tors K and K_0 be given by (3.61a) and (3.61b), respectively. By $H^s(0, 2\pi)$
we denote the Sobolev spaces of order s (see Section A.4 of Appendix A).*

(a) *The operators K and K_0 can be extended to isomorphisms from
$H^{s-1}(0, 2\pi)$ onto $H^s(0, 2\pi)$ for every $s \in \mathbb{R}$.*

(b) *The operator K_0 is coercive from $H^{-1/2}(0, 2\pi)$ into $H^{1/2}(0, 2\pi)$.*

(c) *The operator $C = K - K_0$ is compact from $H^{s-1}(0, 2\pi)$ into $H^s(0, 2\pi)$
for every $s \in \mathbb{R}$.*

Proof: Let $\psi \in L^2(0, 2\pi)$. Then ψ has the representation

$$\psi(t) = \sum_{n\in\mathbb{Z}} \alpha_n e^{int} \quad \text{with} \quad \sum_{n\in\mathbb{Z}} |\alpha_n|^2 < \infty.$$

From (3.65a) and (3.65b), we have that

$$K_0\psi(t) = \alpha_0 + \sum_{n\neq 0} \frac{1}{|n|} \alpha_n e^{int}$$

and thus for any $s \in \mathbb{R}$:

$$\|K_0\psi\|_{H^s}^2 = |\alpha_0|^2 + \sum_{n\neq 0}(1+n^2)^s \frac{1}{n^2}|\alpha_n|^2$$

$$(\psi, K_0\psi) = |\alpha_0|^2 + \sum_{n\neq 0}\frac{1}{|n|}|\alpha_n|^2$$

$$\geq \sum_{n\in\mathbb{Z}}(1+n^2)^{-1/2}|\alpha_n|^2 = \|\psi\|_{H^{-1/2}}^2.$$

From the elementary estimate

$$(1+n^2)^{s-1} \leq \frac{(1+n^2)^s}{n^2} \leq \frac{(1+n^2)^s}{\frac{1}{2}(1+n^2)} = 2(1+n^2)^{s-1}, \quad n \neq 0,$$

we see that K_0 can be extended to an isomorphism from $H^{s-1}(0, 2\pi)$
onto $H^s(0, 2\pi)$ and is coercive for $s = 1/2$. The operator C is bounded

from $H^r(0, 2\pi)$ into $H^s(0, 2\pi)$ for all $r, s \in \mathbb{R}$ by Theorem A.45 of Appendix A. This proves part (c) and that $K = K_0 + C$ is bounded from $H^{s-1}(0, 2\pi)$ into $H^s(0, 2\pi)$. It remains to show that K is also an isomorphism from $H^{s-1}(0, 2\pi)$ onto $H^s(0, 2\pi)$. From the Riesz theory (Theorem A.34), it is sufficient to prove injectivity. Let $\psi \in H^{s-1}(0, 2\pi)$ with $K\psi = 0$. From $K_0\psi = -C\psi$ and the mapping properties of C, we conclude that $K_0\psi \in H^r(0, 2\pi)$ for all $r \in \mathbb{R}$, i.e., $\psi \in H^r(0, 2\pi)$ for all $r \in \mathbb{R}$. In particular, this implies that ψ is continuous and the transformed function $\varphi(\gamma(t)) = \psi(t) / |\dot{\gamma}(t)|$ satisfies Symm's equation (3.56) for $f = 0$. Application of Theorem 3.16 yields $\varphi = 0$. $\square$

We are now in a position to apply all of the Galerkin methods of the previous section. We have seen that the convergence results require estimates of the condition numbers of K on the finite-dimensional spaces X_n. These estimates are sometimes called the *stability property* (see [112]).

Lemma 3.19
Let $r \geq s$. Then there exists $c > 0$ such that

$$\|\psi_n\|_{L^2} \leq cn \|K\psi_n\|_{L^2} \quad \text{for all } \psi_n \in X_n, \qquad (3.66a)$$

$$\|\psi_n\|_{H^r} \leq cn^{r-s} \|\psi_n\|_{H^s} \quad \text{for all } \psi_n \in X_n, \qquad (3.66b)$$

and all $n \in \mathbb{N}$.

Proof: Let $\psi_n(t) = \sum_{|j| \leq n} \alpha_j \exp(ijt) \in X_n$. Then

$$\|K_0\psi_n\|_{L^2}^2 = 2\pi \left[|\alpha_0|^2 + \sum_{\substack{|j| \leq n \\ j \neq 0}} \frac{1}{j^2} |\alpha_j|^2 \right] \geq \frac{1}{n^2} \|\psi_n\|_{L^2}^2, \qquad (3.67)$$

which proves the estimate (3.66a) for K_0. The estimate for K follows from the observations that $K = \left(K K_0^{-1} \right) K_0$ and that $K K_0^{-1}$ is an isomorphism in $L^2(0, 2\pi)$ by Theorem 3.18, part (a).

The proof of (3.66b) is very simple and is left to the reader (see Problem 3.4). $\square$

Combining these estimates with the convergence results of the previous section, we have shown the following.

Theorem 3.20
Let $\psi \in H^r(0, 2\pi)$ be the unique solution of (3.57), i.e.,

$$K\psi(t) := -\frac{1}{\pi} \int_0^{2\pi} \psi(s) \ln |\gamma(t) - \gamma(s)| \, ds = g(t) := f(\gamma(t)),$$

for $t \in [0, 2\pi]$ and some $g \in H^{r+1}(0, 2\pi)$ for $r \geq 0$. Let $g^\delta \in L^2(0, 2\pi)$ with $\|g^\delta - g\|_{L^2} \leq \delta$ and X_n defined by (3.62).

(a) Let $\psi_n^\delta \in X_n$ be the least squares solution, i.e., the solution of

$$\left(K\psi_n^\delta, K\varphi_n \right) = \left(g^\delta, K\varphi_n \right) \quad \text{for all } \varphi_n \in X_n \tag{3.68a}$$

     or

(b) Let $\psi_n^\delta = K\tilde{\psi}_n^\delta$ with $\tilde{\psi}_n^\delta \in X_n$ be the dual least squares solution, i.e., $\tilde{\psi}_n^\delta$ solves

$$\left(K\tilde{\psi}_n^\delta, K\varphi_n \right) = \left(g^\delta, \varphi_n \right) \quad \text{for all } \varphi_n \in X_n \tag{3.68b}$$

     or

(c) Let $\psi_n^\delta \in X_n$ be the Bubnov–Galerkin solution, i.e., the solution of

$$\left(K\psi_n^\delta, \varphi_n \right) = \left(g^\delta, \varphi_n \right) \quad \text{for all } \varphi_n \in X_n. \tag{3.68c}$$

Then there exists $c > 0$ with

$$\|\psi_n^\delta - \psi\|_{L^2} \leq c\left(n\delta + \frac{1}{n^r} \|\psi\|_{H^r} \right) \tag{3.69}$$

for all $n \in \mathbb{N}$.

Proof: We apply Theorems 3.10, 3.11, and 3.15 (with $V = H^{1/2}(0, 2\pi)$ and $V^* = H^{-1/2}(0, 2\pi)$) and use the estimates (see Lemma 3.19)

$$\sigma_n = \max\{\|\varphi_n\|_{L^2} : \varphi_n \in X_n, \|K\varphi_n\|_{L^2} = 1\} \leq cn, \tag{3.70a}$$
$$\rho_n = \max\{\|\varphi_n\|_{L^2} : \varphi_n \in X_n, \|\varphi_n\|_{H^{-1/2}} = 1\} \leq c\sqrt{n}, \tag{3.70b}$$

and

$$\min\{\|\psi - \varphi_n\|_{L^2} : \varphi_n \in X_n\} \leq \|\psi - P_n\psi\|_{L^2} \leq \frac{1}{n^r} \|\psi\|_{H^r}, \tag{3.70c}$$

where $P_n\psi = \sum_{|j| \leq n} \alpha_j \psi_j$ denotes the orthogonal projection of the element $\psi = \sum_{n \in \mathbb{Z}} \alpha_j \psi_j$ onto X_n. For the simple proof of the second inequality in (3.70c), we refer to Problem 3.4. □

It is interesting to note that different error estimates hold for discrete perturbations of the right-hand side. Let us denote by β_k the right-hand sides of (3.68a), (3.68b), or (3.68c), respectively, for $\varphi_k(t) = \exp(ikt)$. Assume that β is perturbed by a vector $\beta^\delta \in \mathbb{C}^{2n+1}$ with $|\beta - \beta^\delta| \leq \delta$. We have to compute r_n of (3.34), (3.40), and (3.52), respectively. Since the functions $\exp(ikt)$, $k = -n, \ldots, n$, are orthogonal, we compute r_n for (3.40) and (3.52) by

$$\max\left\{\sqrt{\sum_{j=-n}^{n} |\rho_j|^2} : \left\|\sum_{j=-n}^{n} \rho_j e^{ij\cdot}\right\|_{L^2} = 1\right\} = \frac{1}{\sqrt{2\pi}}.$$

For the least squares method, however, we have to compute

$$
\begin{aligned}
r_n^2 &= \max\left\{\sum_{j=-n}^{n} |\rho_j|^2 : \left\|\sum_{j=-n}^{n} \rho_j K_0 e^{ij\cdot}\right\|_{L^2} = 1\right\} \\
&= \max\left\{\sum_{j=-n}^{n} |\rho_j|^2 : 2\pi\left(|\rho_0|^2 + \sum_{j\neq 0} \frac{1}{j^2} |\rho_j|^2\right) = 1\right\} \\
&= \frac{n^2}{2\pi},
\end{aligned}
$$

i.e., for discrete perturbations of the right-hand side, the estimate (3.69) is asymptotically the same for the dual least squares method and the Bubnov–Galerkin method, while for the least squares method it has to be replaced by

$$\|\psi_n^\delta - \psi\|_{L^2} \leq c\left(n^2\delta + \frac{1}{n^r}\|\psi\|_{H^r}\right). \tag{3.71}$$

The error estimates (3.69) are optimal under the a priori information $\psi \in H^r(0, 2\pi)$ and $\|\psi\|_{H^r} \leq 1$. This is seen by choosing $n \sim (1/\delta)^{1/(r+1)}$, which gives the asymptotic estimate

$$\|\psi_{n(\delta)}^\delta - \psi\|_{L^2} \leq c\,\delta^{r/(r+1)}.$$

This is optimal by Problem 3.5.

From the preceding analysis, it is clear that the convergence property

$$\min\{\|\psi - \varphi_n\|_{H^s} : \varphi_n \in X_n\} \leq c\left(\frac{1}{n}\right)^{r-s}\|\psi\|_{H^r}, \quad \psi \in H^r(0, 2\pi),$$

and the stability property

$$\|\varphi_n\|_{H^r} \leq cn^{r-s}\|\varphi_n\|_{H^s}, \quad \varphi_n \in X_n,$$

for $r \geq s$ and $n \in \mathbb{N}$ are the essential tools in the proofs. For regions Ω with nonsmooth boundaries, finite element spaces for X_n are more suitable. They satisfy these conditions for a certain range of values of r and s (depending on the smoothness of the solution and the order of the finite elements). We

refer to [45, 48, 111, 112, 113, 220] for more details and boundary value
problems for more complicated partial differential equations.

We refer to Problem 3.6 and Section 3.5, where the Galerkin methods
are explicitly compared for special cases of Symm's equation.

For further literature on Symm's and related integral equations, we refer
to [9, 10, 18, 62, 181, 204, 227].

3.4   Collocation Methods

We have seen that collocation methods are subsumed under the general
theory of projection methods through the use of interpolation operators.
This requires the space Y to be a reproducing kernel Hilbert space, i.e., a
Hilbert space in which all the evaluation functionals $y \mapsto y(t)$ for $y \in Y$
and $t \in [a, b]$ are bounded.

Instead of presenting a general theory as in [162], we avoid the explicit
introduction of reproducing kernel Hilbert spaces and investigate only two
special, but important, cases in detail. First, we will study the minimum
norm collocation method. It will turn out that this is a special case of a
least squares method and can be treated by the methods of the previous
section. In Subsection 3.4.2, we will investigate a second collocation method
for the important example of Symm's equation. We will derive a complete
and satisfactory error analysis for two choices of ansatz functions.

First, we formulate the general collocation method again and derive an
error estimate in the presence of discrete perturbations of the right-hand
side.

Let X be a Hilbert space over the field $\mathbb{K}$, $X_n \subset X$ be finite-dimensional
subspaces with $\dim X_n = n$, and $a \leq t_1 < \cdots < t_n \leq b$ be the collocation
points. Let $K : X \to C[a, b]$ be bounded and one-to-one. Let $Kx = y$, and
assume that the collocation equations

$$K x_n(t_i) = y(t_i), \quad i = 1, \ldots, n, \tag{3.72}$$

are uniquely solvable in X_n for every right-hand side. Choosing a basis
$\{\hat{x}_j : j = 1, \ldots, n\}$ of X_n, we rewrite this as a system $A\alpha = \beta$, where
$x_n = \sum_{j=1}^{n} \alpha_j \hat{x}_j$ and

$$A_{ij} = K\hat{x}_j(t_i), \quad \beta_i = y(t_i). \tag{3.73}$$

The following main theorem is the analog of Theorem 3.9 for collocation
methods. We restrict ourselves to the important case of discrete perturbations of the right-hand side. Continuous perturbations could also be han-
dled but are not of particular interest since point evaluation is no longer

possible when the right-hand side is perturbed in the L^2-sense. This would require stronger norms in the range space and leads to the concept of reproducing kernel Hilbert spaces (see [130]).

Theorem 3.21
Let $\{t_1^{(n)}, \ldots, t_n^{(n)}\} \subset [a, b]$, $n \in \mathbb{N}$, be a sequence of collocation points. Assume that $\bigcup_{n \in \mathbb{N}} X_n$ is dense in X and that the collocation method converges. Let $x_n^\delta = \sum_{j=1}^n \alpha_j^\delta \hat{x}_j \in X_n$, where α^δ solves $A\alpha^\delta = \beta^\delta$. Here, $\beta^\delta \in \mathbb{K}^n$ satisfies $|\beta - \beta^\delta| \leq \delta$ where $|\cdot|$ again denotes the Euclidean norm in $\mathbb{K}$. Then the following error estimate holds:

$$\left\| x_n^\delta - x \right\|_{L^2} \leq c \left(\frac{a_n}{\lambda_n} \delta + \inf\{\|x - z_n\| : z_n \in X_n\} \right), \tag{3.74}$$

where

$$a_n = \max\left\{ \left\| \sum_{j=1}^n \rho_j \hat{x}_j \right\| : \sum_{j=1}^n |\rho_j|^2 = 1 \right\} \tag{3.75}$$

and λ_n denotes the smallest singular value of A.

Proof: Again we write $\left\| x_n^\delta - x \right\| \leq \left\| x_n^\delta - x_n \right\| + \left\| x_n - x \right\|$, where $x_n = R_n y$ solves the collocation equation for β instead of β^δ. The second term is estimated by Theorem 3.7. We estimate the first term by

$$\left\| x_n^\delta - x_n \right\| \leq a_n \left| \alpha^\delta - \alpha \right| = a_n \left| A^{-1}(\beta^\delta - \beta) \right|$$
$$\leq a_n \left| A^{-1} \right|_2 \left| \beta^\delta - \beta \right| \leq \frac{a_n}{\lambda_n} \delta. \qquad \square$$

Again we remark that $a_n = 1$ if $\{\hat{x}_j : j = 1, \ldots, n\}$ forms an orthonormal system in X.

3.4.1 Minimum Norm Collocation

Again, let $K : X \to C[a, b]$ be a linear, bounded, and injective operator from the Hilbert space X into the space $C[a, b]$ of continuous functions on $[a, b]$. We assume that there exists a unique solution of $Kx = y$. Let $a \leq t_1 < \cdots < t_n \leq b$ be the set of collocation points. Solving the equations (3.72) in X is certainly not enough to specify the solution x_n uniquely. An obvious choice is to determine $x_n \in L^2(a, b)$ from the set of solutions of (3.72) that has a minimal L^2-norm among all solutions.

Definition 3.22
$x_n \in X$ is called the moment solution of (3.72) with respect to the collocation points $a \leq t_1 < \cdots < t_n \leq b$ if x_n satisfies (3.72) and

$$\|x_n\|_{L^2} = \min\{\|z_n\|_{L^2} : z_n \in L^2(a, b) \text{ satisfies } (3.72)\}.$$

We can interpret this moment solution as a least squares solution. Since $z \mapsto Kz(t_i)$ is bounded from X into $\mathbb{K}$, the Riesz–Fischer Theorem A.22 yields the existence of $k_i \in X$ with $Kz(t_i) = (k_i, z)$ for all $z \in X$ and $i = 1, \ldots, n$. If, for example, K is the integral operator

$$Kz(t) = \int_a^b k(t, s) \, z(s) \, ds, \quad t \in [a, b], \ z \in L^2(a, b),$$

with real-valued kernel k then $k_i \in L^2(a, b)$ is explicitly given by $k_i(s) = k(t_i, s)$. We rewrite the moment equation (3.72) in the form

$$(k_i, x_n) = y(t_i) = (k_i, x), \quad i = 1, \ldots, n.$$

The minimum norm solution x_n of the set of equations is characterized by the projection theorem (see Theorem A.13 of Appendix A) and is given by the solution of (3.72) in the space $X_n := \operatorname{span}\{k_j : j = 1, \ldots, n\}$.

Now we define the Hilbert space Y by $Y := K(X)$ with inner product

$$(y, z)_Y := (K^{-1}y, K^{-1}z) \quad \text{for } y, z \in K(X).$$

We omit the simple proof of the following lemma.

Lemma 3.23
Y is a Hilbert space that is continuously embedded in $C[a, b]$. Furthermore, K is an isomorphism from X onto Y.

Now we can rewrite (3.72) in the form.

$$(Kk_i, Kx_n)_Y = (Kk_i, y)_Y, \quad i = 1, \ldots, n.$$

Comparing this equation with (3.30a), we observe that (3.72) is the Galerkin equation for the least squares method with respect to X_n. Thus we have shown that the moment solution can be interpreted as the least squares solution for the operator $K : X \to Y$. Application of Theorem 3.10 yields the following theorem.

Theorem 3.24
Let K be one-to-one and $\{k_j : j = 1, \ldots, n\}$ be linearly independent where $k_j \in X$ are such that $Kz(t_j) = (k_j, z)$ for all $z \in X$, $j = 1, \ldots, n$. Then there exists one and only one moment solution x_n of (3.72). x_n is given by

$$x_n = \sum_{j=1}^{n} \alpha_j \, k_j, \tag{3.76}$$

where $\alpha \in \mathbb{K}^n$ solves the linear system $A\alpha = \beta$ with

$$A_{ij} = Kk_j(t_i) = (k_i, k_j) \quad and \quad \beta_i = y(t_i). \tag{3.77}$$

Let $\{t_1^{(n)}, \ldots, t_n^{(n)}\} \subset [a, b]$, $n \in \mathbb{N}$, be a sequence of collocation points such that $\bigcup_{n \in \mathbb{N}} X_n$ is dense in X where

$$X_n := \text{span}\{k_j^{(n)} : j = 1, \ldots, n\}.$$

Then the moment method converges, i.e., the moment solution $x_n \in X_n$ of (3.72) converges to the solution $x \in L^2(a, b)$ of (3.71) in X. If $x_n^\delta = \sum_{j=1}^n \alpha_j^\delta k_j^{(n)}$, where $\alpha^\delta \in \mathbb{K}^n$ solves $A\alpha^\delta = \beta^\delta$ with $|\beta - \beta^\delta| \leq \delta$, then the following error estimate holds:

$$\|x - x_n^\delta\| \leq \frac{a_n}{\lambda_n} \delta + c \min\{\|x - z_n\|_{L^2} : z_n \in X_n\}, \tag{3.78}$$

where

$$a_n = \max\left\{\left\|\sum_{j=1}^n \rho_j k_j^{(n)}\right\|_{L^2} : \sum_{j=1}^n |\rho_j|^2 = 1\right\} \tag{3.79}$$

and where λ_n denotes the smallest singular value of A.

Proof: The definition of $\|\cdot\|_Y$ implies that $\sigma_n = 1$, where σ_n is given by (3.31). Assumption (3.32) for the convergence of the least squares method is obviously satisfied since

$$\min_{z_n \in X_n} \{\|x - z_n\| + \sigma_n \|K(x - z_n)\|_Y\} \leq \|x\| + \sigma_n \|x\| = 2\|x\|.$$

Application of Theorem 3.10 yields the assertion. $\qquad\square$

As an example, we again consider numerical differentiation.

Example 3.25
Let K be defined by

$$(Kx)(t) = \int_0^t x(s)\, ds = \int_0^1 k(t, s)\, x(s)\, ds, \quad t \in [0, 1],$$

with $k(t, s) = \begin{cases} 1, & s \leq t, \\ 0, & s > t. \end{cases}$

We choose equidistant nodes, i.e., $t_j = \frac{j}{n}$ for $j = 0, \ldots, n$. The moment method is to minimize $\|x\|_{L^2}^2$ under the restrictions that

$$\int_0^{t_j} x(s)\, ds = y(t_j), \quad j = 1, \ldots, n. \tag{3.80}$$

The solution x_n is piecewise constant since it is a linear combination of the piecewise constant functions $k(t_j, \cdot)$. Therefore, the finite-dimensional space X_n is given by

$$X_n = \{z_n \in L^2(0,1) : z_n|_{(t_{j-1}, t_j)} \text{ constant, } j = 1, \ldots, n\}. \qquad (3.81)$$

As basis functions $\hat{x}_j$ of X_n we choose $\hat{x}_j(s) = k(t_j, s)$.

Then $x_n = \sum_{j=1}^{n} \alpha_j \, k(t_j, \cdot)$ is the moment solution, where α solves $A\alpha = \beta$ with $\beta_i = y(t_i)$ and

$$A_{ij} = \int_0^1 k(t_i, s) \, k(t_j, s) \, ds = \frac{1}{n} \min\{i, j\}.$$

It is not difficult to see that the moment solution is just the one-sided difference quotient

$$x_n(t_1) = \frac{1}{h} y(t_1), \quad x_n(t_j) = \frac{1}{h} \left[y(t_j) - y(t_{j-1}) \right], \quad j = 2, \ldots, n,$$

for $h = 1/n$.

We have to check the assumptions of Theorem 3.24. First, K is one-to-one and $\{k(t_j, \cdot) : j = 1 \ldots, n\}$ are linearly independent. The union $\bigcup_{n \in \mathbb{N}} X_n$ is dense in $L^2(0,1)$ (see Problem 3.7). We have to estimate a_n from (3.79), the smallest eigenvalue λ_n of A, and $\min\{\|x - z_n\|_{L^2} : z_n \in X_n\}$.

Let $\rho \in \mathbb{R}^n$ with $\sum_{j=1}^{n} \rho_j^2 = 1$. Using the Cauchy–Schwarz inequality, we estimate

$$\int_0^1 \left| \sum_{j=1}^{n} \rho_j k(t_j, s) \right|^2 ds \leq \int_0^1 \sum_{j=1}^{n} k(t_j, s)^2 ds = \sum_{j=1}^{n} t_j = \frac{n+1}{2}.$$

Thus $a_n \leq \sqrt{(n+1)/2}$.

It is straightforward to check that the inverse of A is given by the tridiagonal matrix

$$A^{-1} = n \begin{bmatrix} 2 & -1 & & & \\ -1 & 2 & -1 & & \\ & \ddots & \ddots & \ddots & \\ & & -1 & 2 & -1 \\ & & & -1 & 1 \end{bmatrix}.$$

We estimate the largest eigenvalue $\mu_{\max}$ of A^{-1} by the maximum absolute row sum $\mu_{\max} \leq 4\,n$. This is asymptotically sharp since we can give a lower estimate of $\mu_{\max}$ by the trace formula

$$n\,\mu_{\max} \geq \operatorname{trace}(A^{-1}) = \sum_{j=1}^{n} (A^{-1})_{jj} = (2n-1)\,n,$$

i.e., we have an estimate of λ_n of the form

$$\frac{1}{4n} \leq \lambda_n \leq \frac{1}{2n-1}.$$

In Problem 3.7, it is shown that

$$\min\{\|x - z_n\|_{L^2} : z_n \in X_n\} \leq \frac{1}{n}\,\|x'\|_{L^2}.$$

Thus we have proven the following theorem.

Theorem 3.26
The moment method for (3.80) converges. The following error estimate holds:

$$\|x - x_n^{\delta}\|_{L^2} \leq \sqrt{\frac{n+1}{2}}\,\delta + \frac{c}{n}\,\|x'\|_{L^2}$$

if $x \in H^1(0,1)$. Here, δ is the discrete error on the right-hand side, i.e., $\sum_{j=1}^{n}|\beta_j^{\delta} - y(t_j)|^2 \leq \delta^2$ and $x_n^{\delta} = \sum_{j=1}^{n}\alpha_j^{\delta}\hat{x}_j$, where $\alpha^{\delta} \in \mathbb{R}^n$ solves $A\alpha^{\delta} = \beta^{\delta}$.

The choice $X_n = S_1(t_1,\ldots,t_n)$ of linear splines leads to the two-sided difference quotient (see Problem 3.9). We refer to [64, 161, 162] for further reading on moment collocation.

3.4.2 Collocation of Symm's Equation

We will now study the numerical treatment of Symm's equation (3.57), i.e.,

$$K\psi(t) := -\frac{1}{\pi}\int_{0}^{2\pi} \psi(s)\ln|\gamma(t) - \gamma(s)|\,ds = g(t) \qquad (3.82)$$

for $0 \leq t \leq 2\pi$ by collocation methods. The integral operator K from (3.82) is well defined and bounded from $L^2(0,2\pi)$ into $H^1(0,2\pi)$. We assume throughout this subsection that K is one-to-one (see Theorem 3.16). Then we have seen in Theorem 3.18 that equation (3.82) is uniquely solvable in $L^2(0,2\pi)$ for every $g \in H^1(0,2\pi)$, i.e., K is an isomorphism. We define equidistant collocation points by

$$t_k := k\,\frac{\pi}{n} \quad \text{for } k = 0,\ldots,2n-1.$$

There are several choices for the space $X_n \subset L^2(0, 2\pi)$ of basis functions. Before we study particular cases, let $X_n = \text{span}\{\hat{x}_j : j \in J\} \subset L^2(0, 2\pi)$ be arbitrary. $J \subset \mathbb{Z}$ denotes a set of indices with $2n$ elements. We assume that $\hat{x}_j$, $j \in J$, form an orthonormal system in $L^2(0, 2\pi)$.

The collocation equations (3.72) take the form

$$-\frac{1}{\pi} \int_0^{2\pi} \psi_n(s) \ln|\gamma(t_k) - \gamma(s)| \, ds = g(t_k), \quad k = 0, \ldots, 2n-1, \quad (3.83)$$

with $\psi_n \in X_n$. Let $Q_n : H^1(0, 2\pi) \to Y_n$ be the trigonometric interpolation operator into the $2n$-dimensional space

$$Y_n := \left\{ \sum_{m=-n}^{n-1} a_m e^{imt} : a_m \in \mathbb{C} \right\}. \quad (3.84)$$

We recall some approximation properties of the interpolation operator Q_n: $H^1(0, 2\pi) \to Y_n$. First, it is easily checked that Q_n is given by

$$Q_n \psi = \sum_{k=0}^{2n-1} \psi(t_k) \hat{y}_k$$

with Lagrange interpolation basis functions

$$\hat{y}_k(t) = \frac{1}{2n} \sum_{m=-n}^{n-1} e^{im(t-t_k)}, \quad k = 0, \ldots, 2n-1. \quad (3.85)$$

From Lemma A.43 of Appendix A we have the estimates

$$\|\psi - Q_n \psi\|_{L^2} \leq \frac{c}{n} \|\psi\|_{H^1} \quad \text{for all } \psi \in H^1(0, 2\pi), \quad (3.86a)$$

$$\|Q_n \psi\|_{H^1} \leq c \|\psi\|_{H^1} \quad \text{for all } \psi \in H^1(0, 2\pi). \quad (3.86b)$$

Now we can reformulate the collocation equations (3.83) as

$$Q_n K \psi_n = Q_n g \quad \text{with } \psi_n \in X_n. \quad (3.87)$$

We will use the perturbation result of Theorem 3.8 again and split K into the form $K = K_0 + C$ with

$$K_0 \psi(t) := -\frac{1}{2\pi} \int_0^{2\pi} \psi(s) \left[\ln \left(4 \sin^2 \frac{t-s}{2} \right) - 1 \right] ds. \quad (3.88)$$

Now we specify the spaces X_n. As a first example, we choose the orthonormal functions

$$\hat{x}_j(t) = \frac{1}{\sqrt{2\pi}} e^{ijt} \quad \text{for } j = -n, \ldots, n-1. \quad (3.89)$$

We prove the following convergence result.

Theorem 3.27

Let $\hat{x}_j$, $j = -n, \ldots, n-1$, be given by (3.89). The collocation method is convergent, i.e., the solution $\psi_n \in X_n$ of (3.83) converges to the solution $\psi \in L^2(0, 2\pi)$ of (3.82) in $L^2(0, 2\pi)$.

Let the right-hand side of (3.83) be replaced by $\beta^\delta \in \mathbb{C}^{2n}$ with

$$\sum_{k=0}^{2n-1} \left| \beta_k^\delta - g(t_k) \right|^2 \leq \delta^2.$$

Let $\alpha^\delta \in \mathbb{C}^{2n}$ be the solution of $A\alpha^\delta = \beta^\delta$, where $A_{kj} = K\hat{x}_j(t_k)$. Then the following error estimate holds:

$$\left\| \psi_n^\delta - \psi \right\|_{L^2} \leq c\left[\sqrt{n}\,\delta + \min\{ \|\psi - \varphi_n\|_{L^2} : \varphi_n \in X_n \}\right], \qquad (3.90)$$

where

$$\psi_n^\delta(t) = \frac{1}{\sqrt{2\pi}} \sum_{j=-n}^{n-1} \alpha_j^\delta\, e^{ijt}.$$

If $\psi \in H^r(0, 2\pi)$ for some $r > 0$, then

$$\left\| \psi_n^\delta - \psi \right\|_{L^2} \leq c\left[\sqrt{n}\,\delta + \frac{1}{n^r} \|\psi\|_{H^r} \right]. \qquad (3.91)$$

Proof: By the perturbation Theorem 3.8 it is sufficient to prove the result for K_0 instead of K. By (3.65a) and (3.65b), the operator K_0 maps X_n into $Y_n = X_n$. Therefore, the collocation equation (3.87) for K_0 reduces to

$$K_0\psi_n = Q_n g.$$

We want to apply Theorem 3.7 and have to estimate $R_n K_0$ where in this case $R_n = \left(K_0|_{X_n} \right)^{-1} Q_n$. Since $K_0 : L^2(0, 2\pi) \to H^1(0, 2\pi)$ is invertible, we conclude that

$$\|R_n g\|_{L^2} = \|\psi_n\|_{L^2} \leq c_1 \|K_0\psi_n\|_{H^1} = c_1 \|Q_n g\|_{H^1} \leq c_2 \|g\|_{H^1}$$

for all $g \in H^1(0, 2\pi)$, and thus

$$\|R_n K\psi\|_{L^2} \leq c_2 \|K\psi\|_{H^1} \leq c_3 \|\psi\|_{L^2}$$

for all $\psi \in L^2(0, 2\pi)$. Application of Theorem 3.7 yields convergence.

To prove the error estimate (3.90), we want to apply Theorem 3.21 and hence have to estimate the singular values of the matrix B defined by

$$B_{kj} = K_0\hat{x}_j(t_k), \quad k, j = -n, \ldots, n-1,$$

with $\hat{x}_j$ from (3.89). From (3.65a) and (3.65b), we observe that

$$B_{kj} = \frac{1}{\sqrt{2\pi}} \frac{1}{|j|} e^{ijk\frac{\pi}{n}}, \quad k, j = -n, \ldots, n-1,$$

where $1/|j|$ has to be replaced by 1 if $j = 0$. Since the singular values of B are the square roots of the eigenvalues of B^*B, we compute

$$(B^*B)_{\ell j} = \sum_{k=-n}^{n-1} \overline{B_{k\ell}} \, B_{kj} = \frac{1}{2\pi} \frac{1}{|\ell||j|} \sum_{k=-n}^{n-1} e^{ik(j-\ell)\frac{\pi}{n}} = \frac{n}{\pi} \frac{1}{\ell^2} \delta_{\ell j},$$

where again $1/\ell^2$ has to be replaced by 1 for $\ell = 0$. From this, we see that the singular values of B are given by $\sqrt{n/(\pi\ell^2)}$ for $\ell = 1, \ldots, n$. The smallest singular value is $1/\sqrt{n\pi}$. Estimate (3.74) of Theorem 3.21 yields the assertion. (3.91) follows from Theorem A.44.   □

Comparing the estimates (3.91) with the corresponding error estimate (3.69) for the Galerkin methods, it seems that the estimate for the collocation method is better since the error δ is only multiplied by $\sqrt{n}$ instead of n. Let us now compare the errors of the continuous perturbation $\|y - y^\delta\|_{L^2}$ with the discrete perturbation for both methods. To do this, we have to "extend" the discrete vector β^δ to a function $y_n^\delta \in X_n$. For the collocation method, we have to use the interpolation operator Q_n and define $y_n^\delta \in X_n$ by $y_n^\delta = \sum_{j=1}^{2n-1} \beta_j^\delta \hat{y}_j$, where $\hat{y}_j$ are the Lagrange basis functions (3.85). Then $y_n^\delta(t_k) = \beta_k^\delta$, and we estimate

$$\|y_n^\delta - y\|_{L^2} \leq \|y_n^\delta - Q_n y\|_{L^2} + \|Q_n y - y\|_{L^2}.$$

Writing

$$y_n^\delta(t) - Q_n y(t) = \sum_{j=-n}^{n-1} \rho_j e^{ijt},$$

a simple computation shows that

$$\sum_{k=0}^{n-1} |\beta_k^\delta - y(t_k)|^2 = \sum_{k=0}^{n-1} |\beta_k^\delta - Q_n y(t_k)|^2 = \sum_{k=0}^{n-1} \left| \sum_{j=-n}^{n-1} \rho_j e^{ikj\frac{\pi}{n}} \right|^2$$

$$= 2n \sum_{j=-n}^{n-1} |\rho_j|^2 = \frac{n}{\pi} \|y_n^\delta - Q_n y\|_{L^2}^2. \quad (3.92)$$

Therefore, for the collocation method we have to compare the continuous error δ with the discrete error $\delta \sqrt{n/\pi}$. This gives an extra factor of $\sqrt{n}$ in the first terms of (3.90) and (3.91).

For Galerkin methods, however, we define $y_n^\delta(t) = \frac{1}{2\pi} \sum_{j=-n}^{n} \beta_j^\delta \exp(ijt)$. Then $\left(y_n^\delta, e^{ij\cdot}\right)_{L^2} = \beta_j^\delta$. Let P_n be the orthogonal projection onto X_n. In

$$\left\| y_n^\delta - y \right\|_{L^2} \leq \left\| y_n^\delta - P_n y \right\|_{L^2} + \left\| P_n y - y \right\|_{L^2},$$

we estimate the first term as

$$\left\| y_n^\delta - P_n y \right\|_{L^2}^2 = \frac{1}{2\pi} \sum_{j=-n}^{n} \left| \beta_j^\delta - \left(y, e^{ij\cdot}\right) \right|^2.$$

In this case, the continuous and discrete errors are of the same order.

Choosing trigonometric polynomials as basis functions is particularly suitable for smooth boundary data. If $\partial\Omega$ or the right-hand side f of the boundary value problem (3.54) are not smooth, then spaces of piecewise constant functions are more appropriate. We now study the case where the basis functions $\hat{x}_j \in L^2(0, 2\pi)$ are defined by

$$\hat{x}_0(t) = \begin{cases} \sqrt{\frac{n}{\pi}}, & \text{if } t < \frac{\pi}{2n} \text{ or } t > 2\pi - \frac{\pi}{2n}, \\ 0, & \text{if } \frac{\pi}{2n} < t < 2\pi - \frac{\pi}{2n}, \end{cases} \tag{3.93a}$$

$$\hat{x}_j(t) = \begin{cases} \sqrt{\frac{n}{\pi}}, & \text{if } |t - t_j| < \frac{\pi}{2n}, \\ 0, & \text{if } |t - t_j| > \frac{\pi}{2n}, \end{cases} \tag{3.93b}$$

for $j = 1, \ldots, 2n - 1$. Then $\hat{x}_j$, $j = 0, \ldots, 2n - 1$, are also orthonormal in $L^2(0, 2\pi)$. In the following lemma, we collect some approximation properties of the corresponding spaces X_n.

Lemma 3.28
Let $X_n = \operatorname{span}\{\hat{x}_j : j = 0, \ldots, 2n - 1\}$, where $\hat{x}_j$ are defined by (3.93a) and (3.93b). Let $P_n : L^2(0, 2\pi) \to X_n$ be the orthogonal projection operator. Then $\bigcup_{n \in \mathbb{N}} X_n$ is dense in $L^2(0, 2\pi)$ and there exists $c > 0$ with

$$\left\| \psi - P_n\psi \right\|_{L^2} \leq \frac{c}{n} \left\| \psi \right\|_{H^1} \quad \textit{for all } \psi \in H^1(0, 2\pi), \tag{3.94a}$$

$$\left\| K(\psi - P_n\psi) \right\|_{L^2} \leq \frac{c}{n} \left\| \psi \right\|_{L^2} \quad \textit{for all } \psi \in L^2(0, 2\pi). \tag{3.94b}$$

Proof: Estimate (3.94a) is left as an exercise. To prove estimate (3.94b), we use (implicitly) a duality argument:

$$\left\| K(\psi - P_n\psi) \right\|_{L^2} = \sup_{\varphi \neq 0} \frac{\left(K(\psi - P_n\psi), \varphi\right)_{L^2}}{\left\| \varphi \right\|_{L^2}}$$

$$= \sup_{\varphi \neq 0} \frac{\left(\psi - P_n\psi, K\varphi\right)_{L^2}}{\left\| \varphi \right\|_{L^2}}$$

$$
= \sup_{\varphi \neq 0} \frac{\left(\psi, (I - P_n) K \varphi \right)_{L^2}}{\|\varphi\|_{L^2}}
$$

$$
\leq \|\psi\|_{L^2} \sup_{\varphi \neq 0} \frac{\|(I - P_n) K \varphi\|_{L^2}}{\|\varphi\|_{L^2}}
$$

$$
\leq \frac{\tilde{c}}{n} \|\psi\|_{L^2} \sup_{\varphi \neq 0} \frac{\|K\varphi\|_{H^1}}{\|\varphi\|_{L^2}} \leq \frac{c}{n} \|\psi\|_{L^2} . \qquad \square
$$

Before we prove a convergence theorem, we compute the singular values of the matrix B defined by

$$
B_{kj} = K_0 \hat{x}_j(t_k) = -\frac{1}{2\pi} \int_0^{2\pi} \hat{x}_j(s) \left[\ln \left(4 \sin^2 \frac{t_k - s}{2} \right) - 1 \right] ds. \quad (3.95)
$$

Lemma 3.29

B is symmetric and positive definite. The singular values of B coincide with the eigenvalues and are given by

$$
\mu_0 = \sqrt{\frac{n}{\pi}} \quad \text{and} \quad \mu_m = \sqrt{\frac{n}{\pi}} \frac{\sin \frac{m\pi}{2n}}{2n\pi} \sum_{j \in \mathbb{Z}} \frac{1}{\left(\frac{m}{2n} + j \right)^2} \quad (3.96a)
$$

for $m = 1, \ldots, 2n - 1$. Furthermore, there exists $c > 0$ with

$$
\frac{1}{\sqrt{\pi n}} \leq \mu_m \leq c\sqrt{n} \quad \text{for all } m = 0, \ldots, 2n - 1. \quad (3.96b)
$$

We observe that the condition number of B, i.e., the ratio between the largest and smallest singular values, is again bounded by n.

Proof: We write

$$
B_{kj} = -\frac{1}{2\pi} \sqrt{\frac{n}{\pi}} \int_{t_j - \frac{\pi}{2n}}^{t_j + \frac{\pi}{2n}} \left[\ln \left(4 \sin^2 \frac{s - t_k}{2} \right) - 1 \right] ds = b_{j-k}
$$

with

$$
b_\ell = -\frac{1}{2\pi} \sqrt{\frac{n}{\pi}} \int_{t_\ell - \frac{\pi}{2n}}^{t_\ell + \frac{\pi}{2n}} \left[\ln \left(4 \sin^2 \frac{s}{2} \right) - 1 \right] ds,
$$

where we extended the definition of t_ℓ to all $\ell \in \mathbb{Z}$. Therefore, B is circulant and symmetric. The eigenvectors $x^{(m)}$ and eigenvalues μ_m of B are given by

$$
x^{(m)} = \left(e^{imk\frac{\pi}{n}} \right)_{k=0}^{2n-1} \quad \text{and} \quad \mu_m = \sum_{k=0}^{2n-1} b_k \, e^{imk\frac{\pi}{n}},
$$

respectively, for $m = 0, \ldots, 2n - 1$, as is easily checked. We write μ_m in the form

$$\mu_m = -\frac{1}{2\pi} \int_0^{2\pi} \psi_m(s) \left[\ln\left(4 \sin^2 \frac{s}{2}\right) - 1 \right] ds = K_0 \psi_m(0)$$

with

$$\psi_m(s) = \sqrt{\frac{n}{\pi}} \, e^{imk\frac{\pi}{n}} \quad \text{for } |s - t_k| \leq \frac{\pi}{2n}, \ k \in \mathbb{Z}.$$

Let $\psi_m(t) = \sum_{k \in \mathbb{Z}} \rho_{m,k} \exp(ikt)$. Then by (3.65a) and (3.65b) we have

$$\mu_m = \rho_{m,0} + \sum_{k \neq 0} \frac{\rho_{m,k}}{|k|}.$$

Therefore, we have to compute the Fourier coefficients $\rho_{m,k}$ of ψ_m. They are given by

$$\rho_{m,k} = \frac{1}{2\pi} \int_0^{2\pi} \psi_m(s) e^{-iks} ds = \sqrt{\frac{n}{\pi}} \frac{1}{2\pi} \sum_{j=0}^{2n-1} e^{imj\frac{\pi}{n}} \int_{t_j - \frac{\pi}{2n}}^{t_j + \frac{\pi}{2n}} e^{-iks} ds.$$

For $k = 0$, this reduces to

$$\rho_{m,0} = \sqrt{\frac{n}{\pi}} \frac{1}{2n} \sum_{j=0}^{2n-1} e^{imj\frac{\pi}{n}} = \begin{cases} \sqrt{n/\pi} & \text{if } m = 0, \\ 0 & \text{if } m = 1, \ldots, 2n - 1, \end{cases}$$

and for $k \neq 0$ to

$$\rho_{m,k} = \sqrt{\frac{n}{\pi}} \frac{\sin \frac{\pi k}{2n}}{\pi k} \sum_{j=0}^{2n-1} e^{i(m-k)j\frac{\pi}{n}} = \begin{cases} \sqrt{\frac{n}{\pi}} \frac{2n}{\pi k} \sin \frac{\pi k}{2n}, & \text{if } m - k \in 2n\mathbb{Z}, \\ 0, & \text{if } m - k \notin 2n\mathbb{Z}. \end{cases}$$

Thus we have $\mu_0 = \sqrt{n/\pi}$ and

$$\mu_m = \sqrt{\frac{n}{\pi}} \frac{2n}{\pi} \sum_{k - m \in 2n\mathbb{Z}} \frac{\left|\sin \frac{\pi k}{2n}\right|}{k^2} = \sqrt{\frac{n}{\pi}} \frac{\sin \frac{\pi m}{2n}}{2n\pi} \sum_{j \in \mathbb{Z}} \frac{1}{\left(\frac{m}{2n} + j\right)^2}$$

for $m = 1, \ldots, 2n - 1$. This proves (3.96a). Since all eigenvalues are positive, the matrix B is positive definite and the eigenvalues coincide with the singular values. Taking only the first two terms in the series yields

$$\mu_m \geq \sqrt{\frac{n}{\pi}} \frac{\sin \frac{\pi m}{2n}}{2n\pi} \left(\frac{1}{\left(\frac{m}{2n}\right)^2} + \frac{1}{\left(1 - \frac{m}{2n}\right)^2} \right)$$

$$= \sqrt{\frac{n}{\pi}} \frac{1}{2n\pi} \left(\frac{\sin \pi x}{x^2} + \frac{\sin \pi(1 - x)}{(1 - x)^2} \right)$$

with $x = m/(2n) \in (0, 1)$. From the elementary estimate

$$\frac{\sin \pi x}{x^2} + \frac{\sin \pi (1 - x)}{(1 - x)^2} \geq 8, \quad x \in (0, 1),$$

we conclude that

$$\mu_m \geq \frac{4}{\pi} \frac{1}{\sqrt{\pi n}} \geq \frac{1}{\sqrt{\pi n}}$$

for $m = 1, \ldots, 2n - 1$. The upper estimate of (3.96b) is proven analogously.
□

Now we can prove the following convergence result.

Theorem 3.30
Let $\hat{x}_j$, $j = 0, \ldots, 2n - 1$, be defined by (3.93a) and (3.93b). The collocation method is convergent, i.e., the solution $\psi_n \in X_n$ of (3.83) converges to the solution $\psi \in L^2(0, 2\pi)$ of (3.82) in $L^2(0, 2\pi)$.
Let the right-hand side be replaced by $\beta^\delta \in \mathbb{C}^{2n}$ with

$$\sum_{j=0}^{2n-1} \left| \beta_j^\delta - g(t_j) \right|^2 \leq \delta^2.$$

Let $\alpha^\delta \in \mathbb{C}^{2n}$ be the solution of $A\alpha^\delta = \beta^\delta$, where $A_{kj} = K\hat{x}_j(t_k)$. Then the following error estimate holds:

$$\left\| \psi_n^\delta - \psi \right\|_{L^2} \leq c \left[\sqrt{n} \delta + \min\{ \|\psi - \varphi_n\|_{L^2} : \varphi_n \in X_n \} \right], \quad (3.97)$$

where $\psi_n^\delta = \sum_{j=0}^{2n-1} \alpha_j^\delta \hat{x}_j$. If $\psi \in H^1(0, 2\pi)$, then

$$\left\| \psi_n^\delta - \psi \right\|_{L^2} \leq c \left[\sqrt{n} \delta + \frac{1}{n} \|\psi\|_{H^1} \right]. \quad (3.98)$$

Proof: By the perturbation theorem (Theorem 3.8), it is sufficient to prove the result for K_0 instead of K. Again set

$$R_n = \left[Q_n K_0|_{X_n} \right]^{-1} Q_n : H^1(0, 2\pi) \longrightarrow X_n \subset L^2(0, 2\pi),$$

let $\psi \in H^1(0, 2\pi)$, and set $\psi_n = R_n \psi = \sum_{j=0}^{2n-1} \alpha_j \hat{x}_j$. Then $\alpha \in \mathbb{C}^{2n}$ solves $B\alpha = \beta$ with $\beta_k = \psi(t_k)$, and thus by (3.96b)

$$\|\psi_n\|_{L^2} = |\alpha| \leq \left| B^{-1} \right|_2 |\beta| \leq \sqrt{\pi n} \left[\sum_{k=0}^{2n-1} |\psi(t_k)|^2 \right]^{1/2}$$

where $|\cdot|$ again denotes the Euclidean norm in $\mathbb{C}^n$. Using this estimate and (3.92) for $\beta_k^\delta = 0$, we conclude that

$$\|R_n \psi\|_{L^2} = \|\psi_n\|_{L^2} \leq n \|Q_n \psi\|_{L^2} \quad (3.99)$$

for all $\psi \in H^1(0, 2\pi)$. Thus

$$\|R_n K_0 \psi\|_{L^2} \leq n \|Q_n K_0 \psi\|_{L^2}$$

for all $\psi \in L^2(0, 2\pi)$. Now we estimate $\|R_n K_0 \psi\|_{L^2}$ by the L^2-norm of ψ itself.

Let $\tilde{\psi}_n = P_n \psi \in X_n$ be the orthogonal projection of $\psi \in L^2(0, 2\pi)$ in X_n. Then $R_n K_0 \tilde{\psi}_n = \tilde{\psi}_n$ and $\|\tilde{\psi}_n\|_{L^2} \leq \|\psi\|_{L^2}$, and thus

$$\begin{aligned}
\|R_n K_0 \psi - \tilde{\psi}_n\|_{L^2} &= \|R_n K_0 (\psi - \tilde{\psi}_n)\|_{L^2} \leq n \|Q_n K_0 (\psi - \tilde{\psi}_n)\|_{L^2} \\
&\leq n \|Q_n K_0 \psi - K_0 \psi\|_{L^2} + n \|K_0 \psi - K_0 \tilde{\psi}_n\|_{L^2} \\
&\quad + n \|K_0 \tilde{\psi}_n - Q_n K_0 \tilde{\psi}_n)\|_{L^2}.
\end{aligned}$$

Now we use the error estimates (3.86a), (3.94a), and (3.94b) of Lemma 3.28. This yields

$$\begin{aligned}
\|R_n K_0 \psi - \tilde{\psi}_n\|_{L^2} &\leq c_1 \left[\|K_0 \psi\|_{H^1} + \|\psi\|_{L^2} + \|K_0 \tilde{\psi}_n\|_{H^1} \right] \\
&\leq c_2 \left[\|\psi\|_{L^2} + \|\tilde{\psi}_n\|_{L^2} \right] \leq c_3 \|\psi\|_{L^2},
\end{aligned}$$

i.e., $\|R_n K_0 \psi\|_{L^2} \leq c_4 \|\psi\|_{L^2}$ for all $\psi \in L^2(0, 2\pi)$. Therefore, the assumptions of Theorem 3.7 are satisfied. The application of Theorem 3.21 yields the error estimate (3.98).   □

Among the extensive literature on collocation methods for Symm's integral equation and related equations we mention only the work of [7, 46, 47, 110, 115, 196, 197]. Symm's equation has also been numerically treated by quadrature methods; see [69, 132, 194, 195, 205, 206]. For more general problems, we refer to [8, 48].

3.5 Numerical Experiments for Symm's Equation

In this section, we apply all of the previously investigated regularization strategies to Symm's integral equation

$$K\psi(t) := -\frac{1}{\pi} \int_0^{2\pi} \psi(s) \ln |\gamma(t) - \gamma(s)| \, ds = g(t), \quad 0 \leq t \leq 2\pi,$$

where in this example $\gamma(s) = (\cos s, 2 \sin s)$, $0 \leq s \leq 2\pi$, denotes the parametrization of the ellipse with semiaxes 1 and 2. First, we discuss the numerical computation of $K\psi$. We write $K\psi$ in the form (see (3.60))

$$K\psi(t) = -\frac{1}{2\pi} \int_0^{2\pi} \psi(s) \ln\left(4\sin^2\frac{t-s}{2}\right) ds + \int_0^{2\pi} \psi(s) k(t,s) ds,$$

for $0 \le t \le 2\pi$, with the analytic function

$$k(t,s) = -\frac{1}{2\pi} \ln\frac{|\gamma(t)-\gamma(s)|^2}{4\sin^2\frac{t-s}{2}}, \quad t \ne s,$$

$$k(t,t) = -\frac{1}{\pi} \ln|\dot\gamma(t)|, \quad 0 \le t \le 2\pi.$$

We use the trapezoidal rule for periodic functions (see [130]). Let $t_j = j\frac{\pi}{n}$, $j = 0,\ldots,2n-1$. The smooth part is approximated by

$$\int_0^{2\pi} k(t,s)\,\psi(s)\,ds \approx \frac{\pi}{n} \sum_{j=0}^{2n-1} k(t,t_j)\,\psi(t_j), \quad 0 \le t \le 2\pi.$$

For the weakly singular part, we replace ψ by its trigonometric interpolation polynomial $Q_n\psi = \sum_{j=0}^{2n-1} \psi(t_j)\,L_j$ into the $2n$-dimensional space

$$\left\{\sum_{j=0}^n a_j \cos(jt) + \sum_{j=1}^{n-1} b_j \sin(jt) : a_j, b_j \in \mathbb{R}\right\}$$

over $\mathbb{R}$ (see Section A.4 of Appendix A). From (A.32) and Lemma 3.17, we conclude that

$$-\frac{1}{2\pi} \int_0^{2\pi} \psi(s) \ln\left(4\sin^2\frac{t-s}{2}\right) ds \approx -\frac{1}{2\pi} \int_0^{2\pi} (Q_n\psi)(s) \ln\left(4\sin^2\frac{t-s}{2}\right) ds$$

$$= \sum_{j=0}^{2n-1} \psi(t_j)\,R_j(t), \quad 0 \le t \le 2\pi,$$

where

$$R_j(t) = -\frac{1}{2\pi} \int_0^{2\pi} L_j(s) \ln\left(4\sin^2\frac{t-s}{2}\right) ds$$

$$= \frac{1}{n}\left\{\frac{1}{2n}\cos n(t-t_j) + \sum_{m=1}^{n-1}\frac{1}{m}\cos m(t-t_j)\right\}$$

for $j = 0,\ldots,2n-1$. Therefore, the operator K is replaced by

$$K_n\psi(t) := \sum_{j=0}^{2n-1} \psi(t_j)\left[R_j(t) + \frac{\pi}{n}k(t,t_j)\right], \quad 0 \le t \le 2\pi.$$

It is well known (see [130]) that $K_n\psi$ converges uniformly to $K\psi$ for every 2π-periodic continuous function ψ. Furthermore, the error $\|K_n\psi - K\psi\|_\infty$ is exponentially decreasing for analytic functions ψ. For $t = t_k$, $k = 0,\ldots,2n-1$, we have $K_n\psi(t_k) = \sum_{j=0}^{2n-1} A_{kj}\,\psi(t_j)$ with the symmetric matrix

$$A_{kj} := R_{|k-j|} + \frac{\pi}{n}k(t_k, t_j), \quad k, j = 0\ldots, 2n-1,$$

where

$$R_\ell = \frac{1}{n}\left\{\frac{(-1)^\ell}{2n} + \sum_{m=1}^{n-1}\frac{1}{m}\cos\frac{m\ell\pi}{n}\right\}, \quad \ell = 0,\ldots, 2n-1.$$

For the numerical example, we take $\psi(s) = \exp(3\sin s)$, $0 \le s \le 2\pi$, and $g = K\psi$ or, discretized, $\tilde{\psi}_j = \exp(3\sin t_j)$, $j = 0,\ldots,2n-1$, and $\tilde{g} = A\tilde{\psi}$. We take $n = 60$ and add uniformly distributed random noise on the data $\tilde{g}$. All the results show the average of 10 computations. The errors are measured in the discrete norm $|z|_2^2 := \frac{1}{2n}\sum_{j=0}^{2n-1}|z_j|^2$, $z \in \mathbb{C}^{2n}$.

First, we consider *Tikhonov's regularization method* for $\delta = 0.1$, $\delta = 0.01$, $\delta = 0.001$, and $\delta = 0$. We plot the errors $\left|\tilde{\psi}^{\alpha,\delta} - \tilde{\psi}\right|_2$ and $\left|A\tilde{\psi}^{\alpha,\delta} - \tilde{g}\right|_2$ in the solution and the right-hand side, respectively, versus the regularization parameter α.

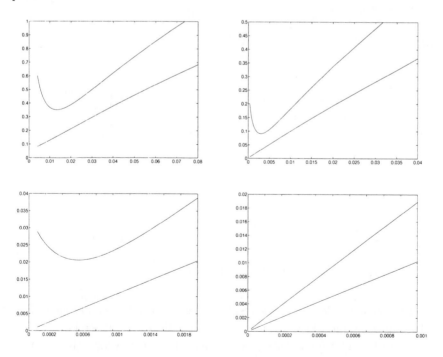

Figure 3.1: Error for Tikhonov's regularization method.

We clearly observe the expected behavior of the errors: For $\delta > 0$ the error in the solution has a well-defined minimum that depends on δ, while the defect always converges to zero as α tends to zero.

The minimal values err_δ of the errors in the solution are approximately 0.351, 0.0909, and 0.0206 for $\delta = 0.1$, 0.01, and 0.001, respectively. From this, we observe the order of convergence: Increasing the error by factor 10 should increase the error by factor $10^{2/3} \approx 4.64$, which roughly agrees with the numerical results where $err_{\delta=0.1}/err_{\delta=0.01} \approx 3.86$ and $err_{\delta=0.01}/err_{\delta=0.001} \approx 4.41$.

The following plots show the results for the *Landweber iteration* with $a = 0.5$ for the same example where again $\delta = 0.1$, $\delta = 0.01$, $\delta = 0.001$, and $\delta = 0$. The errors in the solution and the defects are now plotted versus the iteration number m.

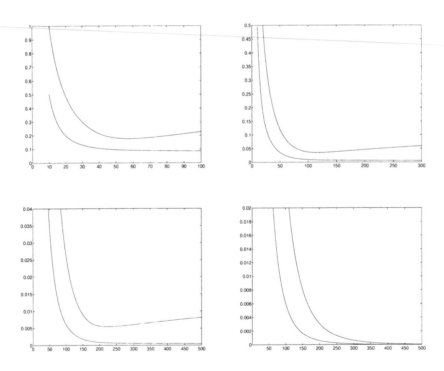

Figure 3.2: Error for Landweber's method ($a = 0.5$).

The following plots show the results for the *conjugate gradient method* for the same example where again $\delta = 0.1$, $\delta = 0.01$, $\delta = 0.001$, and $\delta = 0$. The errors in the solution and the defects are again plotted versus the iteration number m.

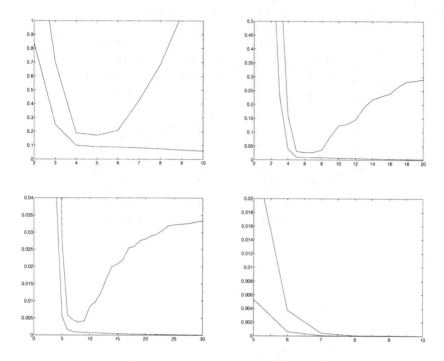

Figure 3.3: Error for the conjugate gradient method.

Here we observe the same behavior as for Tikhonov's method. We note the difference in the results for the Landweber method and the conjugate gradient method. The latter decreases the errors very quickly but is very sensitive to the exact stopping rule, while the Landweber iteration is slow but very stable with respect to the stopping parameter τ. The minimal values are $err_{\delta=0.1} \approx 0.177$, $err_{\delta=0.01} \approx 0.0352$, and $err_{\delta=0.001} \approx 0.0054$ for the Landweber iteration and $err_{\delta=0.1} \approx 0.172$, $err_{\delta=0.01} \approx 0.0266$, and $err_{\delta=0.001} \approx 0.0038$ for the conjugate gradient method. The corresponding factors are considerably larger than $10^{2/3} \approx 4.64$ indicating the optimality of these methods also for smooth solutions (see the remarks following Theorem 2.15).

Next, we compute the same example using some projection methods. First, we list the results for the least squares method and the Bubnov–Galerkin method of Subsections 3.2.1 and 3.2.3. We observe that both methods produce almost the same results, which reflect the estimates of Theorem 3.20. Note that for $\delta = 0$ the error decreases exponentially with m. This reflects the fact that the best approximation $\min\{\|\psi - \varphi_n\| : \varphi_n \in X_n\}$ converges to zero exponentially due to the analyticity of the solution $\psi(s) = \exp(3 \sin s)$ (see [130], Theorem 11.5).

Table 3.1: Least squares method

m	$\delta = 0.1$	$\delta = 0.01$	$\delta = 0.001$	$\delta = 0$
1	38.190	38.190	38.190	38.190
2	15.772	15.769	15.768	15.768
3	5.2791	5.2514	5.2511	5.2511
4	1.6209	1.4562	1.4541	1.4541
5	1.0365	0.3551	$3.433 * 10^{-1}$	$3.432 * 10^{-1}$
6	1.1954	0.1571	$7.190 * 10^{-2}$	$7.045 * 10^{-2}$
10	2.7944	0.2358	$2.742 * 10^{-2}$	$4.075 * 10^{-5}$
12	3.7602	0.3561	$3.187 * 10^{-2}$	$5.713 * 10^{-7}$
15	4.9815	0.4871	$4.977 * 10^{-2}$	$5.570 * 10^{-10}$
20	7.4111	0.7270	$7.300 * 10^{-2}$	$3.530 * 10^{-12}$

Table 3.2: Bubnov–Galerkin method

m	$\delta = 0.1$	$\delta = 0.01$	$\delta = 0.001$	$\delta = 0$
1	38.190	38.190	38.190	38.190
2	15.771	15.769	15.768	15.768
3	5.2752	5.2514	5.2511	5.2511
4	1.6868	1.4565	1.4541	1.4541
5	1.1467	0.3580	$3.434 * 10^{-1}$	$3.432 * 10^{-1}$
6	1.2516	0.1493	$7.168 * 10^{-2}$	$7.045 * 10^{-2}$
10	2.6849	0.2481	$2.881 * 10^{-2}$	$4.075 * 10^{-5}$
12	3.3431	0.3642	$3.652 * 10^{-2}$	$5.713 * 10^{-7}$
15	4.9549	0.4333	$5.719 * 10^{-2}$	$5.570 * 10^{-10}$
20	7.8845	0.7512	$7.452 * 10^{-2}$	$3.519 * 10^{-12}$

Now we turn to the collocation methods of Section 3.4. To implement
the collocation method (3.83) for Symm's integral equation and the basis
functions (3.89), (3.93a), and (3.93b), we have to compute the integrals

$$-\frac{1}{\pi} \int_0^{2\pi} e^{ijs} \ln|\gamma(t_k) - \gamma(s)| \, ds, \qquad (3.100a)$$

$j = -m, \ldots, m - 1, \ k = 0, \ldots, 2m - 1$, and

$$-\frac{1}{\pi} \int_0^{2\pi} \hat{x}_j(s) \ln|\gamma(t_k) - \gamma(s)| \, ds, \quad j, k = 0, \ldots, 2m - 1, \qquad (3.100b)$$

respectively. For the first integral (3.100a), we write using (3.63),

$$-\frac{1}{\pi}\int_0^{2\pi} e^{ijs}\ln|\gamma(t_k)-\gamma(s)|\,ds$$

$$= -\frac{1}{2\pi}\int_0^{2\pi} e^{ijs}\ln\left(4\sin^2\frac{t_k-s}{2}\right)ds - \frac{1}{2\pi}\int_0^{2\pi} e^{ijs}\ln\frac{|\gamma(t_k)-\gamma(s)|^2}{4\sin^2(t_k-s)/2}\,ds$$

$$= \varepsilon_j\, e^{ijt_k} - \frac{1}{2\pi}\int_0^{2\pi} e^{ijs}\ln\frac{|\gamma(t_k)-\gamma(s)|^2}{4\sin^2(t_k-s)/2}\,ds,$$

where $\varepsilon_j = 0$ for $j = 0$ and $\varepsilon_j = 1/|j|$ otherwise. The remaining integral is computed by the trapezoidal rule.

The computation of (3.100b) is more complicated. By definition (3.93a), (3.93b) of $\hat{x}_j$, we have to calculate

$$\int_{t_j-\pi/(2m)}^{t_j+\pi/(2m)} \ln|\gamma(t_k)-\gamma(s)|^2\,ds = \int_{-\pi/(2m)}^{\pi/(2m)} \ln|\gamma(t_k)-\gamma(s+t_j)|^2\,ds.$$

For $j \neq k$, the integrand is analytic, and we use Simpson's rule

$$\int_{-\pi/(2m)}^{\pi/(2m)} g(s)\,ds \approx \sum_{\ell=0}^n w_\ell\, g(s_\ell),$$

where

$$s_\ell = \ell\,\frac{\pi}{mn} - \frac{\pi}{2m}, \qquad w_\ell = \frac{\pi}{3mn}\cdot\begin{cases} 1, & \ell = 0 \text{ or } n, \\ 4, & \ell = 1,3,\ldots,n-1, \\ 2, & \ell = 2,4,\ldots,n-2, \end{cases}$$

$\ell = 0,\ldots,n$. For $j = k$, the integral has a weak singularity at $s = 0$. We split the integrand into

$$\int_{-\pi/(2m)}^{\pi/(2m)} \ln\left(4\sin^2\frac{s}{2}\right)ds + \int_{-\pi/(2m)}^{\pi/(2m)} \ln\frac{|\gamma(t_k)-\gamma(s+t_k)|^2}{4\sin^2(s/2)}\,ds$$

$$= -2\int_{\pi/(2m)}^{\pi} \ln\left(4\sin^2\frac{s}{2}\right)ds + \int_{-\pi/(2m)}^{\pi/(2m)} \ln\frac{|\gamma(t_k)-\gamma(s+t_k)|^2}{4\sin^2(s/2)}\,ds$$

since $\ln(4\sin^2(s/2))$ is even and $\int_0^\pi \ln(4\sin^2(s/2))\,ds = 0$ by (3.63). Both integrals are approximated by Simpson's rule. For the same example as earlier, with 100 integration points for Simpson's rule we obtain the following results.

Table 3.3: Collocation method for basis functions (3.89)

m	$\delta = 0.1$	$\delta = 0.01$	$\delta = 0.001$	$\delta = 0$
1	6.7451	6.7590	6.7573	6.7578
2	1.4133	1.3877	1.3880	1.3879
3	0.3556	$2.791 * 10^{-1}$	$2.770 * 10^{-1}$	$2.769 * 10^{-1}$
4	0.2525	$5.979 * 10^{-2}$	$5.752 * 10^{-2}$	$5.758 * 10^{-2}$
5	0.3096	$3.103 * 10^{-2}$	$1.110 * 10^{-2}$	$1.099 * 10^{-2}$
6	0.3404	$3.486 * 10^{-2}$	$3.753 * 10^{-3}$	$1.905 * 10^{-3}$
10	0.5600	$5.782 * 10^{-2}$	$5.783 * 10^{-3}$	$6.885 * 10^{-7}$
12	0.6974	$6.766 * 10^{-2}$	$6.752 * 10^{-3}$	$8.135 * 10^{-9}$
15	0.8017	$8.371 * 10^{-2}$	$8.586 * 10^{-3}$	$6.436 * 10^{-12}$
20	1.1539	$1.163 * 10^{-1}$	$1.182 * 10^{-2}$	$1.806 * 10^{-13}$

Table 3.4: Collocation method for basis functions (3.93a) and (3.93b)

m	$\delta = 0.1$	$\delta = 0.01$	$\delta = 0.001$	$\delta = 0$
1	6.7461	6.7679	6.7626	6.7625
2	1.3829	1.3562	1.3599	1.3600
3	0.4944	$4.874 * 10^{-1}$	$4.909 * 10^{-1}$	$4.906 * 10^{-1}$
4	0.3225	$1.971 * 10^{-1}$	$2.000 * 10^{-1}$	$2.004 * 10^{-1}$
5	0.3373	$1.649 * 10^{-1}$	$1.615 * 10^{-1}$	$1.617 * 10^{-1}$
6	0.3516	$1.341 * 10^{-1}$	$1.291 * 10^{-1}$	$1.291 * 10^{-1}$
10	0.5558	$8.386 * 10^{-2}$	$6.140 * 10^{-2}$	$6.107 * 10^{-2}$
12	0.6216	$7.716 * 10^{-2}$	$4.516 * 10^{-2}$	$4.498 * 10^{-2}$
15	0.8664	$9.091 * 10^{-2}$	$3.137 * 10^{-2}$	$3.044 * 10^{-2}$
20	1.0959	$1.168 * 10^{-1}$	$2.121 * 10^{-2}$	$1.809 * 10^{-2}$
30	1.7121	$1.688 * 10^{-1}$	$1.862 * 10^{-2}$	$8.669 * 10^{-3}$

The difference for $\delta = 0$ reflects the fact that the best approximation

$$\min\{\|\psi - \varphi_n\| : \varphi_n \in \text{span}\,\{\hat{x}_j : j \in J\}\}$$

converges to zero exponentially for $\hat{x}_j$ defined by (3.89), while it converges to zero only of order $1/n$ for $\hat{x}_j$ defined by (3.93a) and (3.93b) (see Theorem 3.30).

We have seen in this section that the theoretical investigations of the regularization strategies are confirmed by the numerical results for Symm's integral equation.

3.6   The Backus–Gilbert Method

In this section, we will study a different numerical method for "solving" finite moment problems of the following type:

$$\int_a^b k_j(s)\, x(s)\, ds \;=\; y_j, \quad j = 1, \ldots, n. \qquad (3.101)$$

Here, $y_j \in \mathbb{R}$ are any given numbers and $k_j \in L^2(a,b)$ arbitrary given functions. Certainly, we have in mind that $y_j = y(t_j)$ and $k_j = k(t_j, \cdot)$. In the previous section, we studied the moment solution of such problems; see [146, 185]. We saw that the moment solution x_n is a finite linear combination of the functions $\{k_1, \ldots, k_n\}$. Therefore, the moment solution x_n is as smooth as the functions k_j even if the true solution is smoother.

The concept originally proposed by Backus and Gilbert ([11], [12]) does not primarily wish to solve the moment problem but rather wants to determine how well all possible models x can be recovered pointwise.

Define the finite-dimensional operator $K : L^2(a,b) \to \mathbb{R}^n$ by

$$(Kx)_j \;=\; \int_a^b k_j(s)\, x(s)\, ds, \quad j = 1, \ldots, n, \quad x \in L^2(a,b). \qquad (3.102)$$

We try to find a left-inverse S, i.e., a linear operator $S : \mathbb{R}^n \to L^2(a,b)$ such that

$$SKx \;\approx\; x \quad \text{for all } x \in L^2(a,b). \qquad (3.103)$$

Therefore, SKx should be a simultaneous approximation to all possible $x \in L^2(a,b)$. Of course, we have to make clear the meaning of the approximation.

The general form of a linear operator $S : \mathbb{R}^n \to L^2(a,b)$ has to be

$$Sy(t) \;=\; \sum_{j=1}^n y_j\, \varphi_j(t), \quad t \in (a,b), \quad y = (y_j) \in \mathbb{R}^n, \qquad (3.104)$$

for some $\varphi_j \in L^2(a,b)$, which are to be determined from the requirement (3.103):

$$(SKx)(t) \;=\; \sum_{j=1}^n \varphi_j(t) \int_a^b k_j(s)\, x(s)\, ds$$

$$=\; \int_a^b \left[\sum_{j=1}^n k_j(s)\, \varphi_j(t) \right] x(s)\, ds.$$

The requirement $SKx \approx x$ leads to the problem of approximating Dirac's delta distribution $\delta(s-t)$ by linear combinations of the form $\sum_{j=1}^{n} k_j(s)\,\varphi_j(t)$. For example, one can show that the minimum of

$$\int_a^b \int_a^b \left| \sum_{j=1}^{n} k_j(s)\,\varphi_j(t) - \delta(s-t) \right|^2 ds\,dt$$

(in the sense of distributions) is attained at $\varphi(s) = A^{-1}k(s)$, where $k(s) = (k_1(s),\ldots,k_n(s))^\top$ and $A_{ij} = \int_a^b k_i(s)\,k_j(s)\,ds$, $i,j = 1,\ldots,n$. For this minimization criterion, $x = \sum_{j=1}^{n} y_j \varphi_j$ is again the moment solution of Subsection 3.4.1. In [146], it is shown that minimizing with respect to an H^{-s}-norm for $s > 1/2$ leads to projection methods in H^s-spaces. We refer also to [224] for a comparison of several minimization criteria.

The Backus–Gilbert method is based on a pointwise minimization criterion: Treat $t \in [a,b]$ as a fixed parameter and determine the numbers $\varphi_j = \varphi_j(t)$ for $j = 1,\ldots,n$, as the solution of the following minimization problem:

$$\text{minimize} \quad \int_a^b |s - t|^2 \left| \sum_{j=1}^{n} k_j(s)\,\varphi_j \right|^2 ds \qquad (3.105a)$$

subject to $\varphi \in \mathbb{R}^n$ and

$$\int_a^b \sum_{j=1}^{n} k_j(s)\,\varphi_j \, ds = 1. \qquad (3.105b)$$

Using the matrix-vector notation, we rewrite this problem in short form:

$$\text{minimize} \quad \varphi^\top Q(t)\,\varphi \quad \text{subject to} \quad r \cdot \varphi = 1,$$

where

$$Q(t)_{ij} = \int_a^b |s - t|^2 \, k_i(s)\,k_j(s)\,ds, \quad i,j = 1,\ldots,n,$$

$$r_j = \int_a^b k_j(s)\,ds, \quad j = 1,\ldots,n.$$

This is a quadratic minimization problem with one linear equality constraint. We assume that $r \neq 0$ since otherwise the constraint (3.105b) cannot be satisfied. Uniqueness and existence are assured by the following theorem, which also gives a characterization by the Lagrange multiplier rule.

Theorem 3.31

Assume that $\{k_1, \ldots, k_n\}$ are linearly independent. Then the symmetric matrix $Q(t) \in \mathbb{R}^{n \times n}$ is positive definite for every $t \in [a, b]$. The minimization problem (3.105a), (3.105b) is uniquely solvable. $\varphi \in \mathbb{R}^n$ is a solution of (3.105a) and (3.105b) if and only if there exists a number $\lambda \in \mathbb{R}$ (the Lagrange multiplier) such that $(\varphi, \lambda) \in \mathbb{R}^n \times \mathbb{R}$ solves the linear system

$$Q(t)\varphi - \lambda r = 0 \quad and \quad r \cdot \varphi = 1. \tag{3.106}$$

$\lambda = \varphi^\top Q(t)\, \varphi$ is the minimal value of this problem.

Proof: From

$$\varphi^\top Q(t)\, \varphi = \int_a^b |s - t|^2 \left| \sum_{j=1}^n k_j(s)\, \varphi_j \right|^2 ds$$

we conclude first that $\varphi^\top Q(t)\, \varphi \geq 0$ and second that $\varphi^\top Q(t)\, \varphi = 0$ implies that $\sum_{j=1}^n k_j(s)\, \varphi_j = 0$ for almost all $s \in (a, b)$. Since $\{k_j\}$ are linearly independent, $\varphi_j = 0$ for all j follows. Therefore, $Q(t)$ is positive definite. Existence, uniqueness, and equivalence to (3.106) are elementary results from optimization theory; see [221]. □

Definition 3.32

We denote by $\left(\varphi_j(t) \right)_{j=1}^n \in \mathbb{R}^n$ the unique solution $\varphi \in \mathbb{R}^n$ of (3.105a) and (3.105b). The Backus–Gilbert solution x_n *of*

$$\int_a^b k_j(s)\, x_n(s)\, ds = y_j, \quad j = 1, \ldots, n,$$

is defined as

$$x_n(t) = \sum_{j=1}^n y_j\, \varphi_j(t), \quad t \in [a, b]. \tag{3.107}$$

The minimal value $\lambda = \lambda(t) = \varphi(t)^\top Q(t)\varphi(t)$ is called the spread.

We remark that, in general, the Backus–Gilbert solution $x_n = \sum_{j=1}^n y_j\, \varphi_j$ is not a solution of the moment problem, i.e., $\int_a^b k_j(s)\, x_n(s)\, ds \neq y_j$! This is certainly a disadvantage. On the other hand, the solution x is analytic in $[a, b]$ – even for nonsmooth data k_j. We can prove the following lemma.

Lemma 3.33

φ_j and λ are rational functions. More precisely, there exist polynomials $p_j, q \in \mathcal{P}_{2(n-1)}$ and $\rho \in \mathcal{P}_{2n}$ such that $\varphi_j = p_j/q$, $j = 1, \ldots, n$, and $\lambda = \rho/q$. The polynomial q has no zeros in $[a, b]$.

Proof: Obviously, $Q(t) = Q_0 - 2t\,Q_1 + t^2\,Q_2$ with symmetric matrices Q_0, Q_1, Q_2. We search for a polynomial solution $p \in [\mathcal{P}_m]^n$ and $\rho \in \mathcal{P}_{m+2}$ of $Q(t)p(t) - \rho(t)\,r = 0$ with $m = 2(n-1)$. Since the number of equations is $n(m+3) = 2n^2 + n$ and the number of unknowns is $n(m+1) + (m+3) = 2n^2 + n + 1$, there exists a nontrivial solution $p \in [\mathcal{P}_m]^n$ and $\rho \in \mathcal{P}_{m+2}$. If $p(\hat{t}) = 0$ for some $\hat{t} \in [a,b]$, then $\rho(\hat{t}) = 0$ since $r \neq 0$. In this case, we divide the equation by $(t - \hat{t})$. Therefore, we can assume that p has no zero in $[a,b]$.

Now we define $q(t) := r \cdot p(t)$ for $t \in [a,b]$. Then $q \in \mathcal{P}_m$ has no zero in $[a,b]$ since otherwise we would have

$$0 \;=\; \rho(\hat{t})\,r \cdot p(\hat{t}) \;=\; p(\hat{t})^\top Q(\hat{t})\,p(\hat{t});$$

thus $p(\hat{t}) = 0$, a contradiction. Therefore, $\varphi := p/q$ and $\lambda := \rho/q$ solves (3.106). By the uniqueness result, this is the only solution.   □

For the following error estimates, we assume two kinds of a priori information on x depending on the norm of the desired error estimate. Let

$$X_n \;=\; \mathrm{span}\{k_j : j = 1, \ldots, n\}.$$

Theorem 3.34
Let $x \in L^2(a,b)$ be any solution of the finite moment problem (3.101) and $x_n = \sum_{j=1}^{n} y_j \varphi_j$ be the Backus–Gilbert solution. Then the following error estimates hold:

(a) *Assume that x is Lipschitz continuous with constant $\ell > 0$, i.e.,*

$$|x(t) - x(s)| \;\leq\; \ell\,|s - t| \quad \text{for all } s, t \in [a,b].$$

Then

$$|x_n(t) - x(t)| \;\leq\; \ell\,\sqrt{b - a}\,\epsilon_n(t) \tag{3.108}$$

for all $n \in \mathbb{N}$, $t \in [a,b]$, where $\epsilon_n(t)$ is defined by

$$\epsilon_n^2(t) \;:=\; \min\left\{ \int_a^b |s - t|^2\,|z_n(s)|^2\,ds : z_n \in X_n,\ \int_a^b z_n(s)\,ds = 1 \right\}. \tag{3.109}$$

(b) *Let $x \in H^1(a,b)$. Then there exists $c > 0$, independent of x, such that*

$$\|x_n - x\|_{L^2} \;\leq\; c\,\|x'\|_{L^2}\,\|\epsilon_n\|_\infty \quad \text{for all } n \in \mathbb{N}. \tag{3.110}$$

Proof: By the definition of the Backus–Gilbert solution and the constraint on φ, we have

$$x_n(t) - x(t) = \sum_{j=1}^{n} y_j \, \varphi_j(t) \; - \; x(t) \int_a^b \sum_{j=1}^{n} k_j(s) \, \varphi_j(t) \, ds$$

$$= \sum_{j=1}^{n} \int_a^b k_j(s) \left[x(s) - x(t) \right] \varphi_j(t) \, ds.$$

Thus

$$|x_n(t) - x(t)| \leq \int_a^b \left| \sum_{j=1}^{n} k_j(s) \, \varphi_j(t) \right| \, |x(s) - x(t)| \, ds.$$

Now we distinguish between parts (a) and (b):

(a) Let $|x(t) - x(s)| \leq \ell \, |t - s|$. Then, by the Cauchy–Schwarz inequality and the definition of φ_j,

$$|x_n(t) - x(t)| \leq \ell \int_a^b 1 \cdot \left| \sum_{j=1}^{n} k_j(s) \, \varphi_j(t) \right| \, |t - s| \, ds$$

$$\leq \ell \sqrt{b - a} \left[\int_a^b \left| \sum_{j=1}^{n} k_j(s) \, \varphi_j(t) \right|^2 |t - s|^2 \, ds \right]^{1/2}$$

$$= \ell \sqrt{b - a} \, \epsilon_n(t).$$

(b) First, we define the cutoff function λ_δ on $[a, b] \times [a, b]$ by

$$\lambda_\delta(t, s) = \begin{cases} 1, & |t - s| \geq \delta, \\ 0, & |t - s| < \delta. \end{cases} \tag{3.111}$$

Then, by the Cauchy–Schwarz inequality again,

$$\left[\int_a^b \lambda_\delta(t, s) \left| \sum_{j=1}^{n} k_j(s) \, \varphi_j(t) \right| \, |x(s) - x(t)| \, ds \right]^2$$

$$= \left[\int_a^b \left| \sum_{j=1}^{n} k_j(s) \, \varphi_j(t) \, (t - s) \right| \lambda_\delta(t, s) \left| \frac{x(s) - x(t)}{t - s} \right| \, ds \right]^2$$

$$\leq \epsilon_n(t)^2 \int_a^b \lambda_\delta(t, s) \left| \frac{x(s) - x(t)}{s - t} \right|^2 \, ds.$$

Integration with respect to t yields

$$\int_a^b \left[\int_a^b \lambda_\delta(t,s) \left| \sum_{j=1}^n k_j(s)\varphi_j(t) \right| |x(s) - x(t)| \, ds \right]^2 dt$$

$$\leq \|\epsilon_n\|_\infty^2 \int_a^b \int_a^b \left| \frac{x(s) - x(t)}{s - t} \right|^2 \lambda_\delta(t,s) \, ds \, dt.$$

The following technical lemma from the theory of Sobolev spaces yields the assertion.   □

Lemma 3.35

There exists $c > 0$ such that

$$\int_a^b \int_a^b \left| \frac{x(s) - x(t)}{s - t} \right|^2 \lambda_\delta(t,s) \, ds \, dt \leq c \|x'\|_{L^2}^2$$

for all $\delta > 0$ and $x \in H^1(a,b)$. Here, the cutoff function λ_δ is defined by (3.111).

Proof: First, we estimate

$$|x(s) - x(t)|^2 = \left| \int_s^t 1 \cdot x'(\tau) \, d\tau \right|^2 \leq |t - s| \left| \int_s^t |x'(\tau)|^2 \, d\tau \right|$$

and thus, for $s \neq t$,

$$\left| \frac{x(s) - x(t)}{s - t} \right|^2 \leq \frac{1}{|s - t|} \left| \int_s^t |x'(\tau)|^2 \, d\tau \right|.$$

Now we fix $t \in (a,b)$ and write

$$\int_a^b \left| \frac{x(s) - x(t)}{s - t} \right|^2 \lambda_\delta(t,s) \, ds$$

$$\leq \int_a^t \frac{\lambda_\delta(t,s)}{t - s} \int_s^t |x'(\tau)|^2 \, d\tau \, ds + \int_t^b \frac{\lambda_\delta(t,s)}{s - t} \int_t^s |x'(\tau)|^2 \, d\tau \, ds$$

$$= \int_a^b |x'(s)|^2 A_\delta(t,s) \, ds,$$

where

$$A_\delta(t,s) = \begin{cases} \int_a^s \frac{\lambda_\delta(t,\tau)}{|t-\tau|}\, d\tau, & a \le s < t, \\ \int_s^b \frac{\lambda_\delta(t,\tau)}{|t-\tau|}\, d\tau, & t < s \le b, \end{cases} = \begin{cases} \ln\frac{t-a}{\max(\delta,t-s)}, & s \le t, \\ \ln\frac{b-t}{\max(\delta,s-t)}, & s \ge t. \end{cases}$$

Finally, we estimate

$$\int_a^b \int_a^b \frac{|x(s)-x(t)|^2}{|s-t|^2}\, \lambda_\delta(t,s)\, ds\, dt \le \int_a^b |x'(s)|^2 \left(\int_a^b A_\delta(t,s)\, dt \right) ds$$

and

$$\int_a^b A_\delta(t,s)\, dt \le c \quad \text{for all } s \in (a,b) \text{ and } \delta > 0$$

which is seen by elementary integration. $\qquad\square$

From these error estimates, we observe that the rate of convergence depends on the magnitude of ϵ_n, i.e., how well the kernels approximate the delta distribution. Finally, we study the question of convergence for $n \to \infty$.

Theorem 3.36
Let the infinite-dimensional moment problem

$$\int_a^b k_j(s)\, x(s)\, ds = y_j, \quad j = 1, 2 \ldots \,,$$

have a unique solution $x \in L^2(a,b)$. Assume that $\{k_j : j \in \mathbb{N}\}$ is linearly independent and dense in $L^2(a,b)$. Then

$$\|\epsilon_n\|_\infty \longrightarrow 0 \quad \text{for } n \to \infty.$$

Proof: For fixed $t \in [a,b]$ and arbitrary $\delta \in \big(0, (b-a)/2\big)$, we define

$$\tilde{v}(s) := \begin{cases} \frac{1}{|s-t|}, & |s-t| \ge \delta, \\ 0, & |s-t| < \delta, \end{cases} \quad \text{and} \quad v(s) := \left[\int_a^b \tilde{v}(\tau)\, d\tau \right]^{-1} \tilde{v}(s).$$

Then $v \in L^2(a,b)$ and $\int_a^b v(s)\, ds = 1$. Since $\bigcup X_n$ is dense in $L^2(a,b)$, there exists a sequence $\tilde{v}_n \in X_n$ with $\tilde{v}_n \to v$ in $L^2(a,b)$. This implies also that $\int_a^b \tilde{v}_n(s)\, ds \to \int_a^b v(s)\, ds = 1$. Therefore, the functions

$$v_n := \left[\int_a^b \tilde{v}_n(s)\, ds \right]^{-1} \tilde{v}_n \in X_n$$

converge to v in $L^2(a,b)$ and are normalized by $\int_a^b v_n(s)\,ds = 1$. Thus v_n is admissible, and we conclude that

$$\epsilon_n(t)^2 \;\leq\; \int_a^b |s-t|^2\, v_n(s)^2\,ds$$

$$= \int_a^b |s-t|^2\, v(s)^2\,ds \;+\; 2\int_a^b |s-t|^2\, v(s)\big[v_n(s)-v(s)\big]\,ds$$

$$+ \int_a^b |s-t|^2\, \big[v_n(s)-v(s)\big]^2\,ds$$

$$\leq \left[\int_a^b \tilde{v}(s)\,ds\right]^{-2}(b-a)$$

$$+ (b-a)^2\left[2\,\|v\|_{L^2}\,\|v_n-v\|_{L^2} + \|v_n-v\|_{L^2}^2\right].$$

This shows that

$$\limsup_{n\to\infty}\epsilon_n(t) \;\leq\; \sqrt{b-a}\,\left[\int_a^b \tilde{v}(s)\,ds\right]^{-1} \qquad \text{for all } t\in[a,b].$$

Direct computation yields

$$\int_a^b \tilde{v}(s)\,ds \;\geq\; c + |\ln\delta|$$

for some c independent of δ; thus

$$\limsup_{n\to\infty}\epsilon_n(t) \;\leq\; \frac{\sqrt{b-a}}{c+|\ln\delta|} \quad \text{for all } \delta\in\big(0,(b-a)/2\big).$$

This yields pointwise convergence, i.e., $\epsilon_n(t)\to 0$ $(n\to\infty)$ for every $t\in [a,b]$. Since $\epsilon_n(t)$ is monotonic with respect to n, a well-known theorem from classical analysis yields uniform convergence. $\qquad\square$

For further aspects of the Backus–Gilbert method, we refer to [25, 90, 114, 124, 125, 200, 224, 225].

3.7 Problems

3.1 Let $Q_n : C[a,b] \to S_1(t_1,\ldots,t_n)$ be the interpolation operator from Example 3.3. Prove that $\|Q_n\|_\infty = 1$ and derive an estimate of the form

$$\|Q_n x - x\|_\infty \leq c h \|x'\|_\infty$$

for $x \in C^1[a,b]$, where $h = \max\{t_j - t_{j-1} : t = 2, \ldots, n\}$.

3.2 Let $K : X \to X$ be self-adjoint and positive definite and let $y \in X$. Define $\psi(x) = (Kx, x) - 2\operatorname{Re}(y, x)$ for $x \in X$. Prove that $x^* \in X$ is a minimum of ψ if and only if x^* solves $Kx^* = y$.

3.3 Let (V, X, V^*) be a Gelfand triple and $J : V \to X$ the embedding operator. Show that $J^* : X \to V^*$ is one-to-one and that $J^*(X)$ is dense in V^*.

3.4 Define the space X_n by

$$X_n = \left\{ \sum_{|j| \leq n} a_j e^{ijt} : a_j \in \mathbb{C} \right\}$$

and let $P_n : L^2(0, 2\pi) \to X_n$ be the orthogonal projection operator. Prove that for $r \geq s$ there exists $c > 0$ such that

$$\|\psi_n\|_{H^r} \leq c n^{r-s} \|\psi_n\|_{H^s} \quad \text{for all } \psi_n \in X_n,$$

$$\|P_n \psi - \psi\|_{H^s} \leq c \frac{1}{n^{r-s}} \|\psi\|_{H^r} \quad \text{for all } \psi \in H^r(0, 2\pi).$$

3.5 Show that the worst-case error of Symm's equation under the information $\|\psi\|_{H^s} \leq E$ for some $s > 0$ is given by

$$\mathcal{F}(\delta, E, \|\cdot\|_{H^s}) \leq c \delta^{s/(s+1)}.$$

3.6 Let $\Omega \subset \mathbb{R}^2$ be the disc of radius $a = \exp(-1/2)$. Then $\psi = 1$ is the unique solution of Symm's integral equation (3.57) for $f = 1$. Compute explicitly the errors of the least squares solution, the dual least squares solution, and the Bubnov–Galerkin solution as in Section 3.3, and verify that the error estimates of Theorem 3.20 are asymptotically sharp.

3.7 Let $t_k = k/n$, $k = 1, \ldots, n$, be equidistant collocation points. Let X_n be the space of piecewise constant functions as in (3.81) and $P_n : L^2(0, 1) \to X_n$ be the orthogonal projection operator. Prove that $\bigcup X_n$ is dense in $L^2(0, 1)$ and

$$\|x - P_n x\|_{L^2} \leq \frac{1}{n} \|x'\|_{L^2}$$

for all $x \in H^1(0, 1)$ (see Problem 3.1).

3.8 Show that the moment solution can also be interpreted as the solution of a dual least squares method.

3.9 Consider moment collocation of the equation

$$\int_0^t x(s)\,ds \;=\; y(t), \quad t \in [0,1],$$

in the space $X_n = S_1(t_1, \ldots, t_n)$ of linear splines. Show that the moment solution x_n coincides with the two-sided difference quotient, i.e.,

$$x_n(t_j) \;=\; \frac{1}{2h}\left[y(t_{j+1} + h) - y(t_{j-1} - h)\right],$$

where $h = 1/n$. Derive an error estimate for $\left\|x_n^\delta - x\right\|_{L^2}$ as in Example 3.25.

4

Inverse Eigenvalue Problems

4.1   Introduction

Inverse eigenvalue problems are not only interesting in their own right but also have important practical applications. We recall the fundamental paper by Kac [116]. Other applications appear in parameter identification problems for parabolic or hyperbolic differential equations (see [129, 149, 209]) or in grating theory ([123]).

We will study the Sturm–Liouville eigenvalue problem in canonical form. The direct problem is to determine the eigenvalues λ and the corresponding eigenfunctions $u \neq 0$ such that

$$-\frac{d^2u(x)}{dx^2} + q(x)\,u(x) = \lambda\,u(x), \quad 0 \leq x \leq 1, \tag{4.1a}$$

$$u(0) = 0 \quad \text{and} \quad hu'(1) + Hu(1) = 0, \tag{4.1b}$$

where $q \in L^2(0,1)$ and $h, H \in \mathbb{R}$ with $h^2 + H^2 > 0$ are given. In this chapter, we assume that all functions are real-valued. In some applications, e.g., in grating theory, complex-valued functions q are also of practical importance. Essentially all of the results of this chapter hold also for complex-valued q and are proven mainly by the same arguments. We refer to the remarks at the end of each section.

The eigenvalue problem (4.1a), (4.1b) is a special case of the more general eigenvalue problem for w:

$$\frac{d}{dt}\left(p(t)\,\frac{dw(t)}{dt}\right) + [\rho r(t) - g(t)]\,w(t) = 0, \quad t \in [a, b], \tag{4.2a}$$

$$\alpha_a w'(a) + \beta_a\,w(a) = 0, \quad \alpha_b w'(b) + \beta_b w(b) = 0. \tag{4.2b}$$

Here p, r, and g are given functions with $p(t) > 0$ and $r(t) > 0$ for $t \in [a, b]$, and $\alpha_a, \alpha_b, \beta_a, \beta_b \in \mathbb{R}$ are constants with $\alpha_a^2 + \beta_a^2 > 0$ and $\alpha_b^2 + \beta_b^2 > 0$. If we assume, however, that $g \in C[a, b]$ and $p, r \in C^2[a, b]$, then the *Liouville transformation* reduces the eigenvalue problem (4.2a), (4.2b) to the canonical form (4.1a), (4.1b). In particular, we define

$$\sigma(t) := \sqrt{\frac{r(t)}{p(t)}}, \quad f(t) := [p(t)\,r(t)]^{1/4}, \quad L := \int_a^b \sigma(s)\,ds, \tag{4.3}$$

the monotonic function $x : [a, b] \to [0, 1]$ by

$$x(t) := \frac{1}{L}\int_a^t \sigma(s)\,ds, \quad t \in [a, b], \tag{4.4}$$

and the new function $u : [0, 1] \to \mathbb{R}$ by $u(x) := f\big(t(x)\big)\,w\big(t(x)\big)$, $x \in [0, 1]$, where $t = t(x)$ denotes the inverse of $x = x(t)$. Elementary calculations show that u satisfies the differential equation (4.1a) with $\lambda = L^2\rho$ and

$$q(x) = L^2\left[\frac{g(t)}{r(t)} + \frac{f(t)}{r(t)}\left(\frac{p(t)f'(t)}{f(t)^2}\right)'\right]_{t=t(x)}. \tag{4.5}$$

Also, it is easily checked that the boundary conditions (4.2b) are mapped into the boundary conditions

$$h_0 u'(0) + H_0 u(0) = 0 \quad \text{and} \quad h_1 u'(1) + H_1 u(1) = 0 \tag{4.6}$$

with $h_0 = \alpha_a \sigma(a)/(L\,f(a))$ and $H_0 = \beta_a/f(a) - \alpha_a f'(a)/f(a)^2$ and, analogously, h_1, H_1 with a replaced by b.

In this chapter, we will restrict ourselves to the study of the canonical Sturm–Liouville eigenvalue problem (4.1a), (4.1b). In the first part, we study the case $h = 0$ in some detail. At the end of Section 4.3, we briefly discuss the case where $h = 1$. In Section 4.3, we will prove that there exists a countable number of eigenvalues λ_n of this problem and also prove an asymptotic formula. Since q is real-valued, the problem is self-adjoint, and the existence of a countable number of eigenvalues follows from the general spectral theorem of functional analysis (see Appendix A, Theorem A.48).

Since this general theorem provides only the information that the eigenvalues tend to infinity, we need other tools to obtain more information about the rate of convergence. The basic ingredient in the proof of the asymptotic formula will be the asymptotic behavior of the fundamental system of the differential equation (4.1a) as $|\lambda|$ tends to infinity. Although all of the data and the eigenvalues are real-valued, we will use results from complex analysis, in particular Rouché's theorem. This makes it necessary to allow the parameter λ in the fundamental system to be complex-valued. The existence of a fundamental solution and its asymptotics will be the subject of the next section.

Section 4.4 is devoted to the corresponding inverse problem: Given the eigenvalues λ_n, determine the function q. In Section 4.5, we will demonstrate how inverse spectral problems arise in a parameter identification problem for a parabolic initial value problem. Section 4.6, finally, studies numerical procedures for recovering q that have been suggested by Rundell and others (see [148, 188, 189]).

We finish this section with a "negative" result, as seen in Example 4.1.

Example 4.1
Let λ be an eigenvalue and u a corresponding eigenfunction of

$$-u''(x) + q(x)\,u(x) = \lambda\,u(x),\ 0 < x < 1, \quad u(0) = 0,\ u(1) = 0.$$

Then λ is also an eigenvalue with corresponding eigenfunction $v(x) :=$ $u(1 - x)$ of the eigenvalue problem

$$-v''(x) + \tilde{q}(x)\,v(x) = \lambda\,v(x),\ 0 < x < 1, \quad v(0) = 0,\ v(1) = 0,$$

where $\tilde{q}(x) := q(1 - x)$.

This example shows that it is generally impossible to recover the function q unless more information is available. We will see that q can be recovered uniquely, provided we know that it is an even function with respect to $1/2$ or if we know a second spectrum, i.e., a spectrum for a boundary condition different from $u(1) = 0$.

4.2   Construction of a Fundamental System

It is well-known from the theory of linear ordinary differential equations that the following initial value problems are uniquely solvable for every fixed (real- or complex-valued) $q \in C[0, 1]$ and every given $\lambda \in \mathbb{C}$:

$$- u_1'' + q(x)\,u_1 = \lambda\,u_1,\ 0 < x < 1, \quad u_1(0) = 1,\ u_1{}'(0) = 0 \qquad (4.7a)$$

$$-u_2'' + q(x)\, u_2 = \lambda\, u_2, \quad 0 < x < 1, \quad u_2(0) = 0, \ u_2'(0) = 1. \qquad (4.7b)$$

Uniqueness and existence for $q \in L^2(0,1)$ will be shown in Theorem 4.4. The set of functions $\{u_1, u_2\}$ is called a *fundamental system* of the differential equation $-u'' + q\, u = \lambda\, u$ in $(0,1)$. The functions u_1 and u_2 are linearly independent since the *Wronskian determinant* is one:

$$[u_1, u_2] := \det \begin{bmatrix} u_1 & u_2 \\ u_1' & u_2' \end{bmatrix} = u_1 u_2' - u_1' u_2 = 1. \qquad (4.8)$$

This is seen from

$$\frac{d}{dx}[u_1, u_2] = u_1 u_2'' - u_1'' u_2 = u_1 (q - \lambda)\, u_2 - u_2 (q - \lambda)\, u_1 = 0$$

and $[u_1, u_2](0) = 1$. The functions u_1 and u_2 depend on λ and q. We express this dependence often by $u_j = u_j(\cdot, \lambda, q)$, $j = 1, 2$. For $q \in L^2(0,1)$, the solution is not twice continuously differentiable anymore but is only an element of the *Sobolev space*

$$H^2(0,1) := \left\{ u \in C^1[0,1] : u'(x) = \alpha + \int_0^x v(t)\, dt, \ \alpha \in \mathbb{C}, \ v \in L^2(0,1) \right\},$$

see (1.24). We write u'' for v and observe that $u'' \in L^2(0,1)$. The most important example is when $q = 0$. In this case, we can solve (4.7a) and (4.7b) explicitly and have the following.

Example 4.2
Let $q = 0$. Then the solutions of (4.7a) and (4.7b) are given by

$$u_1(x, \lambda, 0) = \cos(\sqrt{\lambda}\, x) \quad \text{and} \quad u_2(x, \lambda, 0) = \frac{\sin(\sqrt{\lambda}\, x)}{\sqrt{\lambda}}, \qquad (4.9)$$

respectively. An arbitrary branch of the square root can be taken since $s \mapsto \cos(sx)$ and $s \mapsto \sin(sx)/s$ are even functions.

We will see that the fundamental solution for any function $q \in L^2(0,1)$ behaves as (4.9) as $|\lambda|$ tends to infinity. For the proof of the next theorem, we need the following technical lemma.

Lemma 4.3
Let $q \in L^2(0,1)$ and $\tilde{k}, k \in C[0,1]$ such that there exists $\mu > 0$ with $|\tilde{k}(\tau)| \leq \exp(\mu\tau)$ and $|k(\tau)| \leq \exp(\mu\tau)$ for all $\tau \in [0,1]$. Let $\tilde{K}, K : C[0,1] \to C[0,1]$ be the Volterra integral operators with kernels $\tilde{k}(x-t)\, q(t)$ and $k(x-t)\, q(t)$, respectively. Then the following estimate holds:

$$\left| \tilde{K}\, K^{n-1} \varphi(x) \right| \leq \|\varphi\|_\infty \, \frac{1}{n!} \, \hat{q}(x)^n \, e^{\mu x}, \quad 0 \leq x \leq 1, \qquad (4.10)$$

for all $\varphi \in C[0,1]$ and all $n \in \mathbb{N}$. Here, $\hat{q}(x) := \int_0^x |q(t)| \, dt$. If $\varphi \in C[0,1]$ satisfies also the estimate $|\varphi(\tau)| \le \exp(\mu\tau)$ for all $\tau \in [0,1]$, then we have

$$\left| \tilde{K} K^{n-1} \varphi(x) \right| \le \frac{1}{n!} \hat{q}(x)^n e^{\mu x}, \quad 0 \le x \le 1, \tag{4.11}$$

for all $n \in \mathbb{N}$.

Proof: We prove the estimates by induction with respect to n.

For $n = 1$, we estimate

$$\left| \tilde{K}\varphi(x) \right| = \left| \int_0^x \tilde{k}(x - t) \, q(t) \, \varphi(t) \, dt \right|$$

$$\le \|\varphi\|_\infty \int_0^x e^{\mu(x-t)} \, |q(t)| \, dt \le \|\varphi\|_\infty \, e^{\mu x} \, \hat{q}(x).$$

Now we assume the validity of (4.10) for n. Since it holds also for $K = \tilde{K}$, we estimate

$$\left| \tilde{K} K^n \varphi(x) \right| \le \int_0^x e^{\mu(x-t)} \, |q(t)| \, |K^n \varphi(t)| \, dt$$

$$\le \|\varphi\|_\infty \frac{1}{n!} e^{\mu x} \int_0^x |q(t)| \, \hat{q}(t)^n \, dt.$$

We compute the last integral by

$$\int_0^x |q(t)| \, \hat{q}(t)^n \, dt = \int_0^x \hat{q}'(t) \, \hat{q}(t)^n \, dt = \frac{1}{n+1} \int_0^x \frac{d}{dt} \left(\hat{q}(t)^{n+1} \right) dt$$

$$= \frac{1}{n+1} \hat{q}(x)^{n+1}.$$

This proves the estimate (4.10) for $n + 1$.

For estimate (4.11), we only change the initial step $n = 1$ into

$$\left| \tilde{K}\varphi(x) \right| \le \int_0^x e^{\mu(x-t)} \, e^{\mu t} \, |q(t)| \, dt \le e^{\mu x} \, \hat{q}(x).$$

The remaining part is proven by the same arguments. □

Now we prove the equivalence of the initial value problems for u_j, $j = 1, 2$, to Volterra integral equations.

Theorem 4.4
Let $q \in L^2(0,1)$ and $\lambda \in \mathbb{C}$. Then we have:

(a) $u_1, u_2 \in H^2(0,1)$ are solutions of (4.7a) and (4.7b), respectively, if and only if $u_1, u_2 \in C[0,1]$ solve the Volterra integral equations:

$$u_1(x) = \cos(\sqrt{\lambda}\,x) + \int_0^x \frac{\sin\sqrt{\lambda}(x-t)}{\sqrt{\lambda}}\, q(t)\, u_1(t)\, dt, \qquad (4.12a)$$

$$u_2(x) = \frac{\sin(\sqrt{\lambda}\,x)}{\sqrt{\lambda}} + \int_0^x \frac{\sin\sqrt{\lambda}(x-t)}{\sqrt{\lambda}}\, q(t)\, u_2(t)\, dt, \qquad (4.12b)$$

respectively, for $0 \le x \le 1$.

(b) The integral equations (4.12a) and (4.12b) and the initial value problems (4.7a) and (4.7b) are uniquely solvable. The solutions can be represented by a Neumann series. Let K denote the integral operator

$$K\varphi(x) := \int_0^x \frac{\sin\sqrt{\lambda}(x-t)}{\sqrt{\lambda}}\, q(t)\, \varphi(t)\, dt, \quad x \in [0,1], \qquad (4.13)$$

and define

$$C(x) := \cos(\sqrt{\lambda}x) \quad and \quad S(x) := \frac{\sin(\sqrt{\lambda}x)}{\sqrt{\lambda}}. \qquad (4.14)$$

Then

$$u_1 = \sum_{n=0}^{\infty} K^n C \quad and \quad u_2 = \sum_{n=0}^{\infty} K^n S. \qquad (4.15)$$

The series converge uniformly with respect to $(x, \lambda, q) \in [0,1] \times \Lambda \times Q$ for all bounded sets $\Lambda \subset \mathbb{C}$ and $Q \subset L^2(0,1)$.

Proof: (a) We will use the following version of partial integration for $f, g \in H^2(0,1)$:

$$\int_a^b [f''(t)\, g(t) - f(t)\, g''(t)]\, dt = [f'(t)\, g(t) - f(t)\, g'(t)]\big|_a^b. \qquad (4.16)$$

We restrict ourselves to the proof for u_1. Let u_1 be a solution of (4.7a). Then

$$\int_0^x S(x-t)\, q(t)\, u_1(t)\, dt = \int_0^x S(x-t)\left[\lambda u_1(t) + u_1''(t)\right] dt$$

$$= \int_0^x u_1(t) \underbrace{\left[\lambda\, S(x - t) + S''(x - t) \right]}_{=0} dt$$

$$+ \left[u_1'(t)\, S(x - t) + u_1(t)\, S'(x - t) \right]\Big|_{t=0}^{t=x}$$

$$= u_1(x) - \cos(\sqrt{\lambda}x).$$

On the other hand, let $u_1 \in C[0,1]$ be a solution of the integral equation (4.12a). Then u_1 is differentiable almost everywhere and

$$u_1'(x) = -\sqrt{\lambda}\, \sin(\sqrt{\lambda}x) + \int_0^x \cos\sqrt{\lambda}(x - t)\, q(t)\, u_1(t)\, dt.$$

From this, we observe that u_1' is continuous and even differentiable almost everywhere and

$$u_1''(x) = -\lambda \cos(\sqrt{\lambda}x) + q(x)\, u_1(x)$$

$$- \int_0^x \sqrt{\lambda}\, \sin\sqrt{\lambda}(x - t)\, q(t)\, u_1(t)\, dt$$

$$= -\lambda\, u_1(x) + q(x)\, u_1(x).$$

This proves the assertion since the right-hand side is in $L^2(0,1)$ and the initial conditions are obviously satisfied.

(b) We observe that all of the functions $k(\tau) = \cos(\sqrt{\lambda}\tau)$, $k(\tau) = \sin(\sqrt{\lambda}\tau)$, and $k(\tau) = \sin(\sqrt{\lambda}\tau)/\sqrt{\lambda}$ for $\tau \in [0,1]$ satisfy the estimate $|k(\tau)| \leq \exp(\mu\tau)$ with $\mu = |\mathrm{Im}\,\sqrt{\lambda}|$. This is obvious for the first two functions. For the third, it follows from

$$\left| \frac{\sin(\sqrt{\lambda}\tau)}{\sqrt{\lambda}} \right| \leq \int_0^\tau \left| \cos(\sqrt{\lambda}s) \right| ds = \int_0^\tau \cosh(\mu s)\, ds \leq e^{\mu\tau}.$$

We have to study the integral operator K with kernel $k(x - t)q(t)$, where $k(\tau) = \sin(\sqrt{\lambda}\tau)/\sqrt{\lambda}$. We apply Lemma 4.3 with $\tilde{K} = K$. Estimate (4.10) yields

$$\|K^n\| \leq \frac{\hat{q}(1)^n}{n!}\, e^\mu < 1$$

for sufficiently large n uniformly for $q \in Q$ and $\lambda \in \Lambda$. Therefore, the Neumann series converges (see Appendix A, Theorem A.29), and part (b) is proven.   □

The integral representation of the previous theorem yields the following asymptotic behavior of the fundamental system by comparing the case for arbitrary q with the case of $q = 0$.

Theorem 4.5

Let $q \in L^2(0,1)$, $\lambda \in \mathbb{C}$, and u_1, u_2 be the fundamental system, i.e., the solutions of the initial value problems (4.7a) and (4.7b), respectively. Then we have for all $x \in [0,1]$:

$$\left| u_1(x) - \cos(\sqrt{\lambda}x) \right| \leq \frac{1}{|\sqrt{\lambda}|} \exp\left(|\text{Im }\sqrt{\lambda}| \, x + \int_0^x |q(t)| \, dt \right), \quad (4.17a)$$

$$\left| u_2(x) - \frac{\sin(\sqrt{\lambda}x)}{\sqrt{\lambda}} \right| \leq \frac{1}{|\lambda|} \exp\left(|\text{Im }\sqrt{\lambda}| \, x + \int_0^x |q(t)| \, dt \right), \quad (4.17b)$$

$$\left| u_1'(x) + \sqrt{\lambda}\sin(\sqrt{\lambda}x) \right| \leq \exp\left(|\text{Im }\sqrt{\lambda}| \, x + \int_0^x |q(t)| \, dt \right), \quad (4.17c)$$

$$\left| u_2'(x) - \cos(\sqrt{\lambda}x) \right| \leq \frac{1}{|\sqrt{\lambda}|} \exp\left(|\text{Im }\sqrt{\lambda}| \, x + \int_0^x |q(t)| \, dt \right). \quad (4.17d)$$

Proof: Again, we use the Neumann series and define $C(\tau) := \cos(\sqrt{\lambda}\tau)$ and $S(\tau) := \sin(\sqrt{\lambda}\tau)/\sqrt{\lambda}$. Let K be the integral operator with kernel $q(t) \sin(\sqrt{\lambda}(x-t))/\sqrt{\lambda}$. Then

$$\left| u_1(x) - \cos(\sqrt{\lambda}x) \right| \leq \sum_{n=1}^{\infty} |K^n C(x)|.$$

Now we set $\tilde{k}(\tau) = \sin(\sqrt{\lambda}\tau)$ and $k(\tau) = \sin(\sqrt{\lambda}\tau)/\sqrt{\lambda}$ and denote by $\tilde{K}$ and K the Volterra integral operators with kernels $\tilde{k}(x-t)$ and $k(x-t)$, respectively. Then $K^n = \frac{1}{\sqrt{\lambda}} \tilde{K} K^{n-1}$ and, by Lemma 4.3, part (b), we conclude that

$$|K^n C(x)| \leq \frac{1}{|\sqrt{\lambda}| \, n!} \left(\int_0^x |q(t)| \, dt \right)^n \exp\left(|\text{Im }\sqrt{\lambda}| \, x \right)$$

for $n \geq 1$. Summation now yields the desired estimate:

$$\left| u_1(x) - \cos(\sqrt{\lambda}x) \right| \leq \frac{1}{|\sqrt{\lambda}|} \exp\left(|\text{Im }\sqrt{\lambda}| \, x + \int_0^x |q(t)| \, dt \right).$$

Since $|S(x)| \leq \frac{1}{|\sqrt{\lambda}|} \exp(|\operatorname{Im} \sqrt{\lambda}| \, x)$, the same arguments prove the estimate (4.17b). Differentiation of the integral equations (4.12a) and (4.12b) yields

$$u_1{}'(x) \; + \; \sqrt{\lambda} \sin(\sqrt{\lambda}\, x) \;\; = \;\; \int_0^x \cos \sqrt{\lambda}(x - t) \, q(t) \, u_1(t) \, dt,$$

$$u_2{}'(x) \; - \; \cos(\sqrt{\lambda}\, x) \;\; = \;\; \int_0^x \cos \sqrt{\lambda}(x - t) \, q(t) \, u_2(t) \, dt.$$

Now we define the operator K as before and $\tilde{K}$ as the one with kernel $q(t) \cos \sqrt{\lambda}(x - t)$. Then

$$u_1{}'(x) \; + \; \sqrt{\lambda} \sin(\sqrt{\lambda}\, x) \;\; = \;\; \tilde{K} \sum_{n=0}^{\infty} K^n C \,,$$

$$u_2{}'(x) \; - \; \cos(\sqrt{\lambda}\, x) \;\; = \;\; \tilde{K} \sum_{n=0}^{\infty} K^n S \,,$$

and we use Lemma 4.3, part (b), again. Summation yields the estimates (4.17c) and (4.17d). □

In the next section, we will need the fact that the eigenfunctions are continuously differentiable with respect to q and λ. We remind the reader of the concept of Fréchet differentiability (F-differentiability) of an operator between Banach spaces X and Y (see Appendix A, Definition A.53). Here we consider the mapping $(\lambda, q) \mapsto u_j(\cdot, \lambda, q)$ from $\mathbb{C} \times L^2(0, 1)$ into $C[0, 1]$ for $j = 1, 2$. We denote these mappings by u_j again and prove the following theorem.

Theorem 4.6
Let $u_j : \mathbb{C} \times L^2(0, 1) \to C[0, 1]$, $j = 1, 2$, be the solution operator of (4.7a) and (4.7b), respectively. Then we have the following:

(a) u_j is continuous.

(b) u_j is continuously F-differentiable for every $(\hat{\lambda}, \hat{q}) \in \mathbb{C} \times L^2(0, 1)$ with partial derivatives

$$\frac{\partial}{\partial \lambda} u_j(\cdot, \hat{\lambda}, \hat{q}) \;=\; u_{j,\lambda}(\cdot, \hat{\lambda}, \hat{q}) \tag{4.18a}$$

and

$$\frac{\partial}{\partial q} u_j(\cdot, \hat{\lambda}, \hat{q}) \, (q) \;=\; u_{j,q}(\cdot, \hat{\lambda}, \hat{q}), \tag{4.18b}$$

where $u_{j,\lambda}(\cdot, \hat{\lambda}, \hat{q})$ and $u_{j,q}(\cdot, \hat{\lambda}, \hat{q})$ are solutions of the following initial boundary value problems for $j = 1, 2$:

$$
\begin{aligned}
-\left(u_{j,\lambda}\right)'' + \left(\hat{q} - \hat{\lambda}\right)u_{j,\lambda} &= u_j(\cdot, \hat{\lambda}, \hat{q}) \quad \text{in } (0, 1) \\
u_{j,\lambda}(0) &= 0, \quad \left(u_{j,\lambda}\right)'(0) = 0, \\
-\left(u_{j,q}\right)'' + \left(\hat{q} - \hat{\lambda}\right)u_{j,q} &= -q\, u_j(\cdot, \hat{\lambda}, \hat{q}) \quad \text{in } (0, 1), \\
u_{j,q}(0) &= 0, \quad \left(u_{j,q}\right)'(0) = 0.
\end{aligned}
\tag{4.19}
$$

(c) Furthermore, for all $x \in [0, 1]$ we have:

$$
\int_0^x u_j(t)^2 dt = [u_{j,\lambda}, u_j](x), \quad j = 1, 2, \tag{4.20a}
$$

$$
\int_0^x u_1(t)\, u_2(t)\, dt = [u_{1,\lambda}, u_2](x) = [u_{2,\lambda}, u_1](x), \tag{4.20b}
$$

$$
-\int_0^x q(t)\, u_j(t)^2 dt = [u_{j,q}, u_j](x), \quad j = 1, 2, \tag{4.20c}
$$

$$
-\int_0^x q(t)\, u_1(t)\, u_2(t)\, dt = [u_{1,q}, u_2](x) = [u_{2,q}, u_1](x), \tag{4.20d}
$$

where $[u, v]$ denotes the Wronskian determinant from (4.8).

Proof: (a), (b): Continuity and differentiability of u_j follow from the integral equations (4.12a) and (4.12b) since the kernel and the right-hand sides depend continuously and differentiably on λ and q. It remains to show the representation of the derivatives in (b). Let $u = u_j$, $j = 1$ or 2. Then

$$
\begin{aligned}
-u''(\cdot, \hat{\lambda} + \varepsilon) &+ \left(\hat{q} - \hat{\lambda} - \varepsilon\right) u(\cdot, \hat{\lambda} + \varepsilon) &= 0, \\
-u''(\cdot, \hat{\lambda}) &+ \left(\hat{q} - \hat{\lambda}\right) u(\cdot, \hat{\lambda}) &= 0;
\end{aligned}
$$

thus

$$
-\frac{1}{\varepsilon}[u(\cdot, \hat{\lambda} + \varepsilon) - u(\cdot, \hat{\lambda})]'' + \left(\hat{q} - \hat{\lambda}\right)\frac{1}{\varepsilon}[u(\cdot, \hat{\lambda} + \varepsilon) - u(\cdot, \hat{\lambda})] = u(\cdot, \hat{\lambda} + \varepsilon).
$$

Furthermore, the homogeneous initial conditions are satisfied for the difference quotient. The right-hand side converges uniformly to $u(\cdot, \hat{\lambda})$ as $\varepsilon \to 0$. Therefore, the difference quotient converges to u_λ uniformly in x. The same arguments yield the result for the derivative with respect to q.

(c) Multiplication of the differential equation for $u_{j,\lambda}$ by u_j and the differential equation for u_j by $u_{j,\lambda}$ and subtraction yields

$$\begin{aligned} u_j^2(x) &= u_j''(x)\, u_{j,\lambda}(x) - u_{j,\lambda}''(x)\, u_j(x) \\ &= \frac{d}{dx}\big(u_j'(x)\, u_{j,\lambda}(x) - u_{j,\lambda}'(x)\, u_j(x)\big). \end{aligned}$$

Integration of this equation and the homogeneous boundary conditions yield the first equation of (4.20a). The proofs for the remaining equations use the same arguments and are left to the reader. □

At no place in this section have we used the assumption that q is real-valued. Therefore, the assertions of Theorems 4.4, 4.5, and 4.6 also hold for complex-valued q.

4.3 Asymptotics of the Eigenvalues and Eigenfunctions

We first restrict ourselves to the Dirichlet problem, i.e., the eigenvalue problem

$$- u''(x) + q(x)\, u(x) = \lambda u(x), \quad 0 < x < 1, \quad u(0) = u(1) = 0. \quad (4.21)$$

Refer to the end of this section for different boundary conditions. Again, let $q \in L^2(0,1)$ be real-valued. We observe that $\lambda \in \mathbb{C}$ is an eigenvalue of this problem if and only if λ is a zero of the function $f(\lambda) := u_2(1, \lambda, q)$. Again, $u_2 = u_2(\cdot, \lambda, q)$ denotes the solution of the initial value problem (4.7b) with initial conditions $u_2(0) = 0$ and $u_2'(0) = 1$. If $u_2(1, \lambda, q) = 0$, then $u_2(\cdot, \lambda, q)$ is the eigenfunction corresponding to the eigenvalue λ. The function f plays exactly the role of the well-known characteristic polynomial for matrices and is therefore called the *characteristic function* of the eigenvalue problem. Theorem 4.6 implies that f is differentiable, i.e., analytic in all of $\mathbb{C}$. This observation makes it possible to use tools from complex analysis. First, we summarize well-known facts about eigenvalues and eigenfunctions for the Sturm–Liouville problem, which can easily be derived from abstract spectral theory (see Theorems A.47 and A.48).

Lemma 4.7
Let $q \in L^2(0,1)$ be real-valued. Then:

(a) *All eigenvalues λ are real.*

(b) There exists a countable number of real eigenvalues λ_j, $j \in \mathbb{N}$, with corresponding eigenfunctions $g_j \in C[0,1]$ such that $\|g_j\|_{L^2} = 1$. The eigenfunctions form a complete orthonormal system in $L^2(0,1)$.

(c) The geometric and algebraic multiplicities of the eigenvalues λ are one, i.e., the eigenspaces are one-dimensional and the zeros of the characteristic function are simple.

Proof: (a) and (b) follow from the fact that the boundary value problem is self-adjoint. We refer to Problem 4.1 for a repetition of the proof.

(c) Let λ be an eigenvalue and u, v be two corresponding eigenfunctions. Choose α, β with $\alpha^2 + \beta^2 > 0$ and such that $\alpha\, u'(0) = \beta\, v'(0)$. The function $w := \alpha u - \beta v$ solves the differential equation and $w(0) = w'(0) = 0$, i.e., w vanishes identically. Therefore, u and v are linearly dependent.

We apply Theorem 4.6, part (c), to show that λ is a simple zero of f. Since $u_2(1,\lambda,q) = 0$, we have from (4.20a) for $j = 2$ that

$$f'(\lambda) \;=\; \frac{\partial}{\partial\lambda} u_2(1,\lambda,q) \;=\; u_{2,\lambda}(1,\lambda,q)$$

$$\;=\; \frac{1}{u_2'(1,\lambda,q)} \int_0^1 u_2(x,\lambda,q)^2 \, dx \;\neq\; 0.$$

This proves part (c). $\square$

We note that there are different ways to normalize the eigenfunctions. Instead of requiring the L^2-norm to be one, we will sometimes normalize them such that $g_j'(0) = 1$. This is possible since $g_j'(0) \neq 0$. Otherwise, the Picard–Lindelöf uniqueness theorem (see Theorem 4.4) would imply that g_j vanishes identically.

Also, we need the following technical result.

Lemma 4.8
Let $z \in \mathbb{C}$ with $|z - n\pi| \geq \pi/4$ for all $n \in \mathbb{Z}$. Then

$$\exp\bigl(|\mathrm{Im}\, z|\bigr) \;<\; 4\, |\sin z|.$$

Proof: Let $\psi(z) = \exp|z_2| / |\sin z|$ for $z = z_1 + iz_2$, $z_1, z_2 \in \mathbb{R}$. We consider two cases:

1st case: $|z_2| > \ln 2/2$. Then

$$\psi(z) \;=\; \frac{2\, e^{|z_2|}}{|e^{iz_1 - z_2} - e^{-iz_1 + z_2}|} \;\leq\; \frac{2\, e^{|z_2|}}{e^{|z_2|} - e^{-|z_2|}} \;=\; \frac{2}{1 - e^{-2|z_2|}} \;<\; 4$$

since $\exp(-2\,|z_2|) < 1/2$.

2nd case: $|z_2| \leq \ln 2/2$. From $|z - n\pi| \geq \pi/4$ for all n, we conclude that $|z_1 - n\pi|^2 \geq \pi^2/16 - z_2^2 \geq \pi^2/16 - (\ln 2)^2/4 \geq \pi^2/64$; thus $|\sin z_1| \geq \sin \frac{\pi}{8}$. With $|\text{Re} \sin z| = |\sin z_1| \, |\cosh z_2| \geq |\sin z_1|$, we conclude that

$$\psi(z) \leq \frac{e^{|z_2|}}{|\text{Re} \sin z|} \leq \frac{\sqrt{2}}{|\sin z_1|} \leq \frac{\sqrt{2}}{|\sin \frac{\pi}{8}|} < 4. \qquad \square$$

Now we prove the "counting lemma," a first crude asymptotic formula for the eigenvalues. As the essential tool in the proof, we use the theorem of Rouché from complex analysis (see [1]), which we state for the convenience of the reader: *Let $U \subset \mathbb{C}$ be a domain and the functions F and G be analytic in $\mathbb{C}$ and $|F(z) - G(z)| < |G(z)|$ for all $z \in \partial U$. Then F and G have the same number of zeros in U.*

Lemma 4.9
Let $q \in L^2(0,1)$ and $N > 2 \exp(\|q\|_{L^1})$ be an integer. Then:

(a) The characteristic function $f(\lambda) := u_2(1, \lambda, q)$ has exactly N zeros in the half-plane

$$H := \{\lambda \in \mathbb{C} : \text{Re} \, \lambda < (N + 1/2)^2 \pi^2\}. \tag{4.22}$$

(b) For every $m > N$ there exists exactly one zero of f in the set

$$U_m := \{\lambda \in \mathbb{C} : |\sqrt{\lambda} - m\pi| < \pi/2\}. \tag{4.23}$$

Here we take the branch with $\text{Re} \sqrt{\lambda} \geq 0$.

(c) There are no other zeros of f in $\mathbb{C}$.

Proof: We are going to apply Rouché's theorem to the function $F(z) = f(z^2) = u_2(1, z^2, q)$ and the corresponding function G of the eigenvalue problem for $q = 0$, i.e., $G(z) := \sin z/z$. For U we take one of the sets W_m or V_R defined by

$$\begin{aligned} W_m &:= \{z \in \mathbb{C} : |z - m\pi| < \pi/2\}, \\ V_R &:= \{z \in \mathbb{C} : |\text{Re} \, z| < (N + 1/2)\pi, \, |\text{Im} \, z| < R\} \end{aligned}$$

for fixed $R > (N + 1/2)\pi$ and want to apply Lemma 4.8:

(i) First let $z \in \partial W_m$: For $n \in \mathbb{Z}$, $n \neq m$, we have $|z - n\pi| \geq |m - n|\pi - |z - m\pi| \geq \pi - \pi/2 > \pi/4$. For $n = m$, we observe that $|z - m\pi| = \pi/2 > \pi/4$. Therefore, we can apply Lemma 4.8 for $z \in \partial W_m$. Furthermore, we note the estimate $|z| \geq m\pi - |z - m\pi| = (m - 1/2)\pi > N\pi > 2N$ for all $z \in \partial W_m$.

(ii) Let $z \in \partial V_R$, $n \in \mathbb{Z}$. Then $|\mathrm{Re}\, z| = (N + 1/2)\pi$ or $|\mathrm{Im}\, z| = R$. In either case, we estimate $|z - n\pi|^2 = (\mathrm{Re}\, z - n\pi)^2 + (\mathrm{Im}\, z)^2 \geq \pi^2/4 > \pi^2/16$. Therefore, we can apply Lemma 4.8 for $z \in \partial V_R$. Furthermore, we have the estimate $|z| \geq (N + 1/2)\pi > 2N$ for all $z \in \partial V_R$.

Application of Theorem 4.5 and Lemma 4.8 yields the following estimate for all $z \in \partial V_R \cup \partial W_m$:

$$\left| F(z) - \frac{\sin z}{z} \right| \leq \frac{1}{|z|^2} \exp\big(|\mathrm{Im}\, z| + \|q\|_{L^1}\big) \leq \frac{4\,|\sin z|}{|z|^2} \frac{N}{2}$$

$$= \frac{2N}{|z|} \left| \frac{\sin z}{z} \right| < \left| \frac{\sin z}{z} \right|.$$

Therefore, F and $G(z) := \sin z / z$ have the same number of zeros in V_R and every W_m. Since the zeros of G are $\pm n\pi$, $n = 1, 2, \ldots$ we conclude that G has exactly $2N$ zeros in V_R and exactly one zero in every W_m. By the theorem of Rouché, this also holds for F.

Now we show that F has no zero outside of $V_R \cup \bigcup_{m>N} W_m$. Again, we apply Lemma 4.8: Let $z \notin V_R \cup \bigcup_{m>N} W_m$. From $z \notin V_R$, we conclude that $|z| = \sqrt{(\mathrm{Re}\, z)^2 + (\mathrm{Im}\, z)^2} \geq (N + 1/2)\pi$. For $n > N$, we have that $|z - n\pi| > \pi/2$ since $z \notin W_n$. For $n \leq N$, we conclude that $|z - n\pi| \geq |z| - n\pi \geq (N + 1/2 - n)\pi \geq \pi/2$. We apply Lemma 4.8 again and use the second triangle inequality. This yields

$$|F(z)| \geq \left| \frac{\sin z}{z} \right| - \frac{1}{|z|^2} \exp\big(|\mathrm{Im}\, z| + \|q\|_{L^1}\big)$$

$$\geq \left| \frac{\sin z}{z} \right| \left[1 - \frac{4 \exp(\|q\|_{L^1})}{|z|} \right]$$

$$\geq \left| \frac{\sin z}{z} \right| \left[1 - \frac{2N}{|z|} \right] > 0$$

since $|z| \geq (N + 1/2)\pi > 2N$. Therefore, we have shown that f has exactly one zero in every U_m, $m > N$, and N zeros in the set

$$H_R := \big\{ \lambda \in \mathbb{C} : 0 < \mathrm{Re}\,\sqrt{\lambda} < (N + 1/2)\pi, \ \big|\mathrm{Im}\,\sqrt{\lambda}\big| < R \big\}$$

and no other zeros. It remains to show that $H_R \subset H$. For $\lambda = |\lambda| \exp(i\theta) \in H_R$, we conclude that $\mathrm{Re}\,\sqrt{\lambda} = \sqrt{|\lambda|} \cos \frac{\theta}{2} < (N + 1/2)\pi$; thus $\mathrm{Re}\,\lambda = |\lambda| \cos\big(2\frac{\theta}{2}\big) \leq |\lambda| \cos^2 \frac{\theta}{2} < (N + 1/2)^2 \pi^2$. $\qquad\square$

This lemma proves again the existence of infinitely many eigenvalues. The arguments are not changed for the case of complex-valued functions q. In this case, the general spectral theory is not applicable anymore since the boundary value problem is not self-adjoint. This lemma also provides more information about the eigenvalue distribution, even for the real-valued case. First, we order the eigenvalues in the form

$$\lambda_1 < \lambda_2 < \lambda_3 < \cdots.$$

Lemma 4.9 implies that

$$\sqrt{\lambda_n} = n\pi + \mathcal{O}(1), \quad \text{i.e.,} \quad \lambda_n = n^2\pi^2 + \mathcal{O}(n). \tag{4.24}$$

For the treatment of the inverse problem, it will be necessary to improve this formula. It is our aim to prove that

$$\lambda_n = n^2\pi^2 + \int_0^1 q(t)\, dt + \tilde{\lambda}_n \quad \text{where} \quad \sum_{n=1}^\infty |\tilde{\lambda}_n|^2 < \infty. \tag{4.25}$$

There are several methods to prove (4.25). We follow the treatment in [177]. The key is to apply the fundamental theorem of calculus to the function $t \mapsto \lambda_n(tq)$ for $t \in [0,1]$, thus connecting the eigenvalues λ_n corresponding to q with the eigenvalues $n^2\pi^2$ corresponding to $q = 0$ by the parameter t. For this approach, we need the differentiability of the eigenvalues with respect to q.

For fixed $n \in \mathbb{N}$, the function $q \mapsto \lambda_n(q)$ from $L^2(0,1)$ into $\mathbb{C}$ is well-defined and Fréchet differentiable by the following theorem.

Theorem 4.10
For every $n \in \mathbb{N}$, the mapping $q \mapsto \lambda_n(q)$ from $L^2(0,1)$ into $\mathbb{C}$ is continuously Fréchet differentiable for every $\hat{q} \in L^2(0,1)$ and

$$\lambda_n'(\hat{q})q = \int_0^1 g_n(x,\hat{q})^2\, q(x)\, dx, \quad q \in L^2(0,1). \tag{4.26}$$

Here,

$$g_n(x,\hat{q}) := \frac{u_2(x,\hat{\lambda}_n,\hat{q})}{\|u_2(\cdot,\hat{\lambda}_n,\hat{q})\|_{L^2}}$$

denotes the L^2-normalized eigenfunction corresponding to $\hat{\lambda}_n := \lambda_n(\hat{q})$.

Proof: We observe that $u_2(1,\hat{\lambda}_n,\hat{q}) = 0$ and will apply the implicit function theorem to the equation

$$u_2(1,\lambda,q) = 0$$

in a neighborhood of $(\hat{\lambda}_n,\hat{q})$. This is possible since the zero $\hat{\lambda}_n$ of $u_2(1,\cdot,\hat{q})$ is simple by Lemma 4.7. The implicit function theorem (see Appendix A,

Theorem A.58) yields the existence of a unique function $\lambda_n = \lambda_n(q)$ such that $u_2(1, \lambda_n(q), q) = 0$ for all q in a neighborhood of $\hat{q}$; we know this already. But it also implies that the function λ_n is continuously differentiable with respect to q and

$$0 = \frac{\partial}{\partial \lambda} u_2(1, \hat{\lambda}_n, \hat{q}) \, \lambda_n'(\hat{q}) q + \frac{\partial}{\partial q} u_2(1, \hat{\lambda}_n, \hat{q}) q \,,$$

i.e., $u_{2,\lambda}(1) \, \lambda_n'(\hat{q}) q + u_{2,q}(1) = 0$. With Theorem 4.6, part (c), we conclude that

$$\lambda_n'(\hat{q})q = -\frac{u_{2,q}(1)}{u_{2,\lambda}(1)} = -\frac{u_{2,q}(1) \, u_2'(1)}{u_{2,\lambda}(1) \, u_2'(1)}$$

$$= -\frac{[u_{2,q}, u_2](1)}{[u_{2,\lambda}, u_2](1)} = \frac{\int_0^1 q(x) \, u_2(x)^2 dx}{\int_0^1 u_2(x)^2 dx} = \int_0^1 g_n(x, \hat{q})^2 q(x) \, dx \,,$$

where we have dropped the arguments $\hat{\lambda}$ and $\hat{q}$. □

Now we are ready to formulate and prove the main theorem which follows.

Theorem 4.11
Let $Q \subset L^2(0,1)$ be bounded, $q \in Q$, and $\lambda_n \in \mathbb{C}$ the corresponding eigenvalues. Then we have

$$\lambda_n = n^2 \pi^2 + \int_0^1 q(t) \, dt - \int_0^1 q(t) \cos(2n\pi t) \, dt + \mathcal{O}(1/n) \qquad (4.27)$$

for $n \to \infty$ uniformly for $q \in Q$. Furthermore, the corresponding eigenfunctions g_n, normalized to $\|g_n\|_{L^2} = 1$, have the following asymptotic behavior:

$$g_n(x) = \sqrt{2} \sin(n\pi x) + \mathcal{O}(1/n) \quad \text{and} \qquad (4.28a)$$
$$g_n'(x) = \sqrt{2} \, n \pi \, \cos(n\pi x) + \mathcal{O}(1) \qquad (4.28b)$$

as $n \to \infty$ uniformly for $x \in [0,1]$ and $q \in Q$.

We observe that the second integral on the right-hand side of (4.27) is the nth Fourier coefficient a_n of q with respect to $\{\cos(2\pi nt) : n = 0, 1, 2, \ldots, \}$. From Fourier theory, it is known that a_n converges to zero, and even more: Parseval's identity yields that $\sum_{n=0}^{\infty} |a_n|^2 < \infty$, i.e., (4.25) is satisfied. If q is smooth enough, e.g., continuously differentiable, then a_n tends to zero faster than $1/n$. In that case, this term disappears in the $\mathcal{O}(1/n)$ expression.

Proof: We split the proof into four parts:

(a) First, we show that $g_n(x) = \sqrt{2}\,\sin(\sqrt{\lambda_n}x) + \mathcal{O}(1/n)$ uniformly for $(x, q) \in [0, 1] \times Q$. By Lemma 4.9, we know that $\sqrt{\lambda_n} = n\pi + \mathcal{O}(1)$, and thus by Theorem 4.5

$$u_2(x, \lambda_n) = \frac{\sin(\sqrt{\lambda_n}x)}{\sqrt{\lambda_n}} + \mathcal{O}(1/n^2).$$

With the formula $2\int_0^1 \sin^2(\alpha t)\,dt = 1 - \sin(2\alpha)/(2\alpha)$, we compute

$$
\begin{aligned}
\int_0^1 u_2(t, \lambda_n)^2 dt &= \frac{1}{\lambda_n} \int_0^1 \sin^2(\sqrt{\lambda_n}t)\,dt + \mathcal{O}(1/n^3) \\
&= \frac{1}{2\lambda_n}\left[1 - \frac{\sin(2\sqrt{\lambda_n})}{2\sqrt{\lambda_n}}\right] + \mathcal{O}(1/n^3) \\
&= \frac{1}{2\lambda_n}\left[1 + \mathcal{O}(1/n)\right].
\end{aligned}
$$

Therefore, we have

$$g_n(x) = \frac{u_2(x, \lambda_n)}{\sqrt{\int_0^1 u_2(t, \lambda_n)^2 dt}} = \sqrt{2}\,\sin(\sqrt{\lambda_n}x) + \mathcal{O}(1/n).$$

(b) Now we show that $\sqrt{\lambda_n} = n\pi + \mathcal{O}(1/n)$ and $g_n(x) = \sqrt{2}\sin(n\pi x) + \mathcal{O}(1/n)$. We apply the fundamental theorem of calculus and use Theorem 4.10:

$$\lambda_n - n^2\pi^2 = \lambda_n(q) - \lambda_n(0) = \int_0^1 \frac{d}{dt}\lambda_n(tq)\,dt \tag{4.29}$$

$$= \int_0^1 \lambda_n'(tq)q\,dt = \int_0^1\int_0^1 g_n(x, tq)^2 q(x)\,dx\,dt = \mathcal{O}(1).$$

This yields $\sqrt{\lambda_n} = n\pi + \mathcal{O}(1/n)$ and, with part (a), the asymptotic form $g_n(x) = \sqrt{2}\,\sin(n\pi x) + \mathcal{O}(1/n)$.

(c) Now the asymptotics of the eigenvalues follow easily from (4.29) by the observation that

$$g_n(x, tq)^2 = 2\sin^2(n\pi x) + \mathcal{O}(1/n) = 1 - \cos(2n\pi x) + \mathcal{O}(1/n),$$

uniformly for $t \in [0, 1]$ and $q \in Q$.

(d) Similarly, we have for the derivatives

$$g_n'(x) \;=\; \frac{u_2'(x, \lambda_n)}{\sqrt{\int_0^1 u_2(t, \lambda_n)^2 dt}} \;=\; \frac{\sqrt{2}\sqrt{\lambda_n}\,\cos(\sqrt{\lambda_n}x) + \mathcal{O}(1)}{\sqrt{1 + \mathcal{O}(1/n)}}$$

$$=\; \sqrt{2}\,n\,\pi\,\cos(n\pi x) \;+\; \mathcal{O}(1). \qquad\qquad \Box$$

Example 4.12

We illustrate Theorem 4.10 by the following two numerical examples:

(a) Let $q_1(x) = \exp\big(\sin(2\pi x)\big)$, $x \in [0,1]$. Then q_1 is analytic and periodic with period 1.

(b) Let $q_2(x) = -5x$ for $0 \le x \le 0.4$ and $q_2(x) = 4$ for $0.4 < x \le 1$. The function q_2 is not continuous.

Plots of the characteristic functions $\lambda \mapsto f(\lambda)$ for q_1, q_2 and $q = 0$, i.e., $\lambda \mapsto \sin\sqrt{\lambda}/\sqrt{\lambda}$ are shown in Figures 4.1 and 4.2.

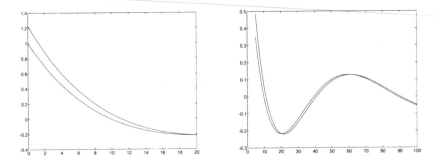

Figure 4.1: Characteristic function of q_1 on $[0,20]$ and $[5,100]$

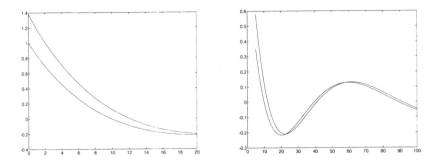

Figure 4.2: Characteristic function of q_2 on $[0,20]$ and $[5,100]$

The Fourier coefficients of q_1 converge to zero of exponential order. The following table shows the eigenvalues λ_n corresponding to q_1, the eigenvalues

$n^2\pi^2$ corresponding to $q = 0$ and the difference

$$c_n := \lambda_n - n^2\pi^2 - \int_0^1 q(x)\,dx \quad \text{for } n = 1, \ldots, 10:$$

λ_n	$n^2\pi^2$	c_n
11.1	9.9	$-2.04 * 10^{-2}$
40.9	39.5	$1.49 * 10^{-1}$
90.1	88.8	$2.73 * 10^{-3}$
159.2	157.9	$-1.91 * 10^{-3}$
248.0	246.7	$7.74 * 10^{-4}$
356.6	354.3	$4.58 * 10^{-4}$
484.9	483.6	$4.58 * 10^{-4}$
632.9	631.7	$4.07 * 10^{-4}$
800.7	799.4	$3.90 * 10^{-4}$
988.2	987.0	$3.83 * 10^{-4}$

We clearly observe the rapid convergence.

Because q_2 is not continuous, the Fourier coefficients converge to zero only slowly. Again, we list the eigenvalues λ_n for q_2, the eigenvalues $n^2\pi^2$ corresponding to $q = 0$, and the differences

$$c_n := \lambda_n - n^2\pi^2 - \int_0^1 q(x)\,dx \quad \text{and}$$

$$d_n := \lambda_n - n^2\pi^2 - \int_0^1 q(x)\,dx + \int_0^1 q(x)\cos(2\pi nx)\,dx$$

for $n = 1, \ldots, 10:$

λ_n	$n^2\pi^2$	c_n	d_n
12.1	9.9	$1.86 * 10^{-1}$	$-1.46 * 10^{-1}$
41.1	39.5	$-3.87 * 10^{-1}$	$8.86 * 10^{-2}$
91.1	88.8	$3.14 * 10^{-1}$	$2.13 * 10^{-2}$
159.8	157.9	$1.61 * 10^{-1}$	$-6.70 * 10^{-3}$
248.8	246.7	$2.07 * 10^{-2}$	$2.07 * 10^{-2}$
357.4	354.3	$8.29 * 10^{-2}$	$-4.24 * 10^{-3}$
484.5	483.6	$-1.25 * 10^{-1}$	$6.17 * 10^{-3}$
633.8	631.7	$1.16 * 10^{-1}$	$3.91 * 10^{-3}$
801.4	799.4	$-6.66 * 10^{-2}$	$-1.38 * 10^{-3}$
989.0	987.0	$5.43 * 10^{-3}$	$5.43 * 10^{-3}$

We will now sketch the modifications necessary for Sturm–Liouville eigenvalue problems of the type

$$-u''(x) + q(x)\, u(x) = \lambda\, u(x), \quad 0 < x < 1, \tag{4.30a}$$

$$u(0) = 0, \quad u'(1) + Hu(1) = 0. \tag{4.30b}$$

Now the eigenvalues are zeros of the characteristic function

$$f(\lambda) \;=\; u_2'(1, \lambda, q) \;+\; H\, u_2(1, \lambda, q), \quad \lambda \in \mathbb{C}. \tag{4.31}$$

For the special case where $q = 0$, we have $u_2(x, \lambda, 0) = \sin(\sqrt{\lambda}x)/\sqrt{\lambda}$. The
characteristic function for this case is then given by

$$g(\lambda) \;=\; \cos\sqrt{\lambda} \;+\; H\,\frac{\sin\sqrt{\lambda}}{\sqrt{\lambda}}.$$

The zeros of f for $q = 0$ and $H = 0$ are $\lambda_n = (n + 1/2)^2\pi^2$, $n = 0, 1, 2, \ldots$
If $H \neq 0$, one has to solve the transcendental equation $z \cot z + H = 0$.
One can show (see Problem 4.4) by an application of the implicit function
theorem in $\mathbb{R}^2$ that the eigenvalues for $q = 0$ behave as

$$\lambda_n \;=\; (n + 1/2)^2\pi^2 \;+\; 2H \;+\; \mathcal{O}(1/n).$$

Lemma 4.7 is also valid since the boundary value problem is again self-
adjoint. The Counting Lemma 4.9 now takes the following form.

Lemma 4.13
Let $q \in L^2(0, 1)$ and $N > 2\exp(\|q\|_{L^1})(1 + |H|)$ be an integer. Then we
have:

(a) The mapping $f(\lambda) := u_2'(1, \lambda, q) + H\, u_2(1, \lambda, q)$ has exactly N zeros
in the half-plane

$$H := \{\lambda \in \mathbb{C} : \operatorname{Re}\lambda < N^2\pi^2\}.$$

(b) f has exactly one zero in every set

$$U_m := \{\lambda \in \mathbb{C} : |\sqrt{\lambda} - (m - 1/2)\pi| < \pi/2\}$$

provided $m > N$.

(c) There are no other zeros of f in $\mathbb{C}$.

For the proof, we refer to Problem 4.3. We can apply the implicit function
theorem to the equation

$$u_2'(1, \lambda_n(q), q) \;+\; H\, u_2(1, \lambda_n(q), q) \;=\; 0$$

since the zeros are again simple. Differentiating this equation with respect to q yields

$$\left[u_{2,\lambda}'(1, \hat{\lambda}_n, \hat{q}) + H u_{2,\lambda}(1, \hat{\lambda}_n, \hat{q}) \right] \lambda_n'(\hat{q})q$$
$$+ \; u_{2,q}'(1, \hat{\lambda}_n, \hat{q}) + H u_{2,q}(1, \hat{\lambda}_n, \hat{q}) = 0.$$

Theorem 4.6 yields

$$\int_0^1 u_2(t)^2 dt = u_{2,\lambda}(1) \underbrace{u_2'(1)}_{=-Hu_2(1)} - u_{2,\lambda}'(1) \, u_2(1)$$

$$= -u_2(1) \left[u_{2,\lambda}'(1) + H u_{2,\lambda}(1) \right],$$

where again we have dropped the arguments $\hat{\lambda}_n$ and $\hat{q}$. Analogously, we compute

$$- \int_0^1 q(t) u_2(t)^2 dt = -u_2(1) \left[u_{2,q}'(1) + H u_{2,q}(1) \right]$$

and thus

$$\lambda_n'(\hat{q})q = -\frac{u_{2,q}'(1) + H u_{2,q}(1)}{u_{2,\lambda}'(1) + H u_{2,\lambda}(1)} = \frac{\int_0^1 q(t) u_2(t)^2 dt}{\int_0^1 u_2(t)^2 dt}.$$

This has the same form as before. We continue as in the case of the Dirichlet boundary condition and arrive at Theorem 4.14.

Theorem 4.14
Let $Q \subset L^2(0, 1)$ be bounded, $q \in Q$, and $H \in \mathbb{R}$. The eigenvalues λ_n have the asymptotic form

$$\lambda_n = \left(n + \frac{1}{2} \right)^2 \pi^2 + 2H + \int_0^1 q(t)\, dt - \int_0^1 q(t)\, \cos(2n + 1)\pi t\, dt + \mathcal{O}(1/n)$$

$$(4.32)$$

as n tends to infinity, uniformly in $q \in Q$. For the L^2-normalized eigenfunctions, we have

$$g_n(x) = \sqrt{2} \sin(n + 1/2)\pi x + \mathcal{O}(1/n) \quad and \quad (4.33a)$$
$$g_n'(x) = \sqrt{2}\,(n + 1/2)\,\pi \cos(n + 1/2)\pi x + \mathcal{O}(1) \quad (4.33b)$$

uniformly for $x \in [0, 1]$ and $q \in Q$.

As mentioned at the beginning of this section, there are other ways to prove the asymptotic formulas for the eigenvalues and eigenfunctions that avoid Lemma 4.9 and the differentiability of λ_n with respect to q. But the proof in, e.g., [228], seems to yield only the asymptotic behavior

$$\lambda_n = m_n^2 \pi^2 + \int_0^1 q(t)\,dt + \mathcal{O}(1/n)$$

instead of (4.27). Here, (m_n) denotes some sequence of natural numbers.

Before we turn to the inverse problem, we make some remarks concerning the case where q is complex-valued. Now the eigenvalue problems are no longer self-adjoint, and the general spectral theory is not applicable anymore. With respect to Lemma 4.7, it is still easy to show that the eigenfunctions corresponding to different eigenvalues are linearly independent and that the geometric multiplicities are still one. The Counting Lemma 4.9 is valid without restrictions. From this, we observe also that the algebraic multiplicities of λ_n are one, at least for $n > N$. Thus, the remaining arguments of this section are valid if we restrict ourselves to the eigenvalues λ_n with $n > N$. Therefore, the asymptotic formulas (4.27), (4.28a), (4.28b), (4.32), (4.33a), and (4.33b) hold equally well for complex-valued q.

4.4   Some Hyperbolic Problems

As a preparation for the following sections, in particular Sections 4.5 and 4.7, we study some initial value problems for the two-dimensional linear hyperbolic partial differential equation

$$\frac{\partial^2 W(x,t)}{\partial x^2} - \frac{\partial^2 W(x,t)}{\partial t^2} + a(x,t)\,W(x,t) = 0,$$

where the coefficient a has the special form $a(x,t) = p(t) - q(x)$. It is well known that the method of characteristics reduces initial value problems to Volterra integral equations of the second kind, which can be studied in spaces of continuous functions. This approach naturally leads to solution concepts for nonsmooth coefficients and boundary data. We will summarize the results in three theorems. In each of them we formulate first the results for the case of smooth coefficients and then for the nonsmooth case. We remark that it is not our aim to relax the solution concept to the weakest possible case but rather to relax the assumptions only to the extent that will be needed in Sections 4.5 and 4.7.

Although most of the problems – at least for smooth data – are subjects of elementary courses on partial differential equations, we include the complete proofs for the convenience of the reader.

Before we begin with the statements of the theorems, we recall some function spaces:

$$C_0[0,1] := \{f \in C[0,1] : f(0) = 0\},$$
$$C_0^j[0,1] := C^j[0,1] \cap C_0[0,1], \quad j = 1,2,$$
$$H^1(0,1) := \left\{f \in C[0,1] : f(x) = \alpha + \int_0^x g(t)\,dt, \ \alpha \in \mathbb{R}, \ g \in L^2(0,1)\right\},$$
$$H_0^1(0,1) := H^1(0,1) \cap C_0[0,1]$$

and equip them with their canonical norms:

$$\|f\|_\infty := \max_{0 \le x \le 1} |f(x)| \quad \text{in } C_0[0,1],$$
$$\|f\|_{C^j} := \max_{\ell=1,\ldots,j} \max_{0 \le x \le 1} \left|f^{(\ell)}(x)\right| \quad \text{in } C_0^j[0,1],$$
$$\|f\|_{H^1} := \sqrt{\|f\|_{L^2}^2 + \|f'\|_{L^2}^2} \quad \text{in } H^1(0,1) \text{ and } H_0^1(0,1).$$

Furthermore, define the triangular regions $\Delta_0 \subset \mathbb{R}^2$ and $\Delta \subset \mathbb{R}^2$ by

$$\Delta_0 := \{(x,t) \in \mathbb{R}^2 : 0 < t < x < 1\}, \tag{4.34a}$$
$$\Delta := \{(x,t) \in \mathbb{R}^2 : |t| < x < 1\}. \tag{4.34b}$$

We begin with an initial value problem, sometimes called the *Goursat problem*.

Theorem 4.15

(a) Let $p,q \in C[0,1]$ and $f \in C^2[0,1]$ with $f(0) = 0$. Then there exists a unique solution $W \in C^2(\overline{\Delta_0})$ of the following hyperbolic initial value problem:

$$\frac{\partial^2 W(x,t)}{\partial x^2} - \frac{\partial^2 W(x,t)}{\partial t^2} + (p(t) - q(x))\,W(x,t) = 0 \quad \text{in } \Delta_0,$$
$$\tag{4.35a}$$

$$W(x,x) = f(x), \quad 0 \le x \le 1, \tag{4.35b}$$
$$W(x,0) = 0, \quad 0 \le x \le 1. \tag{4.35c}$$

(b) The solution operator $(p,q,f) \mapsto W$ has an extension to a bounded operator from $L^2(0,1) \times L^2(0,1) \times C_0[0,1]$ into $C(\overline{\Delta_0})$.

(c) The operator $(p,q,f) \mapsto (W(1,\cdot), W_x(1,\cdot))$ has an extension to a bounded operator from $L^2(0,1) \times L^2(0,1) \times H_0^1(0,1)$ into $H^1(0,1) \times L^2(0,1)$. Here and in the following, we denote by W_x the partial derivative with respect to x.

Proof: First, we extend the problem to the larger region Δ and study the problem

$$\frac{\partial^2 W(x,t)}{\partial x^2} - \frac{\partial^2 W(x,t)}{\partial t^2} + a(x,t)\,W(x,t) \;=\; 0 \quad \text{in } \Delta, \qquad (4.36a)$$

$$W(x,x) \;=\; f(x), \quad 0 \le x \le 1, \qquad (4.36b)$$
$$W(x,-x) \;=\; -f(x), \quad 0 \le x \le 1, \qquad (4.36c)$$

where we have set $a(x,t) := p(|t|) - q(x)$ for $(x,t) \in \Delta$.

To treat problem (4.36a)–(4.36c), we make the change of variables

$$x = \xi + \eta, \quad t = \xi - \eta.$$

Then $(x,t) \in \Delta$ if and only if $(\xi,\eta) \in D$, where

$$D := \big\{ (\xi,\eta) \in (0,1) \times (0,1) : \eta + \xi < 1 \big\}. \qquad (4.37)$$

We set $w(\xi,\eta) := W(\xi + \eta, \xi - \eta)$ for $(\xi,\eta) \in D$. Then W solves problem (4.36a)–(4.36c) if and only if w solves the hyperbolic problem

$$\frac{\partial^2 w(\xi,\eta)}{\partial \xi\,\partial \eta} \;=\; \underbrace{-a(\xi + \eta, \xi - \eta)}_{=:\tilde{a}(\xi,\eta)}\, w(\xi,\eta), \quad (\xi,\eta) \in D, \qquad (4.38a)$$

$$w(\xi,0) \;=\; f(\xi) \quad \text{for } \xi \in [0,1], \qquad (4.38b)$$
$$w(0,\eta) \;=\; -f(\eta) \quad \text{for } \eta \in [0,1]. \qquad (4.38c)$$

Now let w be a solution of (4.38a)–(4.38c). We integrate the differential equation twice and use the initial conditions. Then w solves the integral equation

$$w(\xi,\eta) \;=\; \int_0^\eta \int_0^\xi \tilde{a}(\xi',\eta')\,w(\xi',\eta')\,d\xi'\,d\eta' \;-\; f(\eta) \;+\; f(\xi), \qquad (4.39)$$

for $(\xi,\eta) \in D$. This is a Volterra integral equation in two dimensions. We use the standard method to solve this equation by successive iteration in $C(\overline{D})$. Let A be the Volterra integral operator defined by the integral on the right-hand side of (4.39). By induction with respect to $n \in \mathbb{N}$, it can easily be seen that

$$\big| (A^n w)(\xi,\eta) \big| \;\le\; \|w\|_\infty\, \|\tilde{a}\|_{L^2}^n\, \frac{1}{n!}\, (\xi\eta)^{n/2}, \quad n = 1, 2, \ldots;$$

thus $\|A^n w\|_\infty \leq \|w\|_\infty \|\tilde{a}\|_{L^2}^n \frac{1}{n!}$. Therefore, $\|A^n\|_\infty < 1$ for sufficiently large n and the Neumann series converges (see Appendix A, Theorem A.29). This proves that there exists a unique solution $w \in C(\overline{D})$ of (4.39). From our arguments, uniqueness also holds for (4.36a)–(4.36c).

Now we prove that $w \in C^2(\overline{D})$. We differentiate (4.39) with respect to ξ, which gives

$$w_\xi(\xi, \eta) = \int_0^\eta \left[q(\xi + \eta') - p(|\xi - \eta'|) \right] w(\xi, \eta') \, d\eta' + f'(\xi)$$

$$= \int_\xi^{\xi+\eta} q(y) \, w(\xi, y - \xi) \, dy - \int_{\xi-\eta}^\xi p(|y|) \, w(\xi, \xi - y) \, dy + f'(\xi)$$

and analogously for w_η. This form can be differentiated again. Thus $w \in C^2(\overline{D})$, and we have shown that W is the unique solution of (4.36a)–(4.36c).

Since $a(x, \cdot)$ is an even function and the initial data are odd functions with respect to t, we conclude from the uniqueness result that the solution $W(x, \cdot)$ is also odd. In particular, this implies that $W(x, 0) = 0$ for all $x \in [0, 1]$, which proves that W solves problem (4.35a)–(4.35c) and finishes part (a).

Part (b) follows immediately from the integral equation (4.39) since the integral operator $A : C(\overline{D}) \to C(\overline{D})$ depends continuously on the kernel $\tilde{a} \in L^2(D)$.

For part (c), we observe that

$$W(1, 2\xi - 1) = w(\xi, 1 - \xi) \quad \text{and}$$
$$W_x((1, 2\xi - 1) = \frac{1}{2} w_\xi(\xi, 1 - \xi) + \frac{1}{2} w_\eta(\xi, 1 - \xi).$$

Then the boundedness of the operator follows again by differentiating the integral equation (4.39). By Theorem A.28, there exists a bounded extension of this operator from $L^2(0, 1) \times L^2(0, 1) \times H_0^1(0, 1)$ into $H^1(0, 1) \times L^2(0, 1)$. This ends the proof. $\qquad\square$

If $p, q \in L^2(0, 1)$ and $f \in C_0[0, 1]$, we call the solution

$$W(x, t) = w\left(\frac{1}{2}(x + t), \frac{1}{2}(x - t)\right),$$

where $w \in C(\overline{\Delta})$ solves the integral equation (4.39), a *weak solution* of the Goursat problem (4.35a)–(4.35c). We observe that for every weak solution W there exist sequences $(p_n), (q_n) \subset C[0, 1]$ and $(f_n) \subset C^2[0, 1]$ with $f_n(0) = 0$ and $\|p_n - p\|_{L^2} \to 0$, $\|q_n - q\|_{L^2} \to 0$ and $\|f_n - f\|_\infty \to 0$ such that the solutions $W_n \in C^2(\overline{\Delta_0})$ of (4.35a)–(4.35c) corresponding to p_n, q_n, and f_n converge uniformly to W.

For the special case $p = q = 0$, the integral equation (4.39) reduces to the well-known solution formula

$$W(x,t) = f\left(\frac{1}{2}(x+t)\right) - f\left(\frac{1}{2}(x-t)\right).$$

The next theorem studies a Cauchy problem for the same hyperbolic differential equation.

Theorem 4.16

(a) Let $f \in C^2[0,1]$, $g \in C^1[0,1]$ with $f(0) = f''(0) = g(0) = 0$, and $p, q \in C[0,1]$ and $F \in C(\overline{\Delta_0})$. Then there exists a unique solution $W \in C^2(\overline{\Delta_0})$ of the Cauchy problem

$$\frac{\partial^2 W(x,t)}{\partial x^2} - \frac{\partial^2 W(x,t)}{\partial t^2} + (p(t) - q(x))\, W(x,t) = F(x,t) \quad in\ \Delta_0,$$

$$\tag{4.40a}$$

$$W(1,t) = f(t) \quad for\ 0 \le t \le 1, \tag{4.40b}$$

$$\frac{\partial}{\partial x} W(1,t) = g(t) \quad for\ 0 \le t \le 1. \tag{4.40c}$$

(b) Furthermore, the solution operator $(p, q, F, f, g) \mapsto W$ has an extension to a bounded operator from $L^2(0,1) \times L^2(0,1) \times L^2(\Delta_0) \times H_0^1(0,1) \times L^2(0,1)$ into $C(\overline{\Delta_0})$.

Proof: As in the proof of Theorem 4.15, we set $a(x,t) := p(|t|) - q(x)$ for $(x,t) \in \Delta$ and extend F to an even function on Δ by $F(x,-t) = F(x,t)$ for $(x,t) \in \Delta_0$. We also extend f and g to odd functions on $[-1,1]$ by $f(-t) = -f(t)$ and $g(-t) = -g(t)$ for $t \in [0,1]$. Then $F \in C(\overline{\Delta})$, $f \in C^2[-1,1]$, and $g \in C^1[-1,1]$. We again make the change of variables

$$x = \xi + \eta, \quad t = \xi - \eta, \quad w(\xi,\eta) = W(\xi+\eta, \xi-\eta) \quad for\ (\xi,\eta) \in D,$$

where D is given by (4.37). Then W solves (4.40a)–(4.40c) if and only if w solves

$$\frac{\partial^2 w(\xi,\eta)}{\partial \xi\, \partial \eta} = \tilde{a}(\xi,\eta)\, w(\xi,\eta) + \tilde{F}(\xi,\eta), \quad (\xi,\eta) \in D,$$

where $\tilde{F}(\xi,\eta) = F(\xi+\eta, \xi-\eta)$ and $\tilde{a}(\xi,\eta) = -a(\xi+\eta, \xi-\eta)$. The Cauchy conditions (4.40b) and (4.40c) transform into

$$w(\xi, 1-\xi) = f(2\xi - 1) \quad and \quad w_\xi(\xi, 1-\xi) + w_\eta(\xi, 1-\xi) = 2\, g(2\xi - 1)$$

for $0 \le \xi \le 1$. Differentiating the first equation and solving for w_ξ and w_η yields

$$w_\xi(\xi, 1-\xi) = g(2\xi-1) + f'(2\xi-1) \quad and \quad w_\eta(\xi, 1-\xi) = g(2\xi-1) - f'(2\xi-1)$$

for $0 \leq \xi \leq 1$. Integration of the differential equation with respect to ξ from ξ to $1 - \eta$ yields

$$\frac{\partial w(\xi, \eta)}{\partial \eta} = -\int_{\xi}^{1-\eta} \left[\tilde{a}(\xi', \eta) \, w(\xi', \eta) + \tilde{F}(\xi' + \eta) \right] d\xi' + g(1 - 2\eta) - f'(1 - 2\eta).$$

Now we integrate this equation with respect to η from η to $1 - \xi$ and arrive at

$$w(\xi, \eta) = \int_{\eta}^{1-\xi} \int_{\xi}^{1-\eta'} \left[\tilde{a}(\xi', \eta') \, w(\xi', \eta') + \tilde{F}(\xi', \eta') \right] d\xi' \, d\eta' \qquad (4.41)$$

$$- \int_{\eta}^{1-\xi} g(1 - 2\eta') \, d\eta' \; + \; \frac{1}{2} f(2\xi - 1) \; + \; \frac{1}{2} f(1 - 2\eta)$$

for $(\xi, \eta) \in D$. This is again a Volterra integral equation in two variables. Let A denote the integral operator

$$Aw(\xi, \eta) = \int_{\eta}^{1-\xi} \int_{\xi}^{1-\eta'} \tilde{a}(\xi', \eta') \, w(\xi', \eta') \, d\xi' \, d\eta', \quad (\xi, \eta) \in D.$$

By induction, it is easily seen that

$$|A^n w(\xi, \eta)| \leq \|w\|_{\infty} \|\tilde{a}\|_{L^2}^n \frac{1}{\sqrt{(2n)!}} (1 - \xi - \eta)^n$$

for all $(\xi, \eta) \in D$ and $n \in \mathbb{N}$; thus

$$\|A^n w\|_{\infty} \leq \|w\|_{\infty} \|\tilde{a}\|_{L^2}^n \frac{1}{\sqrt{(2n)!}}$$

for all $n \in \mathbb{N}$. For sufficiently large n, we conclude that $\|A^n\|_{\infty} < 1$, which again implies that (4.41) is uniquely solvable in $C(\overline{D})$. Now we argue exactly as in the proof of Theorem 4.15. $\square$

For the special case $p = q = 0$ and $F = 0$, the integral equation (4.41) reduces to the well-known d'Alembert formula

$$W(x, t) = -\frac{1}{2} \int_{t-(1-x)}^{t+(1-x)} g(\tau) \, d\tau \; + \; \frac{1}{2} f(t + (1 - x)) \; + \; \frac{1}{2} f(t - (1 - x)).$$

Finally, the third theorem studies a quite unusual coupled system for a pair (W, r) of functions. We will treat this system with the same methods as above.

Theorem 4.17

(a) Let $p, q \in C[0,1]$, $F \in C(\overline{\Delta_0})$, $f \in C^2[0,1]$, and $g \in C^1[0,1]$ such that $f(0) = f''(0) = g(0) = 0$. Then there exists a unique pair of functions $(W, r) \in C^2(\overline{\Delta_0}) \times C[0,1]$ with

$$\frac{\partial^2 W(x,t)}{\partial x^2} - \frac{\partial^2 W(x,t)}{\partial t^2} + \big(p(t) - q(x)\big) W(x,t) = F(x,t)\, r(x) \quad \text{in } \Delta_0,$$
$$\text{(4.42a)}$$

$$W(x,x) = \frac{1}{2} \int_0^x r(t)\, dt, \quad 0 \le x \le 1, \tag{4.42b}$$

$$W(x,0) = 0, \quad 0 \le x \le 1, \tag{4.42c}$$

and

$$W(1,t) = f(t) \quad \text{and} \quad \frac{\partial}{\partial x} W(1,t) = g(t) \quad \text{for all } t \in [0,1]. \tag{4.42d}$$

(b) Furthermore, the solution operator $(p, q, F, f, g) \mapsto (W, r)$ has an extension to a bounded operator from $L^2(0,1) \times L^2(0,1) \times C(\overline{\Delta_0}) \times H_0^1(0,1) \times L^2(0,1)$ into $C(\overline{\Delta_0}) \times L^2(0,1)$.

Proof: We apply the same arguments as in the proofs of Theorems 4.15 and 4.16. We define $a(x,t) = p(|t|) - q(x)$ and extend $F(x, \cdot)$ to an even function and f and g to odd functions. We again make the change of variables $x = \xi + \eta$ and $t = \xi - \eta$ and set $\tilde{a}(\xi, \eta) = -a(\xi + \eta, \xi - \eta)$ and $\tilde{F}(\xi, \eta) = F(\xi + \eta, \xi - \eta)$. In Theorem 4.16, we have shown that the solution W of the Cauchy problem (4.42a) and (4.42d) is equivalent to the integral equation

$$w(\xi, \eta) = \int_\eta^{1-\xi} \int_\xi^{1-\eta'} \big[\tilde{a}(\xi', \eta')\, w(\xi', \eta') + \tilde{F}(\xi', \eta')\, r(\xi' + \eta')\big]\, d\xi'\, d\eta'$$

$$- \int_\eta^{1-\xi} g(1 - 2\eta')\, d\eta' + \frac{1}{2} f(2\xi - 1) + \frac{1}{2} f(1 - 2\eta) \quad \text{(4.43a)}$$

for $w(\xi, \eta) = W(\xi + \eta, \xi - \eta)$ (see equation (4.41)). From this and the initial condition (4.42b), we derive a second integral equation. We set $\eta = 0$ in (4.43a), differentiate, and substitute (4.42b). This yields the following Volterra equation after an obvious change of variables:

$$\frac{1}{2} r(x) = - \int_x^1 \big[\tilde{a}(x, y - x)\, w(x, y - x) + r(y)\, \tilde{F}(x, y - x)\big]\, dy$$

$$+ g(2x - 1) + f'(2x - 1). \tag{4.43b}$$

Assume that there exists a solution $(w, r) \in C(\overline{D}) \times C[0, 1]$ of (4.43a) and (4.43b). Then w is differentiable and $\frac{d}{dx} W(x, x) = \frac{d}{dx} w(x, 0) = \frac{1}{2} r(x)$. Now, since $a(x, \cdot)$ and $F(x, \cdot)$ are even and f and g are odd functions, we conclude that $W(x, \cdot)$ is also odd. In particular, $W(x, 0) = 0$ for all $x \in [0, 1]$. This implies $W(0, 0) = 0$ and thus $W(x, x) = \frac{1}{2} \int_0^x r(t) \, dt$. Therefore, we have shown that every solution of equations (4.43a) and (4.43b) satisfies (4.42a)–(4.42d) and vice versa.

Now we will sketch the proof that the system (4.43a), (4.43b) is uniquely solvable for $(w, r) \in C(\overline{D}) \times L^2(0, 1)$. We write the system in the form

$$\begin{pmatrix} w \\ r \end{pmatrix} = A \begin{pmatrix} w \\ r \end{pmatrix} + R$$

with obvious meanings of $A = \begin{pmatrix} A_{11} & A_{12} \\ A_{21} & A_{22} \end{pmatrix}$ and $R = \begin{pmatrix} R_1 \\ R_2 \end{pmatrix} \in C(\overline{D}) \times L^2(0, 1)$. Then A is well defined from $C(\overline{D}) \times L^2(0, 1)$ into itself and depends continuously on $\tilde{a} \in L^2(\Delta)$ and $\tilde{F} \in C(\overline{D})$. We observe that (4.43b) is a Volterra equation for r if w is kept fixed. From Appendix A, Example A.30, we can represent the solution r in the form

$$r(x) = (A_{21} w + R_2)(x) + \int_x^1 \tilde{b}(x, y) (A_{21} w + R_2)(y) \, dy$$

$$= Lw(x) + h(x),$$

where the operator L and function h depend continuously on $\tilde{F}$, $\tilde{a}$, f, and g. Using the explicit expression of $A_{21} w$ yields an estimate of the form

$$|Lw(x)| \leq \tilde{c} \max\{|w(y, z - y)| : y \leq z \leq 1, \, x \leq y \leq 1\} \quad \text{for } x \in [0, 1].$$

Now we substitute $r = Lw + h$ into (4.43a), which yields

$$w(\xi, \eta) = \int_\eta^{1-\xi} \int_\xi^{1-\eta'} \left[\tilde{a}(\xi', \eta') w(\xi', \eta') + \tilde{F}(\xi', \eta') Lw(\xi' + \eta') \right] d\xi' \, d\eta'$$

$$+ \tilde{R}(\xi, \eta)$$

for some function $\tilde{R}$ depending continuously on the data. Let B be the integral operator on the right-hand side and $c := \|\tilde{a}\|_{L^2} + \tilde{c}\|\tilde{F}\|_{L^2}$. By induction, one shows again that

$$|B^n w(\xi, \eta)| \leq \frac{c^n}{\sqrt{(2n)!}} \|w\|_\infty (1 - \xi - \eta)^n \quad \text{in } D.$$

This again implies that $\|B^n\|_\infty$ tends to zero as n tends to infinity. The contraction mapping theorem (see Appendix A, Theorem A.29) yields existence and uniqueness of the system of integral equations in $C(\overline{D}) \times L^2(0,1)$. The regularity of w and p for part (a) and the extension in part (b) are proven analogously as in the proof of Theorem 4.15.   □

4.5   The Inverse Problem

Now we study the inverse spectral problem. This is, given the eigenvalues λ_n of the Sturm–Liouville eigenvalue problem

$$- u''(x) + q(x)\, u(x) = \lambda u(x), \ 0 < x < 1, \quad u(0) = 0, \ u(1) = 0, \quad (4.44)$$

determine the function q. We saw in Example 4.1 that the knowledge of the spectrum $\{\lambda_n : n \in \mathbb{N}\}$ is, in general, not sufficient to determine q uniquely. We need more information, such as a second spectrum μ_n of an eigenvalue problem of the form

$$- v''(x) + q(x)\, v(x) = \mu\, v(x), \quad v(0) = 0, \ v'(1) + Hv(1) = 0, \quad (4.45)$$

or some knowledge about the eigenfunctions.

The basic tool in the uniqueness proof for this inverse problem is the use of the *Gelfand–Levitan–Marchenko integral operator* (see [77]). This integral operator maps solutions of initial value problems for the equation $-u'' + qu = \lambda u$ onto solutions for the equation $-u'' + pu = \lambda u$ and, most importantly, does not depend on λ. It turns out that the kernel of this operator is the solution of the hyperbolic boundary value problem that was studied in the previous section.

Theorem 4.18
Let $p, q \in L^2(0,1)$, $\lambda \in \mathbb{C}$, and $u, v \in H^2(0,1)$ be solutions of

$$- u''(x) + q(x)\, u(x) = \lambda u(x), \quad 0 < x < 1, \qquad u(0) = 0, \quad (4.46a)$$

$$- v''(x) + p(x)\, v(x) = \lambda v(x), \quad 0 < x < 1, \qquad v(0) = 0, \quad (4.46b)$$

such that $u'(0) = v'(0)$. Also let $K \in C(\overline{\Delta_0})$ be the weak solution of the Goursat problem

$$\frac{\partial^2 K(x,t)}{\partial x^2} - \frac{\partial^2 K(x,t)}{\partial t^2} + \big(p(t) - q(x)\big)\, K(x,t) = 0 \quad in \ \Delta_0, \quad (4.47a)$$

$$K(x,0) = 0, \quad 0 \le x \le 1, \quad (4.47b)$$

$$K(x,x) = \frac{1}{2} \int_0^x \big(q(s) - p(s)\big)\, ds, \quad 0 \le x \le 1, \quad (4.47c)$$

where the triangular region Δ_0 is again defined by

$$\Delta_0 := \{(x,t) \in \mathbb{R}^2 : 0 < t < x < 1\}. \tag{4.48}$$

Then we have

$$u(x) = v(x) + \int_0^x K(x,t)\,v(t)\,dt, \quad 0 \le x \le 1. \tag{4.49}$$

We remark that Theorem 4.15 with $f(x) = \frac{1}{2}\int_0^x (q(s) - p(s))\,ds$ implies that this Goursat problem is uniquely solvable in the weak sense.

Proof: First, let $p,q \in C^1[0,1]$. Then $K \in C^2(\overline{\Delta_0})$ by Theorem 4.15. Define w by the right-hand side of (4.49), i.e.,

$$w(x) := v(x) + \int_0^x K(x,t)\,v(t)\,dt \quad \text{for } 0 \le x \le 1.$$

Then $w(0) = v(0) = 0 = u(0)$ and w is differentiable with

$$w'(x) = v'(x) + K(x,x)v(x) + \int_0^x K_x(x,t)\,v(t)\,dt, \quad 0 < x < 1.$$

Again, we denote by K_x, K_t, etc., the partial derivatives. For $x = 0$, we have $w'(0) = v'(0) = u'(0)$. Furthermore,

$$
\begin{aligned}
w''(x) &= v''(x) + v(x)\frac{d}{dx}K(x,x) + K(x,x)\,v'(x) \\[2mm]
&\quad + K_x(x,x)\,v(x) + \int_0^x K_{xx}(x,t)\,v(t)\,dt \\[2mm]
&= \left[p(x) - \lambda + \frac{d}{dx}K(x,x) + K_x(x,x) \right] v(x) + K(x,x)\,v'(x) \\[2mm]
&\quad + \int_0^x \left[(q(x) - p(t))K(x,t)v(t) + K_{tt}(x,t)\,v(t) \right] dt.
\end{aligned}
$$

Partial integration yields

$$
\begin{aligned}
\int_0^x K_{tt}(x,t)\,v(t)\,dt \\[2mm]
= \int_0^x K(x,t)\,v''(t)\,dt + \left[K_t(x,t)\,v(t) - K(x,t)\,v'(t) \right]_{t=0}^{t=x}
\end{aligned}
$$

$$= \int_0^x \big(p(t) - \lambda\big) K(x,t)\, v(t)\, dt \; + \; K_t(x,x)\, v(x) \; - \; K(x,x)\, v'(x).$$

Therefore, we have

$$w''(x) \;=\; \underbrace{\Big[p(x) - \lambda + \frac{d}{dx} K(x,x) + K_x(x,x) + K_t(x,x) \Big]}_{=2\frac{d}{dx} K(x,x) = q(x) - p(x)} v(x)$$

$$+ \big(q(x) - \lambda\big) \int_0^x K(x,t)\, v(t)\, dt$$

$$= \big(q(x) - \lambda\big) \Big[v(x) + \int_0^x K(x,t) v(t)\, dt \Big] \;=\; \big(q(x) - \lambda\big)\, w(x),$$

i.e., w solves the same initial value problem as u. The Picard–Lindelöf uniqueness theorem for initial boundary value problems yields $w = u$. Thus we have proven the theorem for smooth functions p and q.

Now let $p, q \in L^2(0,1)$. Then we choose functions $(p_n), (q_n) \in C^1[0,1]$ with $p_n \to p$ and $q_n \to q$ in $L^2(0,1)$, respectively. Let K_n be the solution of (4.47a)–(4.47c) for p_n and q_n. We have already shown that

$$u_n(x) \;=\; v_n(x) \;+\; \int_0^x K_n(x,t)\, v_n(t)\, dt, \quad 0 \le x \le 1,$$

for all $n \in \mathbb{N}$, where u_n and v_n solve (4.46a) and (4.46b), respectively, with $u_n'(0) = v_n'(0) = u'(0) = v'(0)$. From the continuous dependence results of Theorems 4.6 and 4.15(b), the functions u_n, v_n, and K_n converge uniformly to u, v, and K, respectively. This proves the assertion of the theorem for $p, q \in L^2(0,1)$.   □

As an example we take $p = 0$ and $v(x) = \sin(\sqrt{\lambda}x)/\sqrt{\lambda}$ and have the following result.

Example 4.19

Let u be a solution of

$$- u''(x) + q(x)\, u(x) = \lambda\, u(x), \quad u(0) = 0, \quad u'(0) = 1. \tag{4.50}$$

Then we have the representation

$$u(x) \;=\; \frac{\sin \sqrt{\lambda}\, x}{\sqrt{\lambda}} \;+\; \int_0^x K(x,t)\, \frac{\sin \sqrt{\lambda}\, t}{\sqrt{\lambda}}\, dt, \quad 0 \le x \le 1, \tag{4.51}$$

where the kernel K solves the following Goursat problem:

$$K_{xx}(x,t) - K_{tt}(x,t) - q(x)K(x,t) = 0 \quad \text{in } \Delta_0, \tag{4.52a}$$

$$K(x,0) = 0, \quad 0 \le x \le 1, \tag{4.52b}$$

$$K(x,x) = \frac{1}{2}\int_0^x q(s)\,ds, \quad 0 \le x \le 1. \tag{4.52c}$$

This example has an application that is interesting in itself but that we will also need in Section 4.7.

Theorem 4.20
Let λ_n be the eigenvalues of one of the eigenvalue problems (4.44) or (4.45). Then the set of functions $\{\sin(\sqrt{\lambda_n}\cdot) : n \in \mathbb{N}\}$ is complete in $L^2(0,1)$. This means that $\int_0^1 h(x)\sin\sqrt{\lambda_n}x\,dx = 0$ for all $n \in \mathbb{N}$ implies that $h = 0$.

Proof: Let $T : L^2(0,1) \to L^2(0,1)$ be the Volterra integral operator of the second kind with kernel K, i.e.,

$$Tv(x) := v(x) + \int_0^x K(x,t)\,v(t)\,dt, \quad x \in (0,1), \ v \in L^2(0,1),$$

where K solves the Goursat problem (4.52a)–(4.52c). Then we know that T is an isomorphism from $L^2(0,1)$ onto itself. Define $v_n(x) := \sin\sqrt{\lambda_n}x$ for $x \in [0,1]$, $n \in \mathbb{N}$. Let u_n be the solution of the initial value problem:

$$-u_n'' + q\,u_n = \lambda_n u_n \text{ in } (0,1), \quad u_n(0) = 0, \ u_n'(0) = 1.$$

u_n is the eigenfunction corresponding to λ_n and, by the preceding example,

$$u_n = \frac{1}{\sqrt{\lambda_n}}Tv_n \quad \text{or} \quad v_n = \sqrt{\lambda_n}\,T^{-1}u_n.$$

Now, if $\int_0^1 h(x)v_n(x)\,dx = 0$ for all $n \in \mathbb{N}$, then

$$0 = \int_0^1 h(x)T^{-1}u_n(x)\,dx = \int_0^1 u_n(x)\left(T^*\right)^{-1}h(x)\,dx \quad \text{for all } n \in \mathbb{N},$$

where T^* denotes the L^2-adjoint of T. Since $\{u_n/\|u_n\|_{L^2} : n \in \mathbb{N}\}$ is complete in $L^2(0,1)$ by Lemma 4.7, we conclude that $\left(T^*\right)^{-1}h = 0$ and thus $h = 0$. $\qquad\square$

Now we can prove the main uniqueness theorem.

Theorem 4.21

Let $H \in \mathbb{R}$, $p, q \in L^2(0, 1)$, and $\lambda_n(p)$, $\lambda_n(q)$ be the eigenvalues of the eigenvalue problem

$$-u'' + ru = \lambda u \text{ in } (0, 1), \quad u(0) = 0, \; u(1) = 0,$$

corresponding to $r = p$ and $r = q$, respectively. Furthermore, let $\mu_n(p)$ and $\mu_n(q)$ be the eigenvalues of

$$-u'' + ru = \mu u \text{ in } (0, 1), \quad u(0) = 0, \; u'(1) + Hu(1) = 0,$$

corresponding to $r = p$ and $r = q$, respectively.

If $\lambda_n(p) = \lambda_n(q)$ and $\mu_n(p) = \mu_n(q)$ for all $n \in \mathbb{N}$, then $p = q$.

Proof: From the asymptotics of the eigenvalues, we conclude that

$$\lambda_n(p) = n^2\pi^2 + \int_0^1 p(t)\, dt + o(1), \quad n \to \infty,$$

$$\lambda_n(q) = n^2\pi^2 + \int_0^1 q(t)\, dt + o(1), \quad n \to \infty,$$

and thus

$$\int_0^1 (p(t) - q(t))\, dt = \lim_{n\to\infty} (\lambda_n(p) - \lambda_n(q)) = 0. \qquad (4.53)$$

Now let K be the solution of the Goursat problem (4.47a)–(4.47c). Then K depends only on p and q and is independent of the eigenvalues $\lambda_n := \lambda_n(p) = \lambda_n(q)$ and $\mu_n := \mu_n(p) = \mu_n(q)$. Furthermore, from (4.53) we conclude that $K(1, 1) = 0$.

Now let u_n, v_n be the eigenfunctions corresponding to $\lambda_n(q)$ and $\lambda_n(p)$, respectively, i.e., solutions of the differential equations

$$-u_n''(x) + q(x)\, u_n(x) = \lambda_n\, u_n(x), \quad -v_n''(x) + p(x)\, v_n(x) = \lambda_n\, v_n(x)$$

for $0 < x < 1$ with homogeneous Dirichlet boundary conditions on both sides. Furthermore, we assume that they are normalized by $u_n'(0) = v_n'(0) = 1$. Then Theorem 4.18 is applicable and yields the relationship

$$u_n(x) = v_n(x) + \int_0^x K(x, t)\, v_n(t)\, dt \quad \text{for } x \in [0, 1], \qquad (4.54)$$

and all $n \in \mathbb{N}$. For $x = 1$, the boundary conditions yield

$$0 = \int_0^1 K(1,t)\, v_n(t)\, dt \quad \text{for all } n \in \mathbb{N}. \tag{4.55}$$

Now we use the fact that the set $\{v_n/\|v_n\|_{L^2} : n \in \mathbb{N}\}$ forms a complete orthonormal system in $L^2(0,1)$. From this, $K(1,t) = 0$ for all $t \in [0,1]$ follows.

Now let $\tilde{u}_n$ and $\tilde{v}_n$ be eigenfunctions corresponding to μ_n and q and p, respectively, with the normalization $\tilde{u}_n'(0) = \tilde{v}_n'(0) = 1$. Again, Theorem 4.18 is applicable and yields the relationship (4.54) for $\tilde{u}_n$ and $\tilde{v}_n$ instead of u_n und v_n, respectively. We differentiate this equation and arrive at

$$
\begin{aligned}
0 &= \tilde{u}_n'(1) - \tilde{v}_n'(1) + H\big[\tilde{u}_n(1) - \tilde{v}_n(1)\big] \\
&= \underbrace{K(1,1)}_{=0}\, \tilde{v}_n(1) + \int_0^1 \left(K_x(1,t) + H\underbrace{K(1,t)}_{=0}\right) \tilde{v}_n(t)\, dt.
\end{aligned}
$$

We conclude that $\int_0^1 K_x(1,t)\,\tilde{v}_n(t)\, dt = 0$ for all $n \in \mathbb{N}$. From this, $K_x(1,t) = 0$ for all $t \in (0,1)$ follows since $\{\tilde{v}_n/\|\tilde{v}_n\|_{L^2}\}$ forms a complete orthonormal system.

Now we apply Theorem 4.16, which yields that K has to vanish identically. In particular, this means that

$$0 = K(x,x) = \frac{1}{2}\int_0^x \big(p(s) - q(s)\big)\, ds \quad \text{for all } x \in (0,1).$$

Differentiating this equation yields that $p = q$. $\qquad\square$

We have seen in Example 4.1 that the knowledge of one spectrum for the Sturm–Liouville differential equation is not enough information to recover the function q uniquely. Instead of knowing the spectrum for a second pair of boundary conditions, we can use other kinds of information, as the following theorem shows.

Theorem 4.22
Let $p, q \in L^2(0,1)$ with eigenvalues $\lambda_n(p)$, $\lambda_n(q)$, and eigenfunctions u_n and v_n, respectively, corresponding to Dirichlet boundary conditions $u(0) = 0$, $u(1) = 0$. Let the eigenvalues coincide, i.e. $\lambda_n(p) = \lambda_n(q)$ for all $n \in \mathbb{N}$. Let one of the following assumptions also be satisfied:

(a) Let p and q be even functions with respect to $1/2$, i.e., $p(1-x) = p(x)$ and $q(1-x) = q(x)$ for all $x \in [0,1]$.

(b) Let the Neumann boundary values coincide, i.e., let

$$\frac{u_n{}'(1)}{u_n{}'(0)} = \frac{v_n{}'(1)}{v_n{}'(0)} \quad \text{for all } n \in \mathbb{N}. \tag{4.56}$$

Then $p = q$.

Proof: (a) The eigenfunctions are also even functions. This follows from the fact that the eigenvalues are simple and that u and $\tilde{u}(x) := u(1-x)$ are eigenfunctions corresponding to the same eigenvalue. Therefore, for every eigenfunction u we have that $u'(1) = -u'(0)$, i.e., $u'(1)/u'(0) = -1$. This reduces the uniqueness question for part (a) to part (b).

(b) Now we normalize the eigenfunctions such that $u_n{}'(0) = v_n{}'(0) = 1$. We follow the first part of the proof of Theorem 4.21. From (4.55), we again conclude that $K(1,t)$ vanishes for all $t \in (0,1)$. The additional assumption (4.56) yields that $u_n{}'(1) = v_n{}'(1)$. We differentiate (4.54), set $x = 1$, and arrive at $\int_0^1 K_x(1,t)v_n(t)\,dt = 0$ for all $n \in \mathbb{N}$. Again, this implies that $K_x(1,\cdot) = 0$, and the proof follows the same lines as the proof of Theorem 4.21. $\square$

4.6   A Parameter Identification Problem

This section and the next chapter are devoted to the important field of parameter identification problems for partial differential equations. In Chapter 5, we will study the inverse scattering problem to determine the refrac-
tive index of a medium from measurements of the scattered field, but in this
section we will consider an application of the inverse Sturm–Liouville eigenvalue problem to the following parabolic initial boundary value problem.
First, we formulate the direct problem:
    Let $T > 0$ and $\Omega_T := (0,1) \times (0,T) \subset \mathbb{R}^2$, $q \in C[0,1]$ and $f \in C^2[0,T]$
be given. Let $f(0) = 0$ and $q(x) \geq 0$ for $x \in [0,1]$. Determine $U \in C(\overline{\Omega_T})$,
which is twice continuously differentiable with respect to x and continu-
ously differentiable with respect to t in Ω_T such that $\partial U/\partial x \in C(\overline{\Omega_T})$
and

$$\frac{\partial U(x,t)}{\partial t} = \frac{\partial^2 U(x,t)}{\partial x^2} - q(x)\,U(x,t) \quad \text{in } \Omega_T, \tag{4.57a}$$

$$U(x,0) = 0, \quad x \in [0,1], \tag{4.57b}$$

$$U(0,t) = 0, \quad \frac{\partial}{\partial x}U(1,t) = f(t), \quad t \in (0,T). \tag{4.57c}$$

From the theory of parabolic initial boundary value problems, it is known that there exists a unique solution of this problem. We will prove uniqueness and refer to [133] or (4.59) for the question of existence.

Theorem 4.23
Let $f = 0$. Then $U = 0$ is the only solution of (4.57a)–(4.57c) in Ω_T.

Proof: Multiply the differential equation (4.57a) by $U(x,t)$ and integrate with respect to x. This yields

$$\frac{1}{2}\frac{d}{dt}\int_0^1 U(x,t)^2\,dx \;=\; \int_0^1 \left[\frac{\partial^2 U(x,t)}{\partial x^2}\,U(x,t) - q(x)\,U(x,t)^2\right] dx.$$

We integrate by parts and use the homogeneous boundary conditions:

$$\frac{1}{2}\frac{d}{dt}\int_0^1 U(x,t)^2\,dx \;=\; -\int_0^1 \left[\left(\frac{\partial U(x,t)}{\partial x}\right)^2 + q(x)\,U(x,t)^2\right] dx \;\leq\; 0.$$

This implies that $t \mapsto \int_0^1 U(x,t)^2 dx$ is nonnegative and monotonicly non-increasing. From $\int_0^1 U(x,0)^2\,dx = 0$, we conclude that $\int_0^1 U(x,t)^2\,dx = 0$ for all t, i.e., $U = 0$. $\square$

Now we turn to the inverse problem. Let f be known and, in addition, $U(1,t)$ for all $0 < t \leq T$. The inverse problem is to determine the coefficient q.

In this section, we are only interested in the question if this provides enough information in principle to recover q uniquely, i.e., we will study the question of uniqueness of the inverse problem. It is our aim to prove the following theorem.

Theorem 4.24
Let U_1, U_2 be solutions of (4.57a)–(4.57c) corresponding to $q = q_1 \geq 0$ and $q = q_2 \geq 0$, respectively, and to the same f. Let $f(0) = 0$, $f'(0) \neq 0$, and $U_1(1,t) = U_2(1,t)$ for all $t \in (0,T)$. Then $q_1 = q_2$ on $[0,1]$.

Proof: Let q and U be q_1 or q_2 and U_1 or U_2, respectively. Let λ_n and g_n, $n \in \mathbb{N}$, be the eigenvalues and eigenfunctions, respectively, of the Sturm–Liouville eigenvalue problem (4.45) for $H = 0$, i.e.,

$$-u''(x) + q(x)\,u(x) = \lambda\,u(x), \quad 0 < x < 1, \qquad u(0) = u'(1) = 0.$$

We assume that the eigenfunctions are normalized by $\|g_n\|_{L^2} = 1$ for all $n \in \mathbb{N}$. Furthermore, we can assume that $g_n(1) > 0$ for all $n \in \mathbb{N}$. We know that $\{g_n : n \in \mathbb{N}\}$ forms a complete orthonormal system in $L^2(0,1)$. Theorem 4.14 implies the asymptotic behavior

$$\lambda_n = (n+1/2)^2 + \hat{q} + \tilde{\lambda}_n \quad \text{with} \quad \sum_{n=1}^{\infty} \tilde{\lambda}_n^2 < \infty, \qquad (4.58a)$$

$$g_n(x) = \sqrt{2} \sin(n+1/2)\pi x + \mathcal{O}(1/n), \qquad (4.58b)$$

where $\hat{q} = \int_0^1 q(x)\,dx$. In the first step, we derive a series expansion for the solution U of the initial boundary value problem (4.57a)–(4.57c). From the completeness of $\{g_n : n \in \mathbb{N}\}$, we have the Fourier expansion

$$U(x,t) = \sum_{n=1}^{\infty} a_n(t)\,g_n(x) \quad \text{with} \quad a_n(t) = \int_0^1 U(x,t)\,g_n(x)\,dx, \ n \in \mathbb{N},$$

where the convergence is understood in the $L^2(0,1)$-sense for every $t \in (0,T]$. We would like to substitute this into the differential equation and the initial and boundary conditions. Since for this formal procedure the interchanging of summation and differentiation is not justified, we suggest a different derivation of a_n. We differentiate a_n and use the partial differential equation (4.57a). This yields

$$
\begin{aligned}
a_n{}'(t) &= \int_0^1 \frac{\partial U(x,t)}{\partial t}\,g_n(x)\,dx = \int_0^1 \left[\frac{\partial^2 U(x,t)}{\partial x^2} - q(x)\,U(x,t)\right]g_n(x)\,dx \\
&= \left[g_n(x)\frac{\partial U(x,t)}{\partial x} - U(x,t)g_n{}'(x)\right]_{x=0}^{x=1} \\
&\quad + \int_0^1 U(x,t)\underbrace{\left[g_n{}''(x) - q(x)\,g_n(x)\right]}_{=-\lambda_n g_n(x)}\,dx \\
&= f(t)\,g_n(1) - \lambda_n a_n(t).
\end{aligned}
$$

With the initial condition $a_n(0) = 0$, the solution is given by

$$a_n(t) = g_n(1) \int_0^t f(\tau)\,e^{-\lambda_n(t-\tau)}\,d\tau,$$

i.e., the solution U of (4.57a)–(4.57c) takes the form

$$U(x,t) = \sum_{n=1}^{\infty} g_n(1)\,g_n(x) \int_0^t f(\tau)\,e^{-\lambda_n(t-\tau)}\,d\tau. \qquad (4.59)$$

From partial integration, we observe that

$$\int_0^t f(\tau) e^{-\lambda_n(t-\tau)} d\tau = \frac{1}{\lambda_n} f(t) - \frac{1}{\lambda_n} \int_0^t f'(\tau) e^{-\lambda_n(t-\tau)} d\tau,$$

and this decays as $1/\lambda_n$. Using this and the asymptotic behavior (4.58a) and (4.58b), we conclude that the series (4.59) converges uniformly in $\overline{\Omega_T}$. For $x = 1$, the representation (4.59) reduces to

$$U(1,t) = \sum_{n=1}^{\infty} g_n(1)^2 \int_0^t f(\tau) e^{-\lambda_n(t-\tau)} d\tau$$

$$= \int_0^t f(\tau) \underbrace{\sum_{n=1}^{\infty} g_n(1)^2 e^{-\lambda_n(t-\tau)}}_{=: A(t-\tau)} d\tau, \quad t \in [0,T].$$

Changing the orders of integration and summation is justified by Lebesgue's theorem of dominated convergence. This is seen from the estimate

$$\sum_{n=1}^{\infty} g_n(1)^2 e^{-\lambda_n s} \le c \sum_{n=1}^{\infty} e^{-n^2 \pi^2 s} \le c \int_0^{\infty} e^{-s^2 \pi^2 s} ds = \frac{c}{2\sqrt{\pi s}}$$

and the fact that the function $s \mapsto 1/\sqrt{s}$ is integrable.

Such a representation holds for $U_1(1, \cdot)$ and $U_2(1, \cdot)$ corresponding to q_1 and q_2, respectively. We denote the dependence on q_1 and q_2 by superscripts (1) and (2), respectively. From $U_1(1, \cdot) = U_2(1, \cdot)$, we conclude that

$$0 = \int_0^t f(\tau) \left[A^{(1)}(t-\tau) - A^{(2)}(t-\tau) \right] d\tau = \int_0^t f(t-\tau) \left[A^{(1)}(\tau) - A^{(2)}(\tau) \right] d\tau,$$

i.e., the function $w := A^{(1)} - A^{(2)}$ solves the homogeneous Volterra integral equation of the *first kind* with kernel $f(t-\tau)$. We differentiate this equation twice and use $f(0) = 0$ and $f'(0) \ne 0$. This yields a Volterra equation of the *second kind* for w:

$$f'(0) w(t) + \int_0^t f''(t - s) w(s) ds = 0, \quad t \in [0,T].$$

Since Volterra equations of the second kind are uniquely solvable (see Example A.30 of Appendix A), this yields $w(t) = 0$ for all t, that is

$$\sum_{n=1}^{\infty} \left[g_n^{(1)}(1) \right]^2 e^{-\lambda_n^{(1)} t} = \sum_{n=1}^{\infty} \left[g_n^{(2)}(1) \right]^2 e^{-\lambda_n^{(2)} t} \quad \text{for all } t \in (0,T).$$

We note that $g_n^{(j)}(1) > 0$ for $j = 1, 2$ by our normalization. Now we can apply a result from the theory of Dirichlet series (see Lemma 4.25) and conclude that $\lambda_n^{(1)} = \lambda_n^{(2)}$ and $g_n^{(1)}(1) = g_n^{(2)}(1)$ for all $n \in \mathbb{N}$. Applying the uniqueness result analogous to Theorem 4.22, part (b), for the boundary conditions $u(0) = 0$ and $u'(1) = 0$ (see Problem 4.5), we conclude that $q_1 = q_2$. $\qquad\square$

It remains to prove the following lemma.

Lemma 4.25

Let λ_n and μ_n be strictly increasing sequences that tend to infinity. Let the series

$$\sum_{n=1}^{\infty} \alpha_n e^{-\lambda_n t} \quad and \quad \sum_{n=1}^{\infty} \beta_n e^{-\mu_n t}$$

converge for every $t \in (0, T]$ and uniformly on some interval $[\delta, T]$. Let the limits coincide, that is

$$\sum_{n=1}^{\infty} \alpha_n e^{-\lambda_n t} = \sum_{n=1}^{\infty} \beta_n e^{-\mu_n t} \quad for\ all\ t \in (0, T].$$

If we also assume that $\alpha_n \neq 0$ and $\beta_n \neq 0$ for all $n \in \mathbb{N}$, then $\alpha_n = \beta_n$ and $\lambda_n = \mu_n$ for all $n \in \mathbb{N}$.

Proof: Assume that $\lambda_1 \neq \mu_1$ or $\alpha_1 \neq \beta_1$. Without loss of generality, we can assume that $\mu_2 > \lambda_1$ (otherwise, we have $\mu_1 < \mu_2 \leq \lambda_1 < \lambda_2$ and can interchange the roles of λ_n and μ_n). Define

$$C_n(t) := \alpha_n e^{-(\lambda_n - \lambda_1)t} - \beta_n e^{-(\mu_n - \lambda_1)t} \quad for\ t \geq \delta.$$

By analytic continuation, we conclude that $\sum_{n=1}^{\infty} C_n(t) = 0$ for all $t \geq \delta$ and that the series converges uniformly on $[\delta, \infty)$. Since

$$C_1(t) = \alpha_1 - \beta_1 e^{-(\mu_1 - \lambda_1)t},$$

there exists $\epsilon > 0$ and $t_1 > \delta$ such that $|C_1(t)| \geq \epsilon$ for all $t \geq t_1$. Choose $n_0 \in \mathbb{N}$ with

$$\left| \sum_{n=1}^{n_0} C_n(t) \right| < \frac{\epsilon}{2} \quad for\ all\ t \geq t_1.$$

Then we conclude that

$$\left| \sum_{n=2}^{n_0} C_n(t) \right| = \left| C_1(t) - \sum_{n=1}^{n_0} C_n(t) \right| \geq |C_1(t)| - \left| \sum_{n=1}^{n_0} C_n(t) \right| \geq \frac{\epsilon}{2}$$

for all $t \geq t_1$. Now we let t tend to infinity. The first finite sum converges to zero, which is a contradiction. Therefore, we have shown that $\lambda_1 = \mu_1$ and $\alpha_1 = \beta_1$. Now we repeat the argument for $n = 2$, etc. This proves the lemma. $\qquad\square$

4.7 Numerical Reconstruction Techniques

In this section, we discuss some numerical algorithms suggested and tested by W. Rundell, P. Sacks, and others. We follow closely the papers [148, 188, 189].

From now on, we will assume knowledge of eigenvalues λ_n and μ_n, $n \in \mathbb{N}$, of the Sturm–Liouville eigenvalue problems (4.44) or (4.45). It is our aim to determine the unknown function q. Usually, only a finite number of eigenvalues is known. Then one cannot expect to recover the total function q but only "some portion" of it (see (4.61)).

The first algorithm we will discuss uses the concept of the characteristic function again. For simplicity, we describe the method only for the case where q is known to be an even function, i.e., $q(1 - x) = q(x)$. Then we know that only one spectrum suffices to recover q (see Theorem 4.22).

Recalling the characteristic function $f(\lambda) = u_2(1, \lambda, q)$ for the problem (4.44), the inverse problem can be written as the problem of solving the equations

$$u_2(1, \lambda_n, q) = 0 \quad \text{for all } n \in \mathbb{N} \tag{4.60}$$

for the function q. If we know only a finite number, say λ_n for $n = 1, \ldots, N$, then we assume that q is of the form

$$q(x; a) = \sum_{n=1}^{N} a_n q_n(x), \quad x \in [0, 1], \tag{4.61}$$

for coefficients $a = (a_1, \ldots, a_N) \in \mathbb{R}^N$ and some linear independent even functions q_n. If q is expected to be smooth and periodic, a good choice for q_n is $q_n(x) = \cos(2\pi(n - 1)x)$, $n = 1, \ldots, N$. Equation (4.60) then reduces to the finite nonlinear system $F(a) = 0$, where $F : \mathbb{R}^N \to \mathbb{R}^N$ is defined by

$$F_n(a) := u_2(1, \lambda_n, q(\cdot; a)) \quad \text{for } a \in \mathbb{R}^N \text{ and } n = 1, \ldots, N.$$

Therefore, all of the well-developed methods for solving systems of nonlinear equations can be used. For example, Newton's method

$$a^{(k+1)} = a^{(k)} - F'(a^{(k)})^{-1} F(a^{(k)}), \quad k = 0, 1, \ldots,$$

is known to be quadratically convergent if $F'(a)^{-1}$ is regular. As we know from Section 4.2, Theorem 4.6, the mapping F is continuously Fréchet differentiable for every $a \in \mathbb{R}^N$. The computation of the derivative is rather expensive, and in general it is not known if $F'(a)^{-1}$ is regular. In [148] it was proven that $F'(a)^{-1}$ is regular for sufficiently small a and is of triangular form for $a = 0$. This observation leads to the *simplified Newton method* of the form

$$a^{(k+1)} = a^{(k)} - F'(0)^{-1} F(a^{(k)}), \quad k = 0, 1, \dots.$$

For further aspects of this method, we refer to [148].

Before we describe a second algorithm, we observe that from the asymptotic form (4.27) of the eigenvalues we have an estimate of $\hat{q} = \int_0^1 q(x)\, dx$. Writing the differential equation in the form

$$-u_n''(x) + \big(q(x) - \hat{q}\big)\, u_n(x) = \big(\lambda_n - \hat{q}\big)\, u_n(x), \quad 0 \leq x \leq 1,$$

we observe that we can assume without loss of generality that $\int_0^1 q(x)\, dx = 0$.

Now we describe an algorithm that follows the idea of the uniqueness Theorem 4.21. The algorithm consists of two steps. First, we recover the Cauchy data $f = K(1, \cdot)$ and $g = K_x(1, \cdot)$ from the two sets of eigenvalues. Then we suggest Newton-type methods to compute q from these Cauchy data.

The starting point is Theorem 4.18 for the case $p = 0$. We have already formulated this special case in Example 4.19. Therefore, let (λ_n, u_n) be the eigenvalues and eigenfunctions of the eigenvalue problem (4.44) normalized such that $u_n'(0) = 1$. The eigenvalues λ_n are assumed to be known. From Example 4.19, we have the representation

$$u_n(x) = \frac{\sin \sqrt{\lambda_n} x}{\sqrt{\lambda_n}} + \int_0^x K(x, t) \frac{\sin \sqrt{\lambda_n} t}{\sqrt{\lambda_n}}\, dt, \quad 0 \leq x \leq 1, \qquad (4.62)$$

where K satisfies the Goursat problem (4.52a)–(4.52c) with $K(1, 1) = \frac{1}{2} \int_0^1 q(t)\, dt = 0$. From (4.62) for $x = 1$, we can compute $K(1, t)$ since, by Theorem 4.20, the functions $v_n(t) = \sin \sqrt{\lambda_n} t$ form a complete system in $L^2(0, 1)$. When we know only a finite number $\lambda_1, \dots, \lambda_N$ of eigenvalues, we suggest representing $K(1, \cdot)$ as a finite sum of the form

$$K(1, t) = \sum_{k=1}^N a_k \sin(k\pi t),$$

arriving at the finite linear system

$$\sum_{k=1}^N a_k \int_0^1 \sin(k\pi t) \sin \sqrt{\lambda_n} t\, dt = -\sin \sqrt{\lambda_n} \quad \text{for } n = 1, \dots, N. \qquad (4.63)$$

The same arguments yield a set of equations for the second boundary condition $u'(1) + H\,u(1) = 0$ in the form

$$\sqrt{\mu_n}\,\cos\sqrt{\mu_n} \;+\; H\,\sin\sqrt{\mu_n} \;+\; \int_0^1 \left(K_x(1,t) + H\,K(1,t)\right)\,\sin\sqrt{\mu_n}t\,dt \;=\; 0,$$

where now μ_n are the corresponding known eigenvalues. The representation

$$K_x(1,t) + H\,K(1,t) \;=\; \sum_{k=1}^N b_k\,\sin(k\pi t)$$

leads to the system

$$\sum_{k=1}^N b_k \int_0^1 \sin(k\pi t)\,\sin\sqrt{\mu_n}t\,dt \;=\; -\sqrt{\mu_n}\,\cos\sqrt{\mu_n} \;-\; H\,\sin\sqrt{\mu_n} \quad (4.64)$$

for $n = 1,\dots,N$. Equations (4.63) and (4.64) are of the same form and we restrict ourselves to the discussion of (4.63). Asymptotically, the matrix $A \in \mathbb{R}^{N \times N}$ defined by $A_{kn} = \int_0^1 \sin(k\pi t)\,\sin\sqrt{\lambda_n}t\,dt$ is just $\frac{1}{2}I$. More precisely, from Parseval's identity

$$\sum_{k=1}^\infty \left| \int_0^1 \psi(t)\,\sin(k\pi t)\,dt \right|^2 \;=\; \frac{1}{2}\int_0^1 |\psi(t)|^2\,dt$$

we conclude that

$$\sum_{k=1}^\infty \left| \int_0^1 \sin(k\pi t)\left[\sin\sqrt{\lambda_n}t \;-\; \sin(n\pi t)\right]dt \right|^2$$

$$=\; \frac{1}{2}\int_0^1 \left|\sin\sqrt{\lambda_n}t \;-\; \sin(n\pi t)\right|^2 dt.$$

The estimate (4.29) yields $\left|\lambda_n - n^2\pi^2\right| \le \tilde{c}\,\|q\|_\infty$ and thus

$$\left|\sqrt{\lambda_n} - n\pi\right| \;\le\; \frac{c}{n}\,\|q\|_\infty,$$

where c is independent of q and n. From this, we conclude that

$$\sum_{k=1}^\infty \left| \int_0^1 \sin(k\pi t)\left[\sin\sqrt{\lambda_n}t \;-\; \sin(n\pi t)\right]dt \right|^2 \;\le\; \frac{1}{2}\left|\sqrt{\lambda_n} - n\pi\right|^2$$

$$\le\; \frac{c^2}{2\,n^2}\,\|q\|_\infty^2.$$

The matrix A is thus diagonally dominant and therefore invertible for sufficiently small $\|q\|_\infty$. Numerical experiments have shown that also for "large" values of q the numerical solution of (4.64) does not cause any problems.

We are now facing the following inverse problem: Given (approximate values of) the Cauchy data $f = K(1, \cdot) \in H_0^1(0, 1)$ and $g = K_x(1, \cdot) \in L^2(0, 1)$, compute $q \in L^2(0, 1)$ such that the solution of the Cauchy problem (4.40a)–(4.40c) for $p = 0$ assumes the boundary data $K(x, x) = \frac{1}{2} \int_0^x q(t)\, dt$ for $x \in [0, 1]$. An alternative way of formulating the inverse problem is to start with the Goursat problem (4.52a)–(4.52c): Compute $q \in L^2(0, 1)$ such that the solution of the initial value problem (4.52a)–(4.52c) has Cauchy data $f(t) = K(1, t)$ and $g(t) = K_x(1, t)$ for $t \in [0, 1]$.

We have studied these coupled systems for K and q in Theorem 4.17. Here we apply it for the case where $p = 0$ and $F = 0$. It has been shown that the pair (K, r) solves the system

$$\frac{\partial^2 K(x, t)}{\partial x^2} - \frac{\partial^2 K(x, t)}{\partial t^2} - q(x)\, K(x, t) = 0 \quad \text{in } \Delta_0,$$

$$K(x, x) = \frac{1}{2} \int_0^x r(t)\, dt, \quad 0 \le x \le 1,$$

$$K(x, 0) = 0, \quad 0 \le x \le 1,$$

and

$$K(1, t) = f(t) \quad \text{and} \quad \frac{\partial}{\partial x} K(1, t) = g(t) \quad \text{for all } t \in [0, 1]$$

if and only if $w(\xi, \eta) = K(\xi + \eta, \xi - \eta)$ and r solve the system of integral equations (4.43a) and (4.43b). For this special choice of p and F, (4.43b) reduces to

$$\frac{1}{2} r(x) = -\int_x^1 q(y)\, K(y, 2x - y)\, dy + g(2x - 1) + f'(2x - 1), \quad (4.65)$$

where we have extended f and g to odd functions on $[-1, 1]$. Denote by $T(q)$ the expression on the right-hand side of (4.65). For the evaluation of $T(q)$, one has to solve the Cauchy problem (4.40a)–(4.40c) for $p = 0$. Note that the solution K, i.e., the kernel $K(y, 2x - y)$ of the integral operator T, also depends on q. The operator T is therefore nonlinear!

The requirement $r = q$ leads to a fixed point equation $q = 2T(q)$ in $L^2(0, 1)$. It was shown in [188] that there exists at most one fixed point

$q \in L^\infty(0, 1)$ of T. Even more, Rundell and Sachs proved that the projected operator $P_M T$ is a contraction on the ball $B_M := \{ q \in L^\infty(0, 1) : \|q\|_\infty \leq M \}$ with respect to some weighted L^∞-norms. Here, P_M denotes the projection onto B_M defined by

$$P_M q(x) = \begin{cases} q(x), & |q(x)| \leq M, \\ M \operatorname{sign} q(x), & |q(x)| > M. \end{cases}$$

Also, they showed the effectiveness of the iteration method $q^{(k+1)} = 2T(q^{(k)})$ by several numerical examples. We observe that for $q^{(0)} = 0$ the first iterate $q^{(1)}$ is simply $q^{(1)}(x) = 2 g(2x - 1) + 2 f'(2x - 1)$, $x \in [0, 1]$. We refer to [188] for more details.

As suggested earlier, an alternative numerical procedure based on the kernel function K is to define the operator S from $L^2(0, 1)$ into $H_0^1(0, 1) \times L^2(0, 1)$ by $S(q) = \big(K(1, \cdot), K_x(1, \cdot)\big)$, where K solves the Goursat problem (4.52a)–(4.52c). This operator is well-defined and bounded by Theorem 4.15, part (c). If $f \in H_0^1(0, 1)$ and $g \in L^2(0, 1)$ are the given Cauchy values $K(1, \cdot)$ and $K_x(1, \cdot)$, respectively, then we have to solve the nonlinear equation $S(q) = (f, g)$. Newton's method does it by the iteration procedure

$$q^{(k+1)} = q^{(k)} - S'(q^{(k)})^{-1} \big[S(q^{(k)}) - (f, g) \big], \quad k = 0, 1, \dots . \quad (4.66)$$

For the implementation, one has to compute the Fréchet derivative of S. Using the Volterra equation (4.39) derived in the proof of Theorem 4.15, it is not difficult to prove that S is Fréchet differentiable and that $S'(q)r = \big(W(1, \cdot), W_x(1, \cdot)\big)$, where W solves the inhomogeneous Goursat problem

$$W_{xx}(x, t) - W_{tt}(x, t) - q(x) W(x, t) = K(x, t) r(x) \quad \text{in } \Delta_0, \quad (4.67a)$$

$$W(x, 0) = 0, \quad 0 \leq x \leq 1, \quad (4.67b)$$

$$W(x, x) = \frac{1}{2} \int_0^x r(t)\, dt, \quad 0 \leq x \leq 1. \quad (4.67c)$$

In Theorem 4.17, we showed that $S'(q)$ is an isomorphism. We reformulate this result.

Theorem 4.26
Let $q \in L^2(0, 1)$ and K be the solution of (4.52a)–(4.52c). For every $f \in H_0^1(0, 1)$ and $g \in L^2(0, 1)$, there exists a unique $r \in L^2(0, 1)$ and a solution W of (4.67a)–(4.67c) with $W(1, \cdot) = f$ and $W_x(1, \cdot) = g$, i.e., $S'(q)$ is an isomorphism.

Implementing Newton's method is quite expensive since in every step one has to solve a coupled system of the form (4.67a)–(4.67c). Rundell and Sachs suggested a simplified Newton method of the form

$$q^{(k+1)} \;=\; q^{(k)} \;-\; S'(0)^{-1} \left[\, S\!\left(q^{(k)}\right) \,-\, (f,g) \,\right], \quad k = 0, 1, \ldots.$$

Since $S(0) = 0$, we can invert the linear operator $S'(0)$ analytically. In particular, we have $S'(0)r = \left(W(1, \cdot), W_x(1, \cdot)\right)$, where W now solves

$$W_{xx}(x, t) \;-\; W_{tt}(x, t) \;=\; 0 \quad \text{in } \Delta_0,$$

$$W(x, 0) \;=\; 0, \quad \text{and} \quad W(x, x) \;=\; \frac{1}{2} \int_0^x r(t)\, dt, \quad 0 \le x \le 1,$$

since also $K = 0$. The solution W of the Cauchy problem

$$W_{xx}(x, t) \;-\; W_{tt}(x, t) \;=\; 0 \quad \text{in } \Delta_0,$$

$$W(1, t) \;=\; f(t), \quad \text{and} \quad W_x(1, t) \;=\; g(t), \quad 0 \le t \le 1,$$

is given by

$$W(x, t) \;=\; -\frac{1}{2} \int_{t-(1-x)}^{t+(1-x)} g(\tau)\, d\tau \;+\; \frac{1}{2} f\!\left(t + (1 - x)\right) \;+\; \frac{1}{2} f\!\left(t - (1 - x)\right),$$

where we have extended f and g to odd functions again. The solution r of $S'(0)r = (f, g)$ is therefore given by

$$r(x) \;=\; 2 \frac{d}{dx} W(x, x) \;=\; 2\, f'(2x - 1) \;+\; 2\, g(2x - 1).$$

In this chapter, we have studied only one particular inverse eigenvalue problem. Similar theoretical results and constructive algorithms can be obtained for other inverse spectral problems; see [3, 14]. For an excellent overview, we refer to the forthcoming lecture notes by W. Rundell [187].

4.8   Problems

4.1 Let $q, f \in C[0, 1]$ and $q(x) \ge 0$ for all $x \in [0, 1]$.

   (a) Show that the following boundary value problem on $[0, 1]$ has at most one solution $u \in C^2[0, 1]$:

$$- u''(x) + q(x)\, u(x) \;=\; f(x), \quad u(0) = u(1) = 0. \quad (4.68)$$

(b) Let v_1 and v_2 be the solutions of the following initial value problems on $[0, 1]$:

$$-v_1''(x) + q(x) v_1(x) = 0, \quad v_1(0) = 0, \ v_1'(0) = 1,$$
$$-v_2''(x) + q(x) v_2(x) = 0, \quad v_2(1) = 0, \ v_2'(1) = 1.$$

Show that the Wronskian $v_1'v_2 - v_2'v_1$ is constant. Define the following function for some $a \in \mathbb{R}$:

$$G(x, y) = \begin{cases} a\, v_1(x)\, v_2(y), & 0 \le x \le y \le 1, \\ a\, v_2(x)\, v_1(y), & 0 \le y < x \le 1. \end{cases}$$

Determine $a \in \mathbb{R}$ such that

$$u(x) := \int_0^1 G(x, y)\, f(y)\, dy, \quad x \in [0, 1],$$

solves (4.68).

The function G is called *Green's function* of the boundary value problem (4.68).

(c) Show that the eigenvalue problem

$$-u''(x) + q(x)\, u(x) = \lambda\, u(x), \quad u(0) = u(1) = 0,$$

is equivalent to the eigenvalue problem for the integral equation

$$\frac{1}{\lambda}\, u(x) = \int_0^1 G(x, y)\, u(y)\, dy, \quad x \in [0, 1].$$

Prove Lemma 4.7, parts (a) and (b) by the general spectral theorem (Theorem A.48 of Appendix A).

(d) How can one treat the case when q changes sign?

4.2 Let $H \in \mathbb{R}$. Prove that the transcendental equation $z \cot z + H = 0$ has a countable number of zeros z_n and that

$$z_n = (n + 1/2)\, \pi + \frac{H}{(n + 1/2)\, \pi} + \mathcal{O}(1/n^2).$$

From this,

$$z_n^2 = (n + 1/2)^2 \pi^2 + 2H + \mathcal{O}(1/n)$$

follows.

4.3 Prove Lemma 4.13.

4.4 Let $q \in C[0, 1]$ be real- or complex-valued and λ_n, g_n be the eigenvalues and L^2-normalized eigenfunctions, respectively, corresponding to q and boundary conditions $u(0) = 0$ and $hu'(1) + Hu(1) = 0$. Show by modifying the proof of Theorem 4.20 that $\{g_n : n \in \mathbb{N}\}$ is complete in $L^2(0, 1)$. This gives – even for real q – a proof different from the one obtained by applying the general spectral theory.

4.5 Consider the eigenvalue problem on $[0, 1]$:

$$-u''(x) + q(x)\,u(x) = \lambda\,u(x), \quad u(0) = u'(1) = 0.$$

Prove the following uniqueness result for the inverse problem: Let (λ_n, u_n) and (μ_n, v_n) be the eigenvalues and eigenfunctions corresponding to p and q, respectively. If $\lambda_n = \mu_n$ for all $n \in \mathbb{N}$ and

$$\frac{u_n(1)}{u_n'(0)} = \frac{v_n(1)}{v_n'(0)} \quad \text{for all } n \in \mathbb{N},$$

then p and q coincide.

5

An Inverse Scattering Problem

5.1 Introduction

We consider acoustic waves that travel in a medium, such as a fluid. Let $v(x,t)$ be the velocity vector of a particle at $x \in \mathbb{R}^3$ and time t. Let $p(x,t)$, $\rho(x,t)$, and $S(x,t)$ denote the pressure, density, and specific entropy, respectively, of the fluid. We assume that no exterior forces act on the fluid. Then the movement of the particle is described by the following equations:

$$\frac{\partial v}{\partial t} + (v \cdot \nabla)v + \gamma v + \frac{1}{\rho}\nabla p = 0 \quad \text{(Euler's equation)}, \tag{5.1a}$$

$$\frac{\partial \rho}{\partial t} + \operatorname{div}(\rho v) = 0 \quad \text{(continuity equation)}, \tag{5.1b}$$

$$f(\rho, S) = p \quad \text{(equation of state)}, \tag{5.1c}$$

$$\frac{\partial S}{\partial t} + v \cdot \nabla S = 0 \quad \text{(adiabatic hypothesis)}, \tag{5.1d}$$

where the function f depends on the fluid. γ is a damping coefficient, which we assume to be piecewise constant. This system is nonlinear in the unknown functions v, ρ, p, and S. Let the *stationary case* be described by $v_0 = 0$, time-independent distributions $\rho = \rho_0(x)$ and $S = S_0(x)$, and constant p_0 such that $p_0 = f(\rho_0(x), S_0(x))$. The *linearization* of this nonlinear system is given by the (directional) derivative of this system at (v_0, p_0, ρ_0, S_0). For deriving the linearization, we set

$$v(x,t) = \varepsilon v_1(x,t) + \mathcal{O}(\varepsilon^2),$$

$$
\begin{aligned}
p(x,t) &= p_0 + \varepsilon\, p_1(x,t) + \mathcal{O}(\varepsilon^2), \\
\rho(x,t) &= \rho_0(x) + \varepsilon\, \rho_1(x,t) + \mathcal{O}(\varepsilon^2), \\
S(x,t) &= S_0(x) + \varepsilon\, S_1(x,t) + \mathcal{O}(\varepsilon^2),
\end{aligned}
$$

and we substitute this into (5.1a), (5.1b), (5.1c), and (5.1d). Ignoring terms with $\mathcal{O}(\varepsilon^2)$ leads to the *linear* system

$$
\frac{\partial v_1}{\partial t} + \gamma\, v_1 + \frac{1}{\rho_0}\nabla p_1 = 0, \tag{5.2a}
$$

$$
\frac{\partial \rho_1}{\partial t} + \mathrm{div}(\rho_0\, v_1) = 0, \tag{5.2b}
$$

$$
\frac{\partial f(\rho_0, S_0)}{\partial \rho}\, \rho_1 + \frac{\partial f(\rho_0, S_0)}{\partial S}\, S_1 = p_1, \tag{5.2c}
$$

$$
\frac{\partial S_1}{\partial t} + v_1 \cdot \nabla S_0 = 0. \tag{5.2d}
$$

First, we eliminate S_1. Since

$$
0 = \nabla f\big(\rho_0(x), S_0(x)\big) = \frac{\partial f(\rho_0, S_0)}{\partial \rho}\nabla \rho_0 + \frac{\partial f(\rho_0, S_0)}{\partial S}\nabla S_0,
$$

we conclude by differentiating (5.2c) with respect to t and using (5.2d)

$$
\frac{\partial p_1}{\partial t} = c(x)^2 \left[\frac{\partial \rho_1}{\partial t} + v_1 \cdot \nabla \rho_0\right], \tag{5.2e}
$$

where the *speed of sound c* is defined by

$$
c(x)^2 := \frac{\partial}{\partial \rho} f\big(\rho_0(x), S_0(x)\big).
$$

Now we eliminate v_1 and ρ_1 from the system. This can be achieved by differentiating (5.2e) with respect to time and using Equations (5.2a) and (5.2b). This leads to the *wave equation* for p_1:

$$
\frac{\partial^2 p_1(x,t)}{\partial t^2} + \gamma\frac{\partial p_1(x,t)}{\partial t} = c(x)^2\, \rho_0(x)\, \mathrm{div}\left[\frac{1}{\rho_0(x)}\nabla p_1(x,t)\right]. \tag{5.3}
$$

Now we assume that terms involving $\nabla \rho_0$ are negligible and that p_1 is time-periodic, i.e., of the form

$$
p_1(x,t) = \mathrm{Re}\left[u(x)\, e^{-i\omega t}\right]
$$

with frequency $\omega > 0$ and a complex-valued function $u = u(x)$ depending only on the spatial variable. Substituting this into the wave equation (5.3) yields the three-dimensional *Helmholtz equation* for u:

$$
\Delta u(x) + \frac{\omega^2}{c(x)^2}\left(1 + i\frac{\gamma}{\omega}\right) u = 0.
$$

In free space, $c = c_0$ is constant and $\gamma = 0$. We define the *wave number* and the *index of refraction* by

$$k := \frac{\omega}{c_0} > 0 \quad \text{and} \quad n(x) := \frac{c_0^2}{c(x)^2}\left(1 + i\frac{\gamma}{\omega}\right). \tag{5.4}$$

The Helmholtz equation then takes the form

$$\Delta u + k^2 n u = 0 \tag{5.5}$$

where n is a complex-valued function with $\operatorname{Re} n(x) \geq 0$ and $\operatorname{Im} n(x) \geq 0$.

This equation holds in every source-free domain in $\mathbb{R}^3$. We assume in this chapter that there exists $a > 0$ such that $c(x) = c_0$ and $\gamma(x) = 0$ for all x with $|x| \geq a$, i.e., $n(x) = 1$ for $|x| \geq a$. This means that the inhomogeneous medium $\{x \in \mathbb{R}^3 : n(x) \neq 1\}$ is bounded and contained in the ball $K(0,a) := \{y \in \mathbb{R}^3 : |y| < a\}$ of radius a. By $K[0,a]$, we denote its closure, i.e., $K[0,a] := \{y \in \mathbb{R}^3 : |y| \leq a\}$. We further assume that sources lie outside the ball $K[0,a]$.

These sources generate "incident" fields u^i, which satisfy the unperturbed Helmholtz equation $\Delta u^i + k^2 u^i = 0$ in $K[0,a]$. In this introduction, we assume that u^i is either a *point source* or a *plane wave*, i.e., the time-dependent incident fields have the form

$$p_1^i(x,t) = \frac{1}{|x-z|}\operatorname{Re} e^{ik|x-z|-i\omega t}, \quad \text{i.e.,} \quad u^i(x) = \frac{e^{ik|x-z|}}{|x-z|},$$

for a source at z with $|z| > a$, or

$$p_1^i(x,t) = \operatorname{Re} e^{ik\hat{\theta}\cdot x - i\omega t}, \quad \text{i.e.,} \quad u^i(x) = e^{ik\hat{\theta}\cdot x},$$

for a unit vector $\hat{\theta} \in \mathbb{R}^3$.

In any case, u^i is a solution of the Helmholtz equation $\Delta u^i + k^2 u^i = 0$ in $K[0,a]$. In the first case, p_1^i describes a *spherical wave* that travels away from the source with velocity c_0. In the second case, p_1^i is a plane wave that travels in the direction $\hat{\theta}$ with velocity c_0.

The incident field will be disturbed by the medium described by the index of refraction n and will produce a "scattered wave" u^s. The total field $u = u^i + u^s$ satisfies the Helmholtz equation $\Delta u + k^2 n u = 0$ outside the sources. Furthermore, we expect the scattered field to behave as a spherical wave far away from the medium. This can be described by the following *radiation condition*

$$\frac{\partial u^s(x)}{\partial r} - ik u^s(x) = \mathcal{O}(1/r^2) \quad \text{as } r = |x| \longrightarrow \infty, \tag{5.6}$$

uniformly in $x/|x| \in S^2$. Here we denote by S^2 the unit sphere in $\mathbb{R}^3$. We have now derived a complete description of the direct scattering problem.

Let the wave number $k > 0$, the index of refraction $n = n(x)$ with $n(x) = 1$ for $|x| \geq a$, and the incident field u^i be given. Determine the total field u that satisfies the Helmholtz equation (5.5) in $\mathbb{R}^3$ outside the source region such that the scattered field $u^s = u - u^i$ satisfies the radiation condition (5.6).

In the inverse problem, one tries to determine the index of refraction n from measurements of the field u outside of $K(0, a)$ for several different incident fields u^i and/or different wave numbers k. The following example shows that the radially symmetric case reduces to an ordinary differential equation.

Example 5.1

Let $n = n(r)$ be radially symmetric, i.e., n is independent of the spherical coordinates. Since, in spherical polar coordinates (r, φ, θ),

$$\Delta = \frac{1}{r^2} \frac{\partial}{\partial r} \left(r^2 \frac{\partial}{\partial r} \right) + \frac{1}{r^2 \sin^2 \theta} \frac{\partial^2}{\partial \varphi^2} + \frac{1}{r^2 \sin \theta} \frac{\partial}{\partial \theta} \left(\sin \theta \frac{\partial}{\partial \theta} \right),$$

the Helmholtz equation for radially symmetric $u = u(r)$ reduces to the following ordinary differential equation of second order:

$$\frac{1}{r^2} \left(r^2 u'(r) \right)' + k^2 n(r) u(r) = 0,$$

i.e.,

$$u''(r) + \frac{2}{r} u'(r) + k^2 n(r) u(r) = 0 \quad \text{for } r > 0. \tag{5.7a}$$

From the theory of linear ordinary differential equations of second order with singular coefficients, we know that in a neighborhood of $r = 0$ there exist two linearly independent solutions, a regular one and one with a singularity at $r = 0$. We construct them by making the substitution $u(r) = v(r)/r$ in (5.7a). This yields the equation

$$v''(r) + k^2 n(r) v(r) = 0 \quad \text{for } r > 0. \tag{5.7b}$$

For the simplest case, where $n(r) = 1$, we readily see that $u_1(r) = \alpha \sin(kr)/r$ and $u_2(r) = \beta \cos(kr)/r$ are two linearly independent solutions. u_1 is regular and u_2 is singular at the origin. Neither of them satisfies the radiation condition. However, the combination $u(r) = \gamma \exp(ikr)/r$ does satisfy the radiation condition since

$$u'(r) - iku(r) = -\gamma \frac{\exp(ikr)}{r^2} = \mathcal{O}\left(\frac{1}{r^2} \right)$$

as is readily seen. For the case of arbitrary n, we construct a fundamental system $\{v_1, v_2\}$ of (5.7b), i.e., v_1 and v_2 satisfy (5.7b) with $v_1(0) = 0$, $v_1'(0) = 1$, and $v_2(0) = 1$, $v_2'(0) = 0$. Then $u_1(r) = v_1(r)/r$ is the regular and $u_2(r) = v_2(r)/r$ is the singular solution.

In the next section, we will rigorously formulate the direct scattering problem and prove the uniqueness and existence of a solution. The basic ingredients for the uniqueness proof are a result by Rellich (see [180]) and a unique continuation principle for solutions of the Helmholtz equation. We will prove neither Rellich's lemma nor the general continuation principle, but rather give a simple proof for a special case of a unique continuation principle that is sufficient for the uniqueness proof of the direct problem. This suggestion was recently made by Hähner (see [92]). We will then show the equivalence of the scattering problem with an integral equation. Existence is then proven by an application of the Riesz Theorem A.34 of Appendix A. Section 5.3 is devoted to the introduction of the far field patterns that describe the scattered fields "far away" from the medium. We collect some results on the far field operator, several of which will be needed in Section 5.5. In Section 5.4, we will prove uniqueness of the inverse problem. Finally, in Section 5.5, we will present three numerical algorithms for solving the inverse scattering problem.

5.2 The Direct Scattering Problem

In this section, we will collect properties of solutions of the Helmholtz equation that will be needed later. We will prove uniqueness and existence of the direct scattering problem and introduce the far field pattern. In the remaining part of this chapter, we will restrict ourselves to scattering problems for *plane incident fields*.

Throughout this chapter, we make the following assumptions: Let $n \in C^2(\mathbb{R}^3)$ and $a > 0$ with $n(x) = 1$ for all $|x| \geq a$. Assume that $\operatorname{Re} n(x) \geq 0$ and $\operatorname{Im} n(x) \geq 0$ for all $x \in \mathbb{R}^3$. Let $k \in \mathbb{R}$, $k > 0$, and $\hat{\theta} \in \mathbb{R}^3$ with $|\hat{\theta}| = 1$. We set $u^i(x) := \exp(ik\hat{\theta} \cdot x)$ for $x \in \mathbb{R}^3$. Then u^i solves the Helmholtz equation

$$\Delta u^i + k^2 u^i = 0 \quad \text{in } \mathbb{R}^3. \tag{5.8}$$

We again formulate the direct scattering problem: Given n, k, $\hat{\theta}$ satisfying the previous assumptions, determine $u \in C^2(\mathbb{R}^3)$ such that

$$\Delta u + k^2 n u = 0 \quad \text{in } \mathbb{R}^3, \tag{5.9}$$

and $u^s := u - u^i$ satisfies the *Sommerfeld radiation condition*

$$\frac{\partial u^s}{\partial r} - ik\,u^s = \mathcal{O}(1/r^2) \quad \text{for } r = |x| \to \infty, \tag{5.10}$$

uniformly in $\frac{x}{|x|} \in S^2$. We need some results from the theory of the Helmholtz equation. We omit some of the proofs and refer to [37, 38] for a detailed investigation of the direct scattering problems. The proof of uniqueness relies on the following very important theorem, which we state without proof.

Lemma 5.2 *(Rellich)*
Let u satisfy the Helmholtz equation $\Delta u + k^2 u = 0$ for $|x| > a$. Assume, furthermore, that

$$\lim_{R \to \infty} \int_{|x|=R} |u(x)|^2 \, ds(x) = 0. \tag{5.11}$$

Then $u = 0$ for $|x| > a$.

For the proof, we refer to [38] (Lemma 2.11).

In particular, the condition (5.11) of this lemma is satisfied if $u(x) = \mathcal{O}(1/r^2)$ uniformly in $x/|x| \in S^2$. Note that the assertion of this lemma does not hold if k is imaginary or $k = 0$.

The second important tool for proving uniqueness is the unique continuation principle. For the uniqueness proof, only a special case is sufficient. We present a simple proof by Hähner (see [92]), which is an application of the following result on periodic differential equations with constant coefficients. This lemma will also be needed in the uniqueness proof for the inverse problem (see Section 5.4). First, we define the cube $Q := [-\pi, \pi]^3 \in \mathbb{R}^3$. We identify $L^2(Q)$ with the space of square integrable functions on Q that are 2π-periodic with respect to all variables, i.e., they satisfy $g(2\pi n + x) = g(x)$ for all $x \in \mathbb{R}^3$ and $n \in \mathbb{Z}^3$.

Lemma 5.3
Let $a \in \mathbb{R}^3$, $\alpha \in \mathbb{R}$ and $\hat{e} = (1, i, 0)^\top \in \mathbb{C}^3$. Then, for every $t > 0$ and every 2π-periodic function $g \in L^2(Q)$, there exists a unique 2π-periodic solution $w = w_t(g) \in L^2(Q)$ of the differential equation

$$\Delta w + (2t\hat{e} - ia) \cdot \nabla w - (it + \alpha)\,w = g \quad \text{in } \mathbb{R}^3. \tag{5.12a}$$

The solution is understood in the variational sense, i.e.,

$$\int_{\mathbb{R}^3} w \left[\Delta \varphi - (2t\hat{e} - ia) \cdot \nabla \varphi - (it + \alpha)\,\varphi\right] dx = \int_{\mathbb{R}^3} g\,\varphi\,dx \tag{5.12b}$$

for every function $\varphi \in C^\infty(\mathbb{R}^3)$ with compact support. Furthermore, the following estimate holds:

$$\|w\|_{L^2(Q)} \leq \frac{1}{t} \|g\|_{L^2(Q)} \quad \text{for all } g \in L^2(Q), \ t > 0. \tag{5.13}$$

In other words, there exists a linear and bounded solution operator

$$L_t : L^2(Q) \to L^2(Q), \quad g \mapsto w_t(g),$$

of (5.12b) with the property $\|L_t\|_{L^2(Q)} \leq 1/t$ for all $t > 0$.

Proof: We expand g into the Fourier series

$$g(x) = \sum_{j \in \mathbb{Z}^3} g_j \, e^{i j \cdot x}, \quad x \in \mathbb{R}^3,$$

with Fourier coefficients

$$g_j = \frac{1}{(2\pi)^3} \int_Q g(y) \, e^{-i j \cdot y} \, dy, \quad j \in \mathbb{Z}^3.$$

The representation $w(x) = \sum_{j \in \mathbb{Z}^3} w_j \exp(i j \cdot x)$ leads to the equation

$$w_j \left[-|j|^2 + i j \cdot (2t\hat{e} - i a) - (it + \alpha) \right] = g_j, \quad j \in \mathbb{Z}^3,$$

for the coefficients w_j. We estimate

$$\left| -|j|^2 + i j \cdot (2t\hat{e} - i a) - (it + \alpha) \right| \geq |\mathrm{Im}[\cdots]| = t|2j_1 - 1| \geq t$$

for all $j \in \mathbb{Z}^3$ and $t > 0$. Therefore, the operator

$$L_t g(x) := \sum_{j \in \mathbb{Z}^3} \frac{g_j}{-|j|^2 + i j \cdot (ia + 2t\hat{e}) - (it + \alpha)} e^{i j \cdot x}, \quad g \in L^2(Q),$$

is well-defined and bounded with $\|L_t\|_{L^2(Q)} \leq 1/t$ for every $t > 0$. By truncating the series and using Green's second theorem, it is easy to see that $w = L_t g$ satisfies the variational equation for trigonometric polynomials g. A density argument yields the assertion for all $g \in L^2(Q)$. $\square$

Now we can give a very simple proof of the following version of a unique continuation principle.

Theorem 5.4

Let $n \in C^2(\mathbb{R}^3)$ with $n(x) = 1$ for $|x| \geq a$ be given. Let $u \in C^2(\mathbb{R}^3)$ be a solution of the Helmholtz equation $\Delta u + k^2 n u = 0$ in $\mathbb{R}^3$ such that $u(x) = 0$ for all $|x| \geq b$ for some $b \geq a$. Then u has to vanish in all of $\mathbb{R}^3$.

Proof: Again define $\hat{e} = (1, i, 0)^{\top} \in \mathbb{C}^3$, $\rho = 2b/\pi$, and the function

$$w(x) := e^{i/2\,x_1 - t\,\hat{e}\cdot x}\,u(\rho x), \quad x \in Q := [-\pi, \pi]^3,$$

for some $t > 0$. Then $w(x) = 0$ for all $|x| \geq \pi/2$, in particular near the boundary of the cube. Extend w to a $2\pi-$periodic function in $\mathbb{R}^3$ by $w(2\pi n + x) := w(x)$ for $x \in Q$ and all $n \in \mathbb{Z}^3$, $n \neq 0$. Then $w \in C^2(\mathbb{R}^2)$, and elementary computations show that w satisfies the differential equation

$$\Delta w(x) + \left(2t\hat{e} - ia\right) \cdot \nabla w(x) - (it + 1/4)\,w(x) = -\rho^2 k^2 \tilde{n}(x)\,w(x)$$

for $x \in \mathbb{R}^3$, where $a = (1, 0, 0)^{\top}$ and $\tilde{n}(2\pi n + x) := n(\rho x)$ for all $x \in [-\pi, \pi]^3$ and $n \in \mathbb{Z}^3$. Application of the previous lemma to this differential equation yields the existence of a linear bounded operator L_t from $L^2(Q)$ into itself with $\|L_t\|_{L^2} \leq 1/t$ such that the differential equation is equivalent to

$$w = -\rho^2 k^2\,L_t(\tilde{n}w).$$

Estimating

$$\|w\|_{L^2(Q)} \leq \frac{\rho^2 k^2}{t}\,\|\tilde{n}w\|_{L^2(Q)} \leq \frac{\rho^2 k^2\,\|n\|_{\infty}}{t}\,\|w\|_{L^2(Q)}$$

yields $w = 0$ for sufficiently large $t > 0$. Thus u also has to vanish.   □

The preceding theorem is a special case of a far more general unique continuation principle, which we formulate without proof here.

Let $u \in C^2(\Omega)$ be a solution of the Helmholtz equation $\Delta u + k^2 n\,u = 0$ in a domain $\Omega \subset \mathbb{R}^3$ (i.e., Ω is open and connected). Furthermore, let $n \in C^2(\Omega)$ and $u(x) = 0$ in a neighborhood of some point $x_0 \in \Omega$. Then $u = 0$ in all of Ω. For a proof, we refer to, e.g., [38]. Now we can prove the following uniqueness theorem.

Theorem 5.5 *(Uniqueness)*
The problem (5.9), (5.10) has at most one solution, i.e., if u is a solution corresponding to $u^i = 0$, then $u = 0$.

Proof: Let $u^i = 0$. The radiation condition (5.10) yields

$$\mathcal{O}(1/R^2) = \int_{|x|=R} \left| \frac{\partial u}{\partial r} - ik\,u \right|^2 ds \tag{5.14}$$

$$= \int_{|x|=R} \left(\left| \frac{\partial u}{\partial r} \right|^2 + k^2\,|u|^2 \right) ds + 2\,k\,\mathrm{Im} \int_{|x|=R} u\,\frac{\partial \bar{u}}{\partial r}\,ds.$$

We can transform the last integral using Green's first theorem:

$$\int_{|x|=R} u \frac{\partial \bar{u}}{\partial r} ds = \int_{|x|<R} \left[u \Delta \bar{u} + |\nabla u|^2 \right] dx = \int_{|x|<R} \left[|\nabla u|^2 - k^2 \bar{n} |u|^2 \right] dx,$$

i.e.,

$$\text{Im} \int_{|x|=R} u \frac{\partial \bar{u}}{\partial r} ds = k^2 \int_{|x|<R} \text{Im}\, n |u|^2 dx \geq 0.$$

We substitute this into (5.14) and let R tend to infinity. This yields

$$0 \leq \limsup_{R \to \infty} \int_{|x|=R} \left(\left| \frac{\partial u}{\partial r} \right|^2 + k^2 |u|^2 \right) ds \leq 0,$$

and thus

$$\int_{|x|=R} |u|^2 ds \longrightarrow 0 \quad \text{as } R \to \infty.$$

Rellich's Lemma 5.2 implies $u = 0$ for $|x| > a$. Finally, the unique continuation principle of Theorem 5.4 yields $u = 0$ in $\mathbb{R}^3$.   □

Now let

$$\Phi(x, y) := \frac{e^{ik|x-y|}}{4\pi |x - y|} \quad \text{for } x, y \in \mathbb{R}^3, \ x \neq y, \tag{5.15}$$

be the *fundamental solution* or *free space Green's function* of the Helmholtz equation. Properties of the fundamental solution are summarized in the following theorem.

Theorem 5.6
$\Phi(\cdot, y)$ *solves the Helmholtz equation* $\Delta u + k^2 u = 0$ *in* $\mathbb{R}^3 \setminus \{y\}$ *for every* $y \in \mathbb{R}^3$. *It satisfies the radiation condition*

$$\frac{x}{|x|} \cdot \nabla_x \Phi(x, y) - ik\, \Phi(x, y) = \mathcal{O}(1/|x|^2)$$

uniformly in $\frac{x}{|x|} \in S^2$ *and* $y \in Y$ *for every bounded subset* $Y \subset \mathbb{R}^3$. *In addition,*

$$\Phi(x, y) = \frac{e^{ik|x|}}{4\pi |x|} e^{-ik\hat{x} \cdot y} + \mathcal{O}(1/|x|^2) \tag{5.16}$$

uniformly in $\hat{x} = \frac{x}{|x|} \in S^2$ *and* $y \in Y$.

The proof is not difficult and is left to the reader.   □

Now we construct volume potentials with this fundamental solution. We need another result, which we state without proof. Before doing this, we remind the reader of the definition of Hölder continuity. A function w : $\Omega \to \mathbb{C}^N$ is called *Hölder continuous* if there exists $c > 0$ and $\alpha \in (0,1]$ such that $|w(x) - w(y)| \leq c\,|x - y|^\alpha$ for all $x, y \in \Omega$. The space $C^{p,\alpha}(\Omega)$ consists of all functions that are p-times continuously differentiable and their pth derivatives are all Hölder continuous with exponent $\alpha \in (0,1]$. These spaces, equipped with norms

$$\|w\|_{C^{p,\alpha}} := \max_{|j| \leq p} \left\| D^j w \right\|_\infty + \max_{|j|=p} \sup_{x \neq y} \frac{\left| D^j w(x) - D^j w(y) \right|}{|x - y|^\alpha},$$

are Banach spaces. Here, $j = (j_1, j_2, j_3) \in \mathbb{N}_0^3$ denotes a multi index,

$$D^j = \frac{\partial^{j_1 + j_2 + j_3}}{\partial x_1^{j_1} \partial x_2^{j_2} \partial x_3^{j_3}},$$

and $|j| = j_1 + j_2 + j_3$. We have the following theorem.

Theorem 5.7
Let $\Omega \subset \mathbb{R}^3$ be a bounded domain. For every $\varphi \in C(\overline{\Omega})$ the volume potential

$$v(x) := \int_\Omega \varphi(y)\, \Phi(x, y)\, dy, \quad x \in \mathbb{R}^3, \tag{5.17a}$$

exists as an improper integral. Furthermore, $v \in C^{1,\alpha}(\mathbb{R}^3)$ and

$$\nabla v(x) = \int_\Omega \varphi(y)\, \nabla_x \Phi(x, y)\, dy \quad \text{for } x \in \mathbb{R}^3. \tag{5.17b}$$

If, in addition, φ is Hölder continuous, i.e., $\varphi \in C^\alpha(\Omega)$ for some $\alpha \in (0,1]$, then $v \in C^{2,\alpha}(\Omega)$ and $\Delta v + k^2 v = -\varphi$ in Ω. Furthermore, there exists $c > 0$ (only dependent on Ω) such that

$$\|v\|_{C^{2,\alpha}(\Omega)} \leq c\, \|\varphi\|_{C^\alpha(\Omega)}. \tag{5.18}$$

If $\varphi \in C^\alpha(\Omega)$ is of compact support, then $v \in C^{2,\alpha}(\mathbb{R}^3)$ and $\Delta v + k^2 v = -\varphi$ in $\mathbb{R}^3$, where φ is extended by zero into all of $\mathbb{R}^3$.

For a proof, we refer to [79].

Now we can transform the scattering problem into a Fredholm integral equation of the second kind. The following theorem will be needed quite often later on.

Theorem 5.8

(a) Let $u \in C^2(\mathbb{R}^3)$ be a solution of the scattering problem (5.9), (5.10). Then $u|_{K[0,a]}$ solves the Lippmann–Schwinger integral equation

$$u(x) = u^i(x) - k^2 \int_{|y|<a} (1 - n(y)) \, \Phi(x,y) \, u(y) \, dy, \qquad (5.19)$$

$x \in K[0,a]$. (b) If, on the other hand, $u \in C(K[0,a])$ is a solution of the integral equation (5.19), then u can be extended by the right-hand side of (5.19) to a solution $u \in C^2(\mathbb{R}^3)$ of the scattering problem (5.9), (5.10).

Proof: (a) Let u be a solution of (5.9), (5.10) and v the right-hand side of (5.19) for $x \in \mathbb{R}^3$. Since $u \in C^2(\mathbb{R}^3)$, we conclude by Theorem 5.7 that $v \in C^{2,\alpha}(\mathbb{R}^3)$ and $\Delta v + k^2 v = k^2(1-n)u = k^2 u + \Delta u$. Therefore, $w = v - u$ satisfies the Helmholtz equation $\Delta w + k^2 w = 0$ in all of $\mathbb{R}^3$. Furthermore,

$$w(x) = [u^i(x) - u(x)] - k^2 \int_{|y|<a} (1 - n(y)) \, \Phi(x,y) \, u(y) \, dy, \quad x \in \mathbb{R}^3,$$

and this satisfies the radiation condition (5.10) by Theorem 5.6. The uniqueness Theorem 5.5 yields that $w = 0$; thus $u = v$. This proves the first part.

(b) Let $u \in C(K[0,a])$ be a solution of (5.19). Extend u by the right-hand side of (5.19) to all of $\mathbb{R}^3$. A first application of Theorem 5.7 yields that $u \in C^{1,\alpha}(\mathbb{R}^3)$. Then, since $n \in C^2(\mathbb{R}^3)$), a second application yields $u \in C^{2,\alpha}(\mathbb{R}^3)$. Furthermore, $\Delta u + k^2 u = k^2(1-n)u$ and thus $\Delta u + k^2 n \, u = 0$ in $\mathbb{R}^3$. The radiation condition (5.10) for the scattered field $u^s = u - u^i$ again follows from Theorem 5.6. $\qquad \square$

As a corollary, we immediately have the following.

Theorem 5.9

Under the given assumptions on k, n, and $\hat\theta$, there exists a unique solution u of the scattering problem (5.9), (5.10) or, equivalently, the integral equation (5.19).

Proof: We apply the Riesz theory (Theorem A.34 of Appendix A) to the integral equation $u = u^i - Tu$, where the operator $T : C(K[0,a]) \to C(K[0,a])$ is defined by

$$(Tu)(x) := k^2 \int_{|y|<a} (1 - n(y)) \, \Phi(x,y) \, u(y) \, dy, \quad x \in K[0,a]. \qquad (5.20)$$

This integral operator is compact since the kernel is weakly singular (Theorem A.33 of Appendix A). Therefore, it is sufficient to prove uniqueness of a solution to (5.19). This follows by Theorems 5.8 and 5.5. $\qquad \square$

As another application of the Lippmann–Schwinger integral equation, we derive the following asymptotic behavior of u.

Theorem 5.10

Let u be the solution of the scattering problem (5.9), (5.10). Then

$$u(x) = u^i(x) + \frac{e^{ik|x|}}{|x|} u_\infty(\hat{x}) + \mathcal{O}(1/|x|^2) \quad as \ |x| \to \infty \qquad (5.21)$$

uniformly in $\hat{x} = x/|x|$, where

$$u_\infty(\hat{x}) = -\frac{k^2}{4\pi} \int_{|y|<a} \left(1 - n(y)\right) e^{-ik\hat{x}\cdot y} u(y)\,dy \quad for \ \hat{x} \in S^2. \qquad (5.22)$$

The function $u_\infty : S^2 \to \mathbb{C}$ is called the far field pattern *or* scattering amplitude *of u. It is analytic on S^2 and determines u^s outside of $K(0,a)$ uniquely, i.e., $u_\infty = 0$ if and only if $u^s(x) = 0$ for $|x| \geq a$.*

Proof: Formulas (5.21) and (5.22) follow directly from the asymptotic behavior of the fundamental solution Φ. The analyticity of u_∞ follows from (5.22). Finally, if $u_\infty = 0$, then an application of Rellich's lemma yields that $u^s = u - u^i = 0$ for all $|x| > a$. $\square$

The existence of a far field pattern, i.e., a function u_∞ with

$$u^s(x) = \frac{e^{ik|x|}}{|x|} u_\infty(\hat{x}) + \mathcal{O}(1/|x|^2) \quad as \ |x| \to \infty,$$

is not restricted to scattering problems. Indeed, Theorem 5.11 assures the existence of the far field pattern for every radiating solution of the Helmholtz equation.

We now draw some other conclusions from the Lippmann–Schwinger integral equation $u + Tu = u^i$. We estimate the norm of the integral operator T of (5.20):

$$|(Tu)(x)| \leq k^2 \|1 - n\|_\infty \|u\|_\infty \max_{|x|\leq a} \int_{|y|<a} |\Phi(x,y)|\,dy \quad for \ x \in K[0,a],$$

i.e.,

$$\|T\|_\infty \leq k^2 \|1 - n\|_\infty \max_{|x|\leq a} \int_{|y|<a} \frac{1}{4\pi|x-y|}\,dy$$

$$= \frac{(ka)^2}{2} \|1 - n\|_\infty ; \qquad (5.23)$$

see Problem 5.4. We conclude that $\|T\|_\infty < 1$, provided $(ka)^2 \|1 - n\|_\infty < 2$. The contraction mapping Theorem A.29 yields uniqueness and existence of a solution of the integral equation (5.19) for $(ka)^2 \|1 - n\|_\infty < 2$. We know this already – even for all values of $(ka)^2 \|1 - n\|_\infty$. But Theorem A.29 also tells us that for $(ka)^2 \|1 - n\|_\infty < 2$ the solution can be represented as a Neumann series in the form

$$u = \sum_{j=0}^{\infty} (-1)^j \, T^j u^i. \tag{5.24}$$

The first two terms of the series are

$$u^b(x) := u^i(x) - k^2 \int_{\mathbb{R}^3} (1 - n(y)) \, u^i(y) \, \Phi(x, y) \, dy, \quad x \in \mathbb{R}^3. \tag{5.25}$$

u^b is called the *Born approximation*. It provides a good approximation to u in $K[0, a]$ for small values of $(ka)^2 \|1 - n\|_\infty$ since

$$\|u - u^b\|_\infty \leq \sum_{j=2}^{\infty} \|T\|_\infty^j \, \|u^i\|_\infty = \|T\|_\infty^2 \, \frac{1}{1 - \|T\|_\infty} \leq \frac{(ka)^4}{2} \|1 - n\|_\infty^2$$

for $(ka)^2 \|1 - n\|_\infty \leq 1$. The far field pattern depends on both the direction $\hat{x} \in S^2$ of observation and the direction $\hat{\theta} \in S^2$ of the incident field u^i. Therefore, we often write $u_\infty(\hat{x}; \hat{\theta})$ to indicate this dependence. For the Born approximation, we see from the asymptotic form (5.16) of $\Phi(x, y)$ that

$$u_\infty^b(\hat{x}; \hat{\theta}) = \frac{k^2}{4\pi} \int_{\mathbb{R}^3} (n(y) - 1) \, e^{ik\hat{\theta} \cdot y} \, e^{-ik\hat{x} \cdot y} \, dy$$

$$= \frac{k^2}{4\pi} \int_{\mathbb{R}^3} (n(y) - 1) \, e^{ik(\hat{\theta} - \hat{x}) \cdot y} \, dy, \tag{5.26}$$

and this is just the Fourier transform of $m := n - 1$:

$$u_\infty^b(\hat{x}; \hat{\theta}) = \frac{k^2}{4\pi} m^\sim (k\hat{x} - k\hat{\theta}), \quad \hat{x}, \hat{\theta} \in S^2, \tag{5.27}$$

where the Fourier transform is defined by

$$f^\sim(x) := \int_{\mathbb{R}^3} f(y) \, e^{-ix \cdot y} dy, \quad x \in \mathbb{R}^3.$$

From this, the *reciprocity principle* follows:

$$u_\infty^b(-\hat{\theta}; -\hat{x}) = u_\infty^b(\hat{x}; \hat{\theta}) \quad \text{for } \hat{x}, \hat{\theta} \in S^2. \tag{5.28}$$

We will see that this relation holds for u_∞ itself. Before we can prove this principle for u_∞, we need the important Green's representation theorem which expresses radiating solutions of the Helmholtz equation in terms of the Dirichlet and Neumann boundary data.

Theorem 5.11 *(Green's representation theorem)*
Let $\Omega \subset \mathbb{R}^3$ be a bounded domain and $\Omega^c := \mathbb{R}^3 \setminus \overline{\Omega}$ its exterior. Let the boundary $\partial\Omega$ be sufficiently smooth so that Gauss' theorem holds. Let the unit normal vector $\nu(x)$ in $x \in \partial\Omega$ be directed into the exterior of Ω.
(a) Let $u \in C^2(\Omega) \cap C^1(\overline{\Omega})$. Then

$$
u(x) = \int_{\partial\Omega} \left[\Phi(x,y) \frac{\partial}{\partial\nu} u(y) - u(y) \frac{\partial}{\partial\nu(y)} \Phi(x,y) \right] ds(y)
$$

$$
- \int_\Omega \Phi(x,y) \left[\Delta u(y) + k^2 u(y) \right] dy, \quad x \in \Omega. \tag{5.29a}
$$

(b) Let $u^s \in C^2(\Omega^c) \cap C^1(\overline{\Omega^c})$ be a solution of the Helmholtz equation $\Delta u^s + k^2 u^s = 0$ in Ω^c, and let u^s satisfy the radiation condition (5.10). Then

$$
\int_{\partial\Omega} \left[\Phi(x,\cdot) \frac{\partial}{\partial\nu} u^s - u^s \frac{\partial}{\partial\nu} \Phi(x,\cdot) \right] ds = \begin{cases} 0, & x \in \Omega, \\ -u^s(x), & x \notin \overline{\Omega}. \end{cases} \tag{5.29b}
$$

The far field pattern of u^s has the representation

$$
u_\infty(\hat{x}) = \frac{1}{4\pi} \int_{\partial\Omega} \left[u^s(y) \frac{\partial}{\partial\nu(y)} e^{-ik\hat{x}\cdot y} - e^{-ik\hat{x}\cdot y} \frac{\partial}{\partial\nu} u^s(y) \right] ds(y) \tag{5.30}
$$

for $\hat{x} \in S^2$.

For a proof, we refer to [37], Theorems 3.1 and 3.3. As a corollary, we prove the following useful lemma.

Lemma 5.12
Let $\Omega \in \mathbb{R}^3$ be a domain that is decomposed into two disjoint subdomains: $\overline{\Omega} = \overline{\Omega}_1 \cup \overline{\Omega}_2$ such that $\Omega_1 \cap \Omega_2 = \emptyset$. Let the boundaries $\partial\Omega_1$ and $\partial\Omega_2$ be smooth (i.e., C^2). Let $u_j \in C^2(\Omega_j) \cap C^1(\overline{\Omega}_j)$ for $j = 1,2$ be solutions of the Helmholtz equation $\Delta u_j + k^2 u_j = 0$ in Ω_j. Furthermore, let $u_1 = u_2$ on Γ and $\frac{\partial u_1}{\partial\nu} = \frac{\partial u_2}{\partial\nu}$ on Γ, where Γ denotes the common boundary $\Gamma := \partial\Omega_1 \cap \partial\Omega_2$. Then the function u, defined by

$$
u(x) = \begin{cases} u_1(x), & x \in \Omega_1, \\ u_2(x), & x \in \Omega_2, \end{cases}
$$

can be extended to an analytic function in Ω that satisfies the Helmholtz equation $\Delta u + k^2 u = 0$ in Ω.

Proof: It follows from Green's representation theorem that u_1 and u_2 are analytic in Ω_1 and Ω_2, respectively. We fix $x_0 \in \Gamma \cap \Omega$ and choose a small ball $K(x_0, \varepsilon)$ that is entirely in Ω. Let $K_j := K(x_0, \varepsilon) \cap \Omega_j$, $j = 1, 2$, and $x \in K_1$. We apply Green's representation theorems to u_1 in K_1 and to u_2 in K_2 to arrive at

$$u_1(x) = \int_{\partial K_1} \left[\Phi(x, y) \frac{\partial}{\partial \nu} u_1(y) - u_1(y) \frac{\partial}{\partial \nu(y)} \Phi(x, y) \right] ds(y), \quad x \in K_1,$$

$$0 = \int_{\partial K_2} \left[\Phi(x, y) \frac{\partial}{\partial \nu} u_2(y) - u_2(y) \frac{\partial}{\partial \nu(y)} \Phi(x, y) \right] ds(y) \quad x \in K_1.$$

We add both equations and note that the contributions on $\Gamma \cap K$ cancel. This yields

$$u_1(x) = \int_{\partial K(x_0, \varepsilon)} \left[\Phi(x, y) \frac{\partial}{\partial \nu} u(y) - u(y) \frac{\partial}{\partial \nu(y)} \Phi(x, y) \right] ds(y), \quad x \in K_1.$$

Interchanging the roles of $j = 1$ and $j = 2$ yields

$$u_2(x) = \int_{\partial K(x_0, \varepsilon)} \left[\Phi(x, y) \frac{\partial}{\partial \nu} u(y) - u(y) \frac{\partial}{\partial \nu(y)} \Phi(x, y) \right] ds(y), \quad x \in K_2.$$

The right-hand sides coincide and are analytic functions in $K(x_0, \varepsilon)$. □

5.3   Properties of the Far Field Patterns

First, we will prove a reciprocity principle for u_∞. It states the (physically obvious) fact that it is the same if we illuminate an object from the direction $\hat{\theta}$ and observe it in the direction $-\hat{x}$ or the other way around: illumination from $\hat{x}$ and observation in $-\hat{\theta}$.

Theorem 5.13 *(Reciprocity principle)*
Let $u_\infty(\hat{x}; \hat{\theta})$ be the far field pattern corresponding to the direction $\hat{x}$ of observation and the direction $\hat{\theta}$ of the incident plane wave. Then

$$u_\infty(\hat{x}; \hat{\theta}) = u_\infty(-\hat{\theta}; -\hat{x}) \quad \text{for all } \hat{x}, \hat{\theta} \in S^2. \tag{5.31}$$

Proof: Application of Green's second theorem to u^i and u^s in the interior and exterior of $\{x \in \mathbb{R}^3 : |x| = a\}$, respectively, yields

$$0 = \int_{|y|=a} \left[u^i(y; \hat{\theta}) \frac{\partial}{\partial \nu} u^i(y; -\hat{x}) - u^i(y; -\hat{x}) \frac{\partial}{\partial \nu} u^i(y; \hat{\theta}) \right] ds(y),$$

$$0 = \int_{|y|=a} \left[u^s(y;\hat{\theta}) \frac{\partial}{\partial \nu} u^s(y;-\hat{x}) - u^s(y;-\hat{x}) \frac{\partial}{\partial \nu} u^s(y;\hat{\theta}) \right] ds(y).$$

(More precisely, to prove the second equation, one applies Green's second theorem to u^s in the region $\{x \in \mathbb{R}^3 : a < |x| < R\}$ with $R > a$ and lets R tend to infinity.)

Now we use the representations (5.30) for the far field patterns $u_\infty(\hat{x};\hat{\theta})$ and $u_\infty(-\hat{\theta};-\hat{x})$:

$$4\pi \, u_\infty(\hat{x};\hat{\theta}) = \int_{|y|=a} \left[u^s(y;\hat{\theta}) \frac{\partial}{\partial \nu} u^i(y;-\hat{x}) - u^i(y;-\hat{x}) \frac{\partial}{\partial \nu} u^s(y;\hat{\theta}) \right] ds$$

$$4\pi u_\infty(-\hat{\theta};-\hat{x}) = \int_{|y|=a} \left[u^s(y;-\hat{x}) \frac{\partial}{\partial \nu} u^i(y;\hat{\theta}) - u^i(y;\hat{\theta}) \frac{\partial}{\partial \nu} u^s(y;-\hat{x}) \right] ds.$$

We now subtract the last of these equations from the sum of the first three. This yields

$$4\pi \left[u_\infty(\hat{x};\hat{\theta}) - u_\infty(-\hat{\theta};-\hat{x}) \right]$$

$$= \int_{|y|=a} \left[u(y;\hat{\theta}) \frac{\partial}{\partial \nu} u(y;-\hat{x}) - u(y;-\hat{x}) \frac{\partial}{\partial \nu} u(y;\hat{\theta}) \right] ds(y).$$

So far, we have not used any information of u inside $K(0,a)$. To use this information, we apply Green's second theorem to the boundary integral and use the Helmholtz equation (5.9). This shows that the integral vanishes. $\square$

The far field patterns $u_\infty(\hat{x};\hat{\theta})$, $\hat{x}, \hat{\theta} \in S^2$, define the integral operator

$$Fg(\hat{x}) = \int_{S^2} u_\infty(\hat{x};\hat{\theta}) \, g(\hat{\theta}) \, ds(\hat{\theta}) \quad \text{for } \hat{x} \in S^2, \qquad (5.32)$$

which we will call the *far field operator*. It is certainly compact in $L^2(S^2)$ and is related to the *scattering operator* $S : L^2(S^2) \to L^2(S^2)$ by

$$S = I + \frac{ik}{2\pi} F.$$

The next results prove some properties of these operators. Some of them will be important in Section 5.5. In what follows, we denote by $(\cdot, \cdot)_{L^2}$ the inner product in $L^2(S^2)$. We begin with a technical lemma (see [39]).

Lemma 5.14

For $g, h \in L^2(S^2)$, define the Herglotz wave functions v^i *and* w^i *by*

$$v^i(x) \;=\; \int_{S^2} e^{ikx\cdot\hat{\theta}} \, g(\hat{\theta}) \, ds(\hat{\theta}), \quad x \in \mathbb{R}^3, \tag{5.33a}$$

$$w^i(x) \;=\; \int_{S^2} e^{ikx\cdot\hat{\theta}} \, h(\hat{\theta}) \, ds(\hat{\theta}), \quad x \in \mathbb{R}^3, \tag{5.33b}$$

respectively. Let v and w be the solutions of the scattering problem (5.9), (5.10) corresponding to incident fields v^i and w^i, respectively. Then

$$ik^2 \int_{K(0,a)} (\operatorname{Im} n) \, v\,\overline{w} \, dx$$

$$= \; 2\pi \big(Fg, h\big)_{L^2} \; - \; 2\pi \big(g, Fh\big)_{L^2} \; - \; ik \big(Fg, Fh\big)_{L^2}. \tag{5.34}$$

Proof: Let $v^s = v - v^i$ and $w^s = w - w^i$ denote the scattered fields with far field patterns v_∞ and w_∞. Then, by linearity, $v_\infty = Fg$ and $w_\infty = Fh$ and

$$2ik^2 \int_{K(0,a)} (\operatorname{Im} n)\, v\,\overline{w}\, dx = \int_{K(0,a)} [v\,\Delta\overline{w} - \overline{w}\,\Delta v]\, dx = \int_{\partial K(0,a)} \left[v\,\frac{\partial \overline{w}}{\partial \nu} - \overline{w}\,\frac{\partial v}{\partial \nu} \right] ds.$$

The integral is split into four parts by decomposing $v = v^i + v^s$ and $w = w^i + w^s$. The integral

$$\int_{\partial K(0,a)} \left[v^i\,\frac{\partial \overline{w}^i}{\partial \nu} - \overline{w}^i\,\frac{\partial v^i}{\partial \nu} \right] ds$$

vanishes by Green's second theorem since v^i and $\overline{w}^i$ are solutions of the Helmholtz equation $\Delta u + k^2 u = 0$. We write

$$\int_{\partial K(0,a)} \left[v^s\,\frac{\partial \overline{w}^s}{\partial \nu} - \overline{w}^s\,\frac{\partial v^s}{\partial \nu} \right] ds \;=\; \int_{|x|=R} \left[v^s\,\frac{\partial \overline{w}^s}{\partial \nu} - \overline{w}^s\,\frac{\partial v^s}{\partial \nu} \right] ds$$

and note that

$$v^s(x)\,\frac{\partial \overline{w^s(x)}}{\partial r} - \overline{w^s(x)}\,\frac{\partial v^s(x)}{\partial r} \;=\; -\frac{2ik}{r^2}\, v_\infty(\hat{x})\,\overline{w_\infty(\hat{x})} \;+\; \mathcal{O}(1/r^3).$$

From this

$$\int_{|x|=R} \left[v^s\,\frac{\partial \overline{w}^s}{\partial \nu} - \overline{w}^s\,\frac{\partial v^s}{\partial \nu} \right] ds \;\longrightarrow\; -2ik \int_{S^2} v_\infty\,\overline{w_\infty}\, ds \quad (R \to \infty)$$

follows. Finally, we use the definition of v^i and w^i and the representation (5.30) to compute

$$
\int\limits_{\partial K(0,a)} \left[v^i \frac{\partial \overline{w}^s}{\partial \nu} - \overline{w}^s \frac{\partial v^i}{\partial \nu} \right] ds
$$

$$
= \int\limits_{S^2} g(\hat{\theta}) \int\limits_{\partial K(0,a)} \left[e^{ikx\cdot\hat{\theta}} \frac{\partial \overline{w^s(x)}}{\partial \nu} - \overline{w^s(x)} \frac{\partial}{\partial \nu} e^{ikx\cdot\hat{\theta}} \right] ds(x)\, ds(\hat{\theta})
$$

$$
= -4\pi \int\limits_{S^2} g(\hat{\theta}) \, \overline{w_\infty(\hat{\theta})} \, d(\hat{\theta}) \; = \; -4\pi \left(g, Fh\right)_{L^2}.
$$

Analogously, we have that

$$
\int\limits_{\partial K(0,a)} \left[v^s \frac{\partial \overline{w}^i}{\partial \nu} - \overline{w}^i \frac{\partial v^s}{\partial \nu} \right] ds \; = \; 4\pi \left(Fg, h\right)_{L^2}.
$$

This ends the proof.   □

We can now give a simple proof of the unitarity of the scattering operator for real valued n.

Theorem 5.15
Let $n \in C^2(\mathbb{R}^3)$ be real valued such that the support of $m = 1-n$ is contained in $K(0,a)$. Then F is normal, i.e., $F^*F = FF^*$, and $S := I + \frac{ik}{2\pi} F$ is unitary, i.e., $S^*S = SS^* = I$.

Proof: The preceding lemma implies that

$$
ik\left(Fg, Fh\right)_{L^2} \; = \; 2\pi\left(Fg, h\right)_{L^2} - 2\pi\left(g, Fh\right)_{L^2} \tag{5.35}
$$

for all $g, h \in L^2(S^2)$. By reciprocity (Theorem 5.13), we conclude that

$$
F^*g(\hat{x}) \; = \; \int\limits_{S^2} \overline{u_\infty(\hat{\theta}; \hat{x})} \, g(\hat{\theta}) \, ds(\hat{\theta}) \; = \; \int\limits_{S^2} \overline{u_\infty(-\hat{x}; -\hat{\theta})} \, g(\hat{\theta}) \, ds(\hat{\theta})
$$

$$
= \; \int\limits_{S^2} \overline{u_\infty(-\hat{x}; \hat{\theta})} \, \overline{g(-\hat{\theta})} \, ds(\hat{\theta})
$$

and thus $F^*g = \overline{R\,F\,R\overline{g}}$, where $Rh(\hat{x}) := h(-\hat{x})$ for $\hat{x} \in S^2$. Noting that $(Rg, Rh)_{L^2} = (g, h)_{L^2} = (\overline{h}, \overline{g})_{L^2}$ for all $g, h \in L^2(S^2)$ and using (5.35) twice, we conclude that

$$
ik\left(F^*h, F^*g\right)_{L^2} \; = \; ik\left(RFR\overline{g}, RFR\overline{h}\right)_{L^2} \; = \; ik\left(FR\overline{g}, FR\overline{h}\right)_{L^2}
$$

$$
\begin{aligned}
&= 2\pi \left(FR\bar{g}, R\bar{h}\right)_{L^2} - 2\pi \left(R\bar{g}, FR\bar{h}\right)_{L^2} \\
&= 2\pi \left(RFR\bar{g}, \bar{h}\right)_{L^2} - 2\pi \left(\bar{g}, RFR\bar{h}\right)_{L^2} \\
&= 2\pi \left(h, F^*g\right)_{L^2} - 2\pi \left(F^*h, g\right)_{L^2} \\
&= 2\pi \left(Fh, g\right)_{L^2} - 2\pi \left(h, Fg\right)_{L^2} \\
&= ik \left(Fh, Fg\right)_{L^2}.
\end{aligned}
$$

This holds for all $g, h \in L^2(S^2)$; thus $F^*F = F F^*$. Finally, from (5.35), we conclude that

$$
-\left(g, ikF^*Fh\right)_{L^2} = 2\pi \left(g, (F^* - F)h\right)_{L^2} \quad \text{for all } g, h \in L^2(S^2),
$$

i.e., $ikF^*F = 2\pi \left(F - F^*\right)$. This formula, together with the normality of F, yields $S^*S = S S^* = I$. $\qquad\square$

It is well known that the eigenvalues of unitary operators all lie on the unit circle in $\mathbb{C}$. From the definition $S = I + \frac{ik}{2\pi}F$, we conclude that the eigenvalues of F lie on the circle $|2\pi i/k - z| = 2\pi/k$ with center $2\pi i/k$ and radius $2\pi/k$. This holds for real-valued indices of refraction n. For further results for absorbing media, i.e., for which n is complex valued, we refer to the original literature [39].

We will now study the far field equation

$$
\int_{S^2} u_\infty (\hat{x}; \hat{\theta}) \, g(\hat{\theta}) \, ds(\hat{\theta}) = f(\hat{x}), \quad \hat{x} \in S^2,
$$

for different right-hand sides f. We will see that its null space is characterized by the following unusual eigenvalue problem. We set for abbreviation $K = K(0, a)$.

Problem: Determine $v, w \in C^2(K) \cap C^1(\overline{K})$ such that

$$
\Delta v + k^2 v = 0 \text{ in } K, \qquad \Delta w + k^2 n w = 0 \text{ in } K, \qquad (5.36a)
$$

$$
v = w \text{ on } \partial K, \qquad \frac{\partial v}{\partial \nu} = \frac{\partial w}{\partial \nu} \text{ on } \partial K. \qquad (5.36b)
$$

We can show the following theorem (see [42, 43, 121]).

Theorem 5.16

(a) $g \in L^2(S^2)$ is a solution of the homogeneous integral equation

$$
\int_{S^2} u_\infty (\hat{x}; \hat{\theta}) \, g(\hat{\theta}) \, ds(\hat{\theta}) = 0, \quad \hat{x} \in S^2, \qquad (5.37)
$$

if and only if there exists $w \in C^2(K) \cap C^1(\overline{K})$ such that (v, w) solve (5.36a), (5.36b), where v is the Herglotz function defined by

$$v(x) = \int_{S^2} e^{ikx \cdot \hat{y}} g(\hat{y}) \, ds(\hat{y}), \quad x \in \mathbb{R}^3. \tag{5.38}$$

(b) Let $z \in K$ be fixed. The integral equation

$$\int_{S^2} u_\infty(\hat{x}; \hat{\theta}) g(\hat{\theta}) \, ds(\hat{\theta}) = e^{-ikz \cdot \hat{x}}, \quad \hat{x} \in S^2, \tag{5.39}$$

of the first kind is solvable in $L^2(S^2)$ if and only if the boundary value problem

$$\Delta v + k^2 v = 0 \text{ in } K, \qquad \Delta w + k^2 n w = 0 \text{ in } K, \tag{5.40a}$$

$$w(x) - v(x) = \frac{\exp(ik \, |x - z|)}{|x - z|} \quad \text{on } \partial K, \tag{5.40b}$$

$$\frac{\partial w(x)}{\partial \nu} - \frac{\partial v(x)}{\partial \nu} = \frac{\partial}{\partial \nu} \frac{\exp(ik \, |x - z|)}{|x - z|} \quad \text{on } \partial K, \tag{5.40c}$$

has a solution (v, w) with v of the form (5.38).

Proof: (a) Let $g \in L^2(S^2)$ be a solution of (5.37) and define v by (5.38). We observe that the left-hand side of (5.37) is a superposition of far field patterns. Therefore, the far field pattern w_∞ of the scattered field w^s that corresponds to the incident field v vanishes. The corresponding total field $w = w^s + v$ satisfies the Helmholtz equation $\Delta w + k^2 n w = 0$ in $\mathbb{R}^3$. By Rellich's lemma (Lemma 5.2), the scattered field $w^s = w - v$ vanishes outside of K. This yields that

$$w - v = 0 \text{ on } \partial K \quad \text{and} \quad \frac{\partial(w - v)}{\partial \nu} = 0 \text{ on } \partial K.$$

This proves the first part.

Now let v be of the form (5.38) and let there exist w such that (v, w) solves the eigenvalue problem (5.36a), (5.36b). We extend w to all of $\mathbb{R}^3$ by setting $w(x) := v(x)$ for $x \notin K$. Then $w \in C^1(\mathbb{R}^3)$ and even $w \in C^2(\mathbb{R}^3)$ by Lemma 5.12. Furthermore, w satisfies the Helmholtz equation $\Delta w + k^2 n w = 0$ in all of $\mathbb{R}^3$. $w - v$ vanishes in the exterior of K and obviously satisfies the radiation condition. Therefore, w is the unique total field corresponding to the incident field v. The far field pattern w_∞ of the corresponding scattered field $w - v$ vanishes. As in the previous part, we see that w is a superposition

$$w(x) = \int_{S^2} u(x; \hat{\theta}) \, g(\hat{\theta}) \, ds(\hat{\theta})$$

of total fields. For the corresponding far field patterns, we conclude that

$$0 = w_\infty(\hat{x}) = \int_{S^2} u_\infty(\hat{x}; \hat{\theta}) \, g(\hat{\theta}) \, ds(\hat{\theta})$$

for all $\hat{x} \in S^2$. This proves part (a).

(b) The proof is very similar to the preceding one. Let $g \in L^2(S^2)$ be a solution of (5.39) and define v as in (5.38). As in part (a), the integral is the far field pattern w_∞ corresponding to the total field w that satisfies the Helmholtz equation. Now w_∞ does not vanish but is equal to the function $\exp(-ikz \cdot x)$. By Theorem 5.6, the only radiating solution of the Helmholtz equation with this far field pattern is the spherical wave $\exp(ik\,|x - z|)/\,|x - z|$. Since z is contained in K, the scattered waves $w - v$ and $\exp(ik\,|x - z|)/\,|x - z|$ have to coincide outside of K. This proves the theorem. $\qquad\square$

As an application, we give conditions under which the range of the far field operator F is dense in $L^2(S^2)$. From $(Fg, h)_{L^2} = (g, F^*h)_{L^2}$ for all $g, h \in L^2(S^2)$, it is seen that the orthogonal complement of the range of F is characterized by the null space of the adjoint F^* of F.

Theorem 5.17
The null space $\{h \in L^2(S^2) : F^*h = 0\}$ consists exactly of those functions $h \in L^2(S^2)$ for which the corresponding Herglotz functions

$$v(x) := \int_{S^2} e^{ikx \cdot \hat{y}} \, \overline{h(-\hat{y})} \, ds(\hat{y}), \quad x \in \mathbb{R}^3,$$

satisfy the interior transmission problem (5.36a), (5.36b) for some $w \in C^2(K) \cap C^1(\overline{K})$.

Proof: By using the reciprocity principle (Theorem 5.13), we conclude that

$$F^*h = 0 \iff \int_{S^2} \overline{u_\infty(\hat{\theta}; \hat{x})} \, h(\hat{\theta}) \, ds(\hat{\theta}) = 0 \quad \text{for all } \hat{x} \in S^2$$

$$\iff \int_{S^2} u_\infty(-\hat{x}; -\hat{\theta}) \, \overline{h(\hat{\theta})} \, ds(\hat{\theta}) = 0 \quad \text{for all } \hat{x} \in S^2$$

$$\iff \int_{S^2} u_\infty(\hat{x}; \hat{\theta}) \, \overline{h(-\hat{\theta})} \, ds(\hat{\theta}) = 0 \quad \text{for all } \hat{x} \in S^2.$$

Application of Theorem 5.16 yields the assertion.   □

In particular, the range of F is dense in $L^2(S^2)$ if the interior transmission problem (5.36a), (5.36b) admits only the trivial solution $v = 0$ and $w = 0$. Denseness results for this and related scattering problems have been derived in, e.g., [35].

We now study the eigenvalue problem (5.36a), (5.36b) for the special case of n being real valued and radially symmetric, i.e., $n = n(r)$. For the general case, we refer the reader to [36]. We call the wave number k an *eigenvalue* of (5.36a), (5.36b) if there exist nontrivial fields (v, w) that satisfy (5.36a), (5.36b). We assume that $n \in C^2[0, \infty)$ such that $n(r) > 1$ for $0 \leq r < 1$ and $n(r) = 1$ for $r \geq 1$. The Helmholtz equations for v and w reduce to ordinary differential equations. In Example 5.1, we proved the representation $v(r) = \alpha \sin(kr)/r$, $0 < r < 1$ for some $\alpha \in \mathbb{R}$. With the substitution $w(r) = y(r)/r$, the equation for y reduces to (see (5.7b))

$$y''(r) + k^2 n(r) y(r) = 0, \quad r > 0, \tag{5.41}$$

with boundary condition $y(0) = 0$ since w has to be regular at 0. (v, w) solves the eigenvalue problem (5.36a), (5.36b) if and only if v and y satisfy the boundary conditions

$$y(1) = w(1) = v(1) = \alpha \sin k$$

and

$$y'(1) = w(1) + w'(1) = v(1) + v'(1) = \alpha k \cos k.$$

We use the *Liouville transformation* (see Section 4.1)

$$s = s(r) := \int_0^r \sqrt{n(t)}\, dt, \quad z(s) := \left[n(r(s))\right]^{1/4} y(r(s))$$

again. Here, $s \mapsto r(s)$ denotes the inverse function of the monotonic function $r \mapsto s(r)$. It transforms the differential equation (5.41) into the following form for z:

$$z''(s) + \left(k^2 - q(s)\right) z(s) = 0 \quad \text{for } 0 < s < \hat{s}, \tag{5.42}$$

with

$$q(s) := \frac{1}{4}\left[n(r)^{-3/4}\left(n(r)^{-5/4} n'(r)\right)'\right]_{r=r(s)}$$

and

$$\hat{s} = s(1) = \int_0^1 \sqrt{n(t)}\, dt > 1.$$

The boundary conditions transform to

$$z(0) = 0, \quad z(\hat{s}) = \alpha \sin k, \quad z'(\hat{s}) = \alpha k \cos k + \frac{1}{4} \alpha \sin k.$$

The quantity k^2 plays the role of λ of the previous chapter. Again, we denote by $u_2 = u_2(s, k^2, q)$ the function of the fundamental system corresponding to (5.42) with $u_2(0) = 0$ and $u_2'(0) = 1$. Then $z = \beta u_2$ for some $\beta \in \mathbb{R}$. In order to satisfy the remaining boundary conditions, α and β have to satisfy the following system of two equations:

$$\alpha \sin k - \beta u_2(\hat{s}) = 0 \quad \text{and} \quad \alpha \left(k \cos k + \frac{1}{4} \sin k \right) - \beta u_2'(\hat{s}) = 0.$$

The determinant is given by

$$f(k) = u_2(\hat{s}) \left[k \cos k + \frac{1}{4} \sin k \right] - u_2'(\hat{s}) \sin k.$$

With the asymptotic form of $u_2(\hat{s})$ for $k \to \infty$ (see Theorem 4.5), we observe that

$$
\begin{aligned}
f(k) &= \frac{\sin(k\hat{s})}{k} k \cos k - \cos(k\hat{s}) \sin k + \mathcal{O}(1/k) \\
&= \sin(k(\hat{s} - 1)) + \mathcal{O}(1/k).
\end{aligned}
$$

From this, we see that this determinant vanishes at infinitely many discrete values of k.

For this special case, we have shown the existence of infinitely many eigenvalues of (5.36a) and (5.36b). The corresponding eigenfunctions v have the form (5.38) since, by the following lemma, we have that

$$v(x) = \alpha \frac{\sin(kr)}{r} = \frac{\alpha k}{4\pi} \int_{S^2} e^{-ikr\hat{x}\cdot\hat{y}} \, ds(\hat{y}), \quad x = r\hat{x} \in \mathbb{R}^3.$$

We observe that in this radially symmetric space the range of the far field operator F is not dense in $L^2(S^2)$ if k is a zero of f.

Lemma 5.18
Let $\rho \in \mathbb{R}$ and $\hat{x} \in S^2$. Then

$$\int_{S^2} e^{i\rho\hat{x}\cdot\hat{y}} ds(\hat{y}) = 4\pi \frac{\sin \rho}{\rho}.$$

Proof: Because the integrand is spherically symmetric, we can assume without loss of generality that $\hat{x}$ is the "north pole", that is $\hat{x} = (0, 0, 1)^\top$.

Then

$$\int_{S^2} e^{i\rho \hat{x} \cdot \hat{y}}\, ds(\hat{y}) \;=\; \int_0^{2\pi} \int_0^{\pi} e^{i\rho \cos\theta}\, \sin\theta\, d\theta\, d\varphi$$

$$\;=\; 2\pi \int_{-1}^{1} e^{i\rho s}\, ds \;=\; 4\pi\, \frac{\sin\rho}{\rho}. \qquad\qquad \square$$

We have seen in this special case that in general we cannot avoid the existence of eigenvalues for real indices of refraction n. In [36, 38] it is shown for the general case that under certain assumptions on n there exist at most a countable number of eigenvalues with infinity as their only possible accumulation point.

For the case when the index of refraction has a nonvanishing imaginary part, i.e., the medium is *absorbing*, there exist no eigenvalues.

Theorem 5.19

If $\mathrm{Im}\, n(x_0) > 0$ for at least one $x_0 \in K(0, a)$, then the eigenvalue problem (5.36a), (5.36b) has no real eigenvalue $k > 0$.

Proof: Let (v, w) be a solution of (5.36a) and (5.36b) corresponding to some $k > 0$. An application of Green's second theorem yields

$$0 \;=\; \int_{\partial K}\left(v\, \frac{\partial \overline{v}}{\partial \nu} - \overline{v}\, \frac{\partial v}{\partial \nu}\right) ds \;=\; \int_{\partial K}\left(w\, \frac{\partial \overline{w}}{\partial \nu} - \overline{w}\, \frac{\partial w}{\partial \nu}\right) ds$$

$$\;=\; \int_K (w\, \Delta \overline{w} - \overline{w}\, \Delta w)\, dx \;=\; 2ik^2 \int_K \mathrm{Im}\, n\, |w|^2 dx.$$

Since $\mathrm{Im}\, n(x) \geq 0$ for all x and $\mathrm{Im}\, n(x) > 0$ in a neighborhood of x_0, we conclude that w has to vanish in a neighborhood of x_0. The general unique continuation principle (see the remark following Theorem 5.4) implies that w vanishes in all of K. Therefore, the Cauchy data $v|_{\partial K}$ and $\partial v/\partial \nu|_{\partial K}$ also vanish on ∂K. Green's representation Theorem 5.11 yields that v vanishes in K. Therefore, k cannot be an eigenvalue. $\qquad\qquad \square$

5.4   Uniqueness of the Inverse Problem

In this section, we want to determine if the knowledge of the far field pattern $u_\infty(\hat{x}; \hat{\theta})$ provides enough information to recover the index of refraction $n = n(x)$. Therefore, let two functions $n_1, n_2 \in C^2(\mathbb{R}^3)$ be given with $n_1(x) = n_2(x) = 1$ for $|x| \geq a$. We assume that the corresponding far field patterns $u_{1,\infty}$ and $u_{2,\infty}$ coincide and we wish to show that n_1 and n_2 also coincide. As a first simple case, we consider the Born approximation again. Let

$$u_{1,\infty}^b(\hat{x}; \hat{\theta}) = u_{2,\infty}^b(\hat{x}; \hat{\theta}) \quad \text{for all } \hat{x} \in S^2 \text{ and some } \hat{\theta} \in S^2.$$

Formula (5.27) implies that $m_1^{\sim}(k\hat{x} - k\hat{\theta}) = m_2^{\sim}(k\hat{x} - k\hat{\theta})$ for all $\hat{x} \in S^2$. Here, $m_j := 1 - n_j$ for $j = 1, 2$. Therefore, the Fourier transforms of m_1 and m_2 coincide on a sphere with center $k\hat{\theta}$ and radius $k > 0$. This, however, is not enough to conclude that m_1 and m_2 coincide.

Let us now assume that

$$u_{1,\infty}^b(\hat{x}; \hat{\theta}) = u_{2,\infty}^b(\hat{x}; \hat{\theta}) \quad \text{for all } \hat{x} \in S^2 \text{ and all } \hat{\theta} \in S^2.$$

Then $m_1^{\sim}(k\hat{x} - k\hat{\theta}) = m_2^{\sim}(k\hat{x} - k\hat{\theta})$ for all $\hat{x}, \hat{\theta} \in S^2$. Therefore, the Fourier transforms coincide on the set $\{k(\hat{x} - \hat{\theta}) : \hat{x}, \hat{\theta} \in S^2\}$, which describes a ball in $\mathbb{R}^3$ with center zero and radius $2k$. Since the Fourier transforms of m_1 and m_2 are analytic functions, the unique continuation principle for analytic functions yields that $m_1^{\sim}$ and $m_2^{\sim}$ coincide on all of $\mathbb{R}^3$ and thus $m_1 = m_2$. Therefore, the knowledge of $\{u_\infty^b(\hat{x}; \hat{\theta}) : \hat{x}, \hat{\theta} \in S^2\}$ is (theoretically) sufficient to recover the index of refraction.

The same arguments also show that the knowledge of $u_\infty^b(\hat{x}; \hat{\theta})$ for all $\hat{x} \in S^2$, some $\hat{\theta} \in S^2$, and all k from an interval of $\mathbb{R}^+$ is sufficient to recover n. We refer to Problem 5.1 for an investigation of this case.

These arguments hold for the Born approximation to the far field pattern. We will now prove an analogous uniqueness theorem for the actual far field pattern, which is due to A. Nachman [159], R. Novikov [171], and A. Ramm [178]. The proof consists of three steps, which we will formulate as lemmata. For the first result, we consider a fixed index of refraction $n \in C^2(\mathbb{R}^3)$ with $n(x) = 1$ for $|x| \geq a$ and show that the span of all total fields that correspond to scattering problems with plane incident fields is dense in the space of solutions of the Helmholtz equation in $K(0, a)$.

Lemma 5.20
Let $n \in C^2(\mathbb{R}^3)$ with $n(x) = 1$ for $|x| \geq a$. Let $u(\cdot; \hat{\theta})$ denote the total field corresponding to the incident field $e^{ik\hat{\theta}\cdot x}$. Let $b > a$ and define the space H by

$$H := \left\{ v \in C^2(K(0, b)) : \Delta v + k^2 n\, v = 0 \text{ in } K(0, b) \right\}. \tag{5.43}$$

Then $\mathrm{span}\{u(\cdot; \hat{\theta})|_{K(0,a)} : \hat{\theta} \in S^2\}$ is dense in $H|_{K(0,a)}$ with respect to the norm in $L^2(K(0, a))$.

Proof: Let $v \in \overline{H}$ with

$$\left(v, u(\cdot; \hat{\theta})\right)_{L^2} := \int_{K(0,a)} v(x)\, \overline{u(x; \hat{\theta})}\, dx = 0 \quad \text{for all } \hat{\theta} \in S^2.$$

The Lippmann–Schwinger equation yields $u = (I + T)^{-1} u^i$; thus

$$0 = \left(v, (I+T)^{-1} u^i(\cdot; \hat{\theta})\right)_{L^2} = \left((I+T^*)^{-1} v, u^i(\cdot; \hat{\theta})\right)_{L^2} \tag{5.44}$$

for all $\hat{\theta} \in S^2$. Set $w := (I + T^*)^{-1} v$. Then $w \in L^2\big(K(0,a)\big)$ and satisfies the "adjoint equation"

$$v(x) = w(x) + k^2\big(1 - \overline{n(x)}\big) \int_{K(0,a)} \overline{\Phi(x,y)}\, w(y)\, dy, \quad x \in K[0,a].$$

Now set

$$\tilde{w}(x) := \int_{K(0,a)} \overline{w(y)}\, \Phi(x,y)\, dy \quad \text{for } x \in \mathbb{R}^3.$$

Then $\tilde{w}$ is a volume potential with L^2-density $\overline{w}$. It is still a solution of the Helmholtz equation $\Delta \tilde{w} + k^2 \tilde{w} = 0$ for $|x| > a$, as is readily seen. Its far field pattern vanishes since

$$\overline{\tilde{w}_\infty(\hat{\theta})} = \frac{1}{4\pi} \int_{K(0,a)} w(y)\, \exp(ik\hat{\theta} \cdot y)\, dy = \frac{1}{4\pi}\left(w, u^i(\cdot; -\hat{\theta})\right)_{L^2} = 0$$

for all $\hat{\theta} \in S^2$. Rellich's lemma implies that $\tilde{w}(x) = 0$ for all $x \notin K[0,a]$. Now let $v_j \in H$ with $v_j \to v$ in $L^2\big(K(0,a)\big)$. Then

$$\int_{K(0,a)} \overline{v}\, v_j\, dx = \int_{K(0,a)} \overline{w}\, v_j\, dx + k^2 \int_{K(0,a)} (1-n)\, \tilde{w}\, v_j\, dx$$

$$= \int_{K(0,a)} \overline{w}\, v_j\, dx + \int_{K(0,a)} \tilde{w}\left[\Delta v_j + k^2 v_j\right] dx.$$

Now we substitute the definition of $\tilde{w}$ and change the orders of integration. This yields

$$\int_{K(0,a)} \overline{v}\, v_j\, dx$$

$$= \int_{K(0,a)} \overline{w(y)} \left[v_j(y) + \int_{K(0,a)} \Phi(x,y)\left[\Delta v_j(x) + k^2 v_j(x)\right] dx\right] dy.$$

Green's representation Theorem 5.11 yields

$$\int\limits_{K(0,a)} \overline{v}\, v_j\, dx \;=\; \int\limits_{K(0,a)} \overline{w(y)} \int\limits_{|x|=a} \left[\Phi(\cdot,y)\,\frac{\partial v_j}{\partial \nu} - v_j\,\frac{\partial \Phi(\cdot,y)}{\partial \nu}\right] ds\, ds(y).$$

Since v_j satisfies the Helmholtz equation $\Delta v_j + k^2 v_j = 0$ for $a < |x| < b$, we transform the inner integral from $\{x : |x| = a\}$ onto $\{x : |x| = c\}$ for some $a < c < b$. Changing the orders of integration again yields that

$$\int\limits_{K(0,a)} \overline{v}\, v_j\, dx \;=\; \int\limits_{|x|=c} \left[\tilde{w}\,\frac{\partial v_j}{\partial \nu} - v_j\,\frac{\partial \tilde{w}}{\partial \nu}\right] ds \;=\; 0$$

since $\tilde{w}$ vanishes outside $K[0,a]$. Letting j tend to infinity yields $v = 0$. $\qquad\square$

The second lemma proves a certain "orthogonality relation" between solutions of the Helmholtz equation with different indices of refraction n_1 and n_2.

Lemma 5.21
Let $n_1, n_2 \in C^2(\mathbb{R}^3)$ be two indices of refraction with $n_1(x) = n_2(x) = 1$ for all $|x| \geq a$ and assume that $u_{1,\infty}(\hat{x}; \hat{\theta}) = u_{2,\infty}(\hat{x}; \hat{\theta})$ for all $\hat{x}, \hat{\theta} \in S^2$. Then

$$\int\limits_{K(0,a)} v_1(x)\, v_2(x)\, [n_1(x) - n_2(x)]\, dx \;=\; 0 \tag{5.45}$$

for all solutions $v_j \in C^2\big(K(0,b)\big)$ of the Helmholtz equation $\Delta v_j + k^2 n_j v_j = 0$, $j = 1, 2$, in $K(0,b)$, where $b > a$.

Proof: Let v_1 be any fixed solution of $\Delta v_1 + k^2 n_1 v_1 = 0$ in $K(0,a)$. By the denseness result of Lemma 5.20 it is sufficient to prove the assertion for $v_2 := u_2(\cdot; \hat{\theta})$ and arbitrary $\hat{\theta} \in S^2$. We set $u = u_1(\cdot, \hat{\theta}) - u_2(\cdot, \hat{\theta})$. Then u satisfies the Helmholtz equation

$$\Delta u \;+\; k^2 n_1 u \;=\; k^2(n_2 - n_1)u_2.$$

We multiply this equation by v_1 and the Helmholtz equation for v_1 by u, subtract the results, and integrate. This yields

$$\int\limits_{K(0,a)} \left(v_1\, \Delta u - u\, \Delta v_1\right) dx \;=\; k^2 \int\limits_{K(0,a)} (n_2 - n_1)\, u_2\, v_1\, dx.$$

Green's second formula applied to the left integral yields

$$k^2 \int\limits_{K(0,a)} (n_2 - n_1)\, u_2\, v_1\, dx \;=\; \int\limits_{\partial K(0,a)} \left(v_1\,\frac{\partial u}{\partial \nu} - u\,\frac{\partial v_1}{\partial \nu}\right) ds.$$

From $u_{1,\infty}(\cdot,\hat{\theta}) = u_{2,\infty}(\cdot,\hat{\theta})$ and Rellich's lemma, it follows that u vanishes outside $K(0,a)$. Therefore, the surface integral vanishes and the assertion is proven. $\qquad\square$

The original proof of the third important "ingredient" of the uniqueness proof was first given in [210]. It is of independent interest and states that the set of all products $v_1 v_2$ of functions v_j that satisfy the Helmholtz equations $\Delta v_j + k^2 n_j v_j = 0$ in some bounded region Ω is dense in $L^2(\Omega)$. As a motivation we consider the case $k = 0$ and prove a result that goes back to Calderón [26].

Lemma 5.22
Let $\Omega \subset \mathbb{R}^3$ be a bounded domain. Then the set

$$\{u_1 u_2 : u_1, u_2 \text{ harmonic in } \Omega\}$$

is dense in $L^2(\Omega)$.

Proof: Let $g \in L^2(\Omega)$ such that

$$\int_\Omega g(x)\, u_1(x)\, u_2(x)\, dx \;=\; 0$$

for all harmonic functions u_1 and u_2. We fix an arbitrary $y \in \mathbb{R}^3$ and determine *complex* vectors $z^1, z^2 \in \mathbb{C}^3$ with the properties

$$z^1 \cdot z^1 \;=\; z^2 \cdot z^2 \;=\; 0 \quad \text{and} \quad z^1 + z^2 \;=\; -iy.$$

Here, $z \cdot z$ denotes the complex product $z \cdot z = \sum_{j=1}^{3} z_j^2 \in \mathbb{C}$. Such vectors z^1 and z^2 exist, as is easily seen by separating these vectors into their real and imaginary parts. Now we define the functions

$$u_1(x) \;:=\; \exp(z^1 \cdot x) \quad \text{and} \quad u_2(x) \;:=\; \exp(z^2 \cdot x).$$

Both functions are harmonic in all of $\mathbb{R}^3$. Therefore,

$$0 \;=\; \int_\Omega g(x)\, u_1(x)\, u_2(x)\, dx \;=\; \int_\Omega g(x)\, e^{(z^1 + z^2) \cdot x}\, dx$$

$$=\; \int_\Omega g(x)\, e^{-iy \cdot x}\, dx.$$

This holds for every real vector $y \in \mathbb{R}^3$. Therefore, the Fourier transform of the function g, extended by zero into all of $\mathbb{R}^3$, vanishes identically. Therefore, $g = 0$, and the lemma is proven. $\qquad\square$

Our situation is more complicated since we have to consider products of solutions of different differential equations with nonconstant coefficients. The idea is to construct solutions u of the Helmholtz equation $\Delta u + k^2 n\, u = 0$ in Ω that behave asymptotically as $\exp(z \cdot x)$. Here we take $n = n_1$ or n_2. The following result is crucial.

Theorem 5.23

Let $K(0, b) \subset \mathbb{R}^3$ be a ball of radius b and $n \in C^2(K(0, b))$ such that $n - 1$ has compact support in $K(0, b)$. Then there exist $T > 0$ and $C > 0$ such that for all $z \in \mathbb{C}^3$ with $z \cdot z = 0$ and $|z| \geq T$ there exists a solution $u_z \in L^2(K(0, b))$ of the differential equation

$$\Delta u_z + k^2 n\, u_z = 0 \quad in\ K(0, b) \tag{5.46}$$

of the form

$$u_z(x) = e^{z \cdot x}\left(1 + v_z(x)\right), \quad x \in K(0, b). \tag{5.47}$$

Furthermore, v_z satisfies the estimate

$$\|v_z\|_{L^2} \leq \frac{C}{|z|} \quad for\ all\ z \in \mathbb{C}^3\ with\ z \cdot z = 0\ and\ |z| \geq T. \tag{5.48}$$

Proof: The proof consists of two parts. First, we will construct v_z only for $z = t\hat{e}$, where $\hat{e} = (1, i, 0)^\top \in \mathbb{C}^3$ and t being sufficiently large. In the second part, we will consider the general case by rotating the geometry.

Let $z = t\hat{e}$ for some $t > 0$. By scaling the functions, we can assume without loss of generality that $K(0, b)$ is contained in the cube $Q = [-\pi, \pi]^3 \subset \mathbb{R}^3$. Furthermore, we will only construct a solution u_z of the Helmholtz equation in the variational sense, i.e., a function $u_z \in L^2(Q)$ with

$$\int\limits_Q u_z \left[\Delta \varphi + k^2 n\, \varphi\right] dx = 0$$

for all $\varphi \in C^\infty(\mathbb{R}^3)$ with support in the interior of Q. By well-known regularity results (Lemma of Weyl; see [103], IV 4.2), any variational solution is also a classical solution in the interior of Q.

We substitute the ansatz

$$u(x) = e^{t\hat{e}\cdot x}\left[1 + \exp(-i/2\, x_1)\, w_t(x)\right]$$

into the Helmholtz equation (5.46). This yields the following differential equation for w_t:

$$\Delta w_t(x) + \left(2t\hat{e} - ia\right)\cdot \nabla w_t(x) - \left(it + 1/4\right) w_t(x)$$
$$= -k^2 n(x)\, w_t(x) - k^2 n(x)\, \exp(i/2\, x_1) \quad in\ Q,$$

where $a = (1, 0, 0)^\top \in \mathbb{R}^3$. We will determine a 2π-periodic solution of this equation. Since this equation has the form of (5.12a) (for $\alpha = 1/4$), we use the solution operator L_t of Lemma 5.3 and write this equation in the form

$$w_t + k^2 L_t(n w_t) = L_t \tilde{n} \quad \text{in } Q, \tag{5.49}$$

where we have set $\tilde{n}(x) = -k^2 n(x) \exp(i/2 \, x_1)$. For large values of t, the operator $K_t : w \mapsto k^2 L_t(nw)$ is a contraction mapping in $L^2(Q)$. This follows from the estimates

$$
\begin{aligned}
\|K_t w\|_{L^2(Q)} &= k^2 \|L_t(nw)\|_{L^2(Q)} \\
&\leq \frac{k^2}{t} \|nw\|_{L^2(Q)} \leq \frac{k^2 \|n\|_\infty}{t} \|w\|_{L^2(Q)},
\end{aligned}
$$

which implies that $\|K_t\|_{L^2(Q)} < 1$ for sufficiently large $t > 0$. For these values of t, there exists a unique solution w_t of (5.49). Since the solution depends continuously on the right-hand side, we conclude that there exists $c > 0$ with

$$\|w_t\|_{L^2(Q)} \leq c \, \|L_t \tilde{n}\|_{L^2(Q)} \leq \frac{ck^2}{t} \|n\|_{L^2(Q)}$$

for all $t \geq T$ and some $T > 0$.

By applying Green's theorem, it is not difficult to see that with this solution w_t the function $u(x) = e^{t \hat{e} \cdot x} [1 + \exp(-i/2 \, x_1) w_t(x)]$ is a solution of (5.46) in the variational sense (see Problem 5.2). This proves the theorem for the special choice $z = t \hat{e}$.

Now let $z \in \mathbb{C}^3$ be arbitrary with $z \cdot z = 0$ and $|z| \geq T$. From this, we observe that $|\operatorname{Re} z| = |\operatorname{Im} z|$ and $(\operatorname{Re} z) \cdot (\operatorname{Im} z) = 0$. We decompose z in the unique form $z = t (\hat{a} + i\hat{b})$ with $\hat{a}, \hat{b} \in S^2$ and $t > 0$ and $\hat{a} \cdot \hat{b} = 0$. We define the cross product $\hat{c} = \hat{a} \times \hat{b}$ and the orthogonal matrix $R = [\hat{a} \, \hat{b} \, \hat{c}] \in \mathbb{R}^{3 \times 3}$. Then $t \, R\hat{e} = z$ and thus $R^\top z = t\hat{e}$. The substitution $x \mapsto Rx$ tranforms the Helmholtz equation (5.46) into

$$\Delta w(x) + k^2 n(Rx) \, w(x) = 0, \quad x \in K(0, b),$$

for $w(x) = v(Rx)$, $x \in K(0, b)$. Application of the first part of this proof yields the existence of a solution w of this equation of the form

$$w(x) = e^{t \hat{e} \cdot x} [1 + \exp(-i/2 \, x_1) w_t(x)],$$

where w_t satisfies $\|w_t\|_{L^2(Q)} \leq C/t$ for $t \geq T$. From $v(x) = w(R^\top x)$, we conclude that

$$
\begin{aligned}
v(x) &= e^{t \hat{e} \cdot R^\top x} [1 + \exp(-i/2 \, \hat{a} \cdot x) w_t(R^\top x)] \\
&= e^{z \cdot x} [1 + \exp(-i/2 \, \hat{a} \cdot x) w_t(R^\top x)],
\end{aligned}
$$

which proves the theorem for this case. □

Now we are able to prove the following.

Theorem 5.24

Let $\Omega \subset \mathbb{R}^3$ be a bounded domain and let $n_1, n_2 \in C^2(\Omega)$ such that $n_1 - 1$ and $n_2 - 1$ have compact support in Ω. Then the set

$$\{u_1 u_2 : u_j \in C^2(\overline{\Omega}) \text{ solves } (5.9) \text{ for } n = n_j, \ j = 1, 2\}$$

of products is dense in $L^2(\Omega)$.

Proof: Choose $b > 0$ such that $\overline{\Omega}$ is contained in the ball $K(0, b)$. Let $g \in L^2(\Omega)$ be such that

$$\int_\Omega g(x)\, u_1(x)\, u_2(x)\, dx \ = \ 0 \tag{5.50}$$

for all solutions $u_j \in C^2(\overline{\Omega})$ of the Helmholtz equation $\Delta u_j + k^2 n_j u_j = 0$ in Ω, $j = 1, 2$. In particular, (5.50) holds for all solutions of the Helmholtz equation in $K(0, b)$.

Fix an arbitrary vector $y \in \mathbb{R}^3 \setminus \{0\}$ and a number $\rho > 0$. Choose a unit vector $\hat{a} \in \mathbb{R}^3$ and a vector $b \in \mathbb{R}^3$ with $|b|^2 = |y|^2 + \rho^2$ such that $\{y, \hat{a}, b\}$ forms an orthogonal system in $\mathbb{R}^3$. Set

$$z^1 := \frac{1}{2}b - \frac{i}{2}(y + \rho\,\hat{a}) \quad \text{and} \quad z^2 := -\frac{1}{2}b - \frac{i}{2}(y - \rho\,\hat{a}).$$

Then $z^j \cdot z^j = |\operatorname{Re} z^j|^2 - |\operatorname{Im} z^j|^2 + 2i \operatorname{Re} z^j \cdot \operatorname{Im} z^j = |b|^2/4 - (|y|^2 + \rho^2)/4 = 0$ and $|z^j|^2 = (|b|^2 + |y|^2 + \rho^2)/4 \geq \rho^2/4$. Furthermore, $z^1 + z^2 = -i\,y$.

Now we apply Theorem 5.23 with z^j to the Helmholtz equations $\Delta u_j + k^2 n_j u_j = 0$ in $K(0, b)$. We substitute the forms (5.47) of u_j into the orthogonality relation (5.45) and arrive at

$$0 \ = \ \int_{K(0,b)} e^{(z^1 + z^2)\cdot x} \left[1 + v_1(x)\right]\left[1 + v_2(x)\right] g(x)\, dx$$

$$= \ \int_{K(0,b)} e^{-i\, y \cdot x} \left[1 + v_1(x) + v_2(x) + v_1(x)\, v_2(x)\right] g(x)\, dx.$$

By Theorem 5.23, there exist constants $T > 0$ and $C > 0$ with

$$\|v_j\|_{L^2} \ \leq \ \frac{C}{|z^j|} \ \leq \ \frac{2C}{\rho}$$

for all $\rho \geq T$. Now we use the Cauchy–Schwarz inequality and let ρ tend to infinity. This yields

$$\int_{K(0,b)} e^{-i\, y \cdot x} \, g(x) \, dx \;=\; 0.$$

Since the vector $y \in \mathbb{R}^3 \setminus \{0\}$ was arbitrary, we conclude that the Fourier transform of g vanishes. This yields $g = 0$. $\qquad\square$

This proof does not work in $\mathbb{R}^2$ because in that case there is no corresponding decomposition of y. To the author's knowledge, there does not exist a proof of the corresponding denseness result in $\mathbb{R}^2$.

As a corollary, we have the following uniqueness theorem.

Theorem 5.25
Let $n_1, n_2 \in C^2(\mathbb{R}^3)$ be two indices of refraction with $n_1(x) = n_2(x) = 1$ for all $|x| \geq a$. Let $u_{1,\infty}$ and $u_{2,\infty}$ be the corresponding far field patterns, and assume that they coincide, i.e., $u_{1,\infty}(\hat{x}; \hat{\theta}) = u_{2,\infty}(\hat{x}; \hat{\theta})$ for all $\hat{x}, \hat{\theta} \in S^2$. Then $n_1 = n_2$.

Proof: We combine the orthogonality relation of Lemma 5.21 with the denseness result of Theorem 5.24, where we choose $\Omega = K(0, b)$. $\qquad\square$

Since it is not known if Theorem 5.24 holds in $\mathbb{R}^2$, this uniqueness proof has no analog in $\mathbb{R}^2$. There are, however, several partial uniqueness results in $\mathbb{R}^2$. We refer to [211] for a review of these results.

5.5   Numerical Methods

In this section, we describe three types of numerical algorithms for the approximate solution of the inverse scattering problem. We assume as before that $n \in C^2(\mathbb{R}^3)$ with $n(x) = 1$ outside some ball $K = K(0, a)$ of radius $a > 0$.

The numerical methods we will describe now are all based on the Lippmann–Schwinger integral equation. We define the volume potential $V\varphi$ with density φ by

$$V\varphi(x) \;:=\; \int_K \frac{e^{ik|x-y|}}{4\pi\,|x-y|} \, \varphi(y)\,dy, \quad x \in K. \tag{5.51}$$

Then the Lippmann–Schwinger equation (5.19) takes the form

$$u \,+\, k^2 V(mu) \;=\; u^i \quad \text{in } K, \tag{5.52}$$

where again $m = 1 - n$ and $u^i(x, \hat{\theta}) = \exp(ik\hat{\theta} \cdot x)$. The far field pattern of $u^s = -k^2 V(mu)$ is given by

$$u_\infty(\hat{x}) = -\frac{k^2}{4\pi} \int_K m(y)\, u(y)\, e^{-ik\hat{x}\cdot y}\, dy, \quad \hat{x} \in S^2. \qquad (5.53)$$

Defining the integral operator $W : L^2(K) \to L^2(S^2)$ by

$$W\psi(\hat{x}) := -\frac{k^2}{4\pi} \int_K \psi(y)\, e^{-ik\hat{x}\cdot y}\, dy, \quad \hat{x} \in S^2, \qquad (5.54)$$

we note that the inverse scattering problem is to solve the system of equations

$$u + k^2 V(mu) = u^i \quad \text{in } K, \qquad (5.55a)$$
$$W(mu) = f \quad \text{on } S^2. \qquad (5.55b)$$

Here, f denotes the measured far field pattern. From the uniqueness results of Section 5.4, we expect that the far field patterns of more than one incident field have to be known. Therefore, from now on, we consider $u^i = u^i(x, \hat{\theta}) = \exp(ik\hat{\theta} \cdot x)$, $u = u(x, \hat{\theta})$, and $f = f(\hat{x}, \hat{\theta})$ to be functions of two variables. The operators V and W from equations (5.51) and (5.54) can be considered as linear and bounded operators

$$V : \quad C(\overline{K} \times S^2) \longrightarrow C(\overline{K} \times S^2), \qquad (5.56a)$$
$$W : \quad C(\overline{K} \times S^2) \longrightarrow L^2(S^2 \times S^2). \qquad (5.56b)$$

In the next subsections, we will discuss three methods for solving the inverse scattering problem, the first two of which are based on the system (5.55a), (5.55b). We will formulate the algorithms and prove convergence results only for the setting in function spaces, although for the practical implementations these algorithms have to be discretized. The methods suggested by Gutman and Klibanov [88, 89] and Kleinman and van den Berg [126] are iteration methods based on the system (5.55a), (5.55b). The first one is a regularized simplified Newton method, the second a modified gradient method. In Subsection 5.5.3, we will describe a different method, which has been proposed by Colton and Monk in several papers (see [40] – [44]).

5.5.1 A Simplified Newton Method

By scaling the problem, we assume throughout this subsection that the support of $m = 1 - n$ is contained in the cube $Q = [-\pi, \pi]^3 \subset \mathbb{R}^3$. We define the nonlinear mapping

$$T : C(Q) \times C(Q \times S^2) \longrightarrow C(Q \times S^2) \times C(S^2 \times S^2) \qquad (5.57a)$$

by

$$T(m, u) := \big(u + k^2 V(mu), \, W(mu) \big), \quad m \in C(Q), \; u \in C\big(Q \times S^2\big).$$
$$(5.57b)$$

Then the inverse problem can be written in the form

$$T(m, u) = (u^i, f).$$

The Newton method is to compute iterations (m_ℓ, u_ℓ), $\ell = 0, 1, 2, \ldots$ by

$$(m_{\ell+1}, u_{\ell+1}) = (m_\ell, u_\ell) - T'(m_\ell, u_\ell)^{-1}\big[T(m_\ell, u_\ell) - (u^i, f)\big] \quad (5.58)$$

for $\ell = 0, 1, 2, \ldots$ Since the components of the mapping T are bilinear, it is not difficult to see that the Fréchet derivative $T'(m, u)$ of T at (m, u) is given by

$$T'(m, u)(\mu, v) = \big(k^2 V(\mu u) + v + k^2 V(mv), \, W(\mu u) + W(mv) \big) \quad (5.59)$$

for $\mu \in C(Q)$ and $v \in C\big(Q \times S^2\big)$.

The simplified Newton method is to replace $T'(m_\ell, u_\ell)$ by some fixed $T'(\hat{m}, \hat{u})$ (see Theorem A.57 of Appendix A). Then it is known that under certain assumptions linear convergence can be expected. We choose $\hat{m} = 0$ and $\hat{u} = u^i$. Then the simplified Newton method sets $m_{\ell+1} = m_\ell + \mu$ and $u_{\ell+1} = u_\ell + v$, where (μ, v) solves $T'(0, u^i)(\mu, v) = (u^i, f) - T(m_\ell, u_\ell)$. Using the characterization of T', we are led to the following algorithm:

(A) Set $m_0 = 0$, $u_0 = u^i$ and $\ell = 0$.

(B) Determine $(\mu, v) \in C(Q) \times C\big(Q \times S^2\big)$ from the system of equations

$$k^2 V(\mu u^i) + v = u^i - u_\ell - k^2 V(m_\ell u_\ell), \quad (5.60a)$$
$$W(\mu u^i) = f - W(m_\ell u_\ell). \quad (5.60b)$$

(C) Set $m_{\ell+1} = m_\ell + \mu$ and $u_{\ell+1} = u_\ell + v$, replace ℓ by $\ell+1$, and continue with step (B).

Solving an equation of the form $W(\mu u^i) = \rho$ means solving the integral equation of the first kind

$$\int_Q \mu(y) \, e^{iky \cdot (\hat{\theta} - \hat{x})} \, dy = -\frac{4\pi}{k^2} \rho(\hat{x}, \hat{\theta}), \quad \hat{x}, \hat{\theta} \in S^2. \quad (5.61)$$

We will approximately solve this equation by a special collocation method. We observe that the left-hand side is essentially the Fourier transform $\mu\tilde{\ }$

of μ evaluated at $\xi = k(\hat{x} - \hat{\theta})$. As in Gutman and Klibanov [89], we define $N \in \mathbb{N}$ to be the largest integer not exceeding $2k/\sqrt{3}$, the set

$$\mathcal{Z}_N := \{j \in \mathbb{Z}^3 : |j_s| \leq N, \ s = 1, 2, 3\}$$

of grid points, and the finite-dimensional space

$$X_N := \left\{ \sum_{j \in \mathcal{Z}_N} a_j \, e^{i j \cdot x} : a_j \in \mathbb{C} \right\}. \tag{5.62}$$

Then, for every $j \in \mathcal{Z}_N$, there exist unit vectors $\hat{x}_j, \hat{\theta}_j \in S^2$ with $j = k(\hat{x}_j - \hat{\theta}_j)$. This is easily seen from the fact that the intersection of S^2 with the sphere of radius 1 and center j/k is not empty. For every $j \in \mathcal{Z}_N$, we fix the unit vectors $\hat{x}_j$ and $\hat{\theta}_j$ such that $j = k(\hat{x}_j - \hat{\theta}_j)$.

We solve (5.61) approximately by substituting $\hat{x}_j$ and $\hat{\theta}_j$ into this equation. This yields

$$\int_Q \mu(y) \, e^{-i j \cdot y} \, dy = -\frac{4\pi}{k^2} \rho(\hat{x}_j, \hat{\theta}_j), \quad j \in \mathcal{Z}_N. \tag{5.63}$$

Since the left-hand sides are just the first Fourier coefficients of μ, the unique solution of (5.63) in X_N is given by $\mu = L_1 \rho$, where the operator $L_1 : C(S^2 \times S^2) \to X_N$ is defined by

$$L_1 \rho(x) = -\frac{1}{2\pi^2 k^2} \sum_{j \in \mathcal{Z}_N} \rho(\hat{x}_j, \hat{\theta}_j) \, e^{i j \cdot x}. \tag{5.64}$$

The regularized algorithm now takes the form

(A_r) Set $m_0 = 0$, $u_0 = u^i$ and $\ell = 0$.

(B_r) Set

$$\mu := L_1 \big[f - W(m_\ell u_\ell) \big] \quad \text{and}$$

$$v := u^i - u_\ell - k^2 V(m_\ell u_\ell) - k^2 V(\mu u^i).$$

(C_r) Set $m_{\ell+1} = m_\ell + \mu$ and $u_{\ell+1} = u_\ell + v$, replace ℓ by $\ell+1$, and continue with step (B_r).

Then we can prove the following (see [89]).

Theorem 5.26
There exists $\varepsilon > 0$ such that, if $m \in C(Q)$ with $\|m\|_\infty \leq \varepsilon$ and $u = u(x, \hat{\theta})$ is the corresponding total field with exact far field pattern $f(\hat{x}, \hat{\theta}) = u_\infty(\hat{x}, \hat{\theta})$,

then the sequence (m_ℓ, u_ℓ) constructed by the regularized algorithm (A_r), (B_r), (C_r) converges to some $(\tilde{m}, \tilde{u}) \in X_N \times C(Q \times S^2)$ that satisfies the scattering problem with index of refraction $\tilde{m}$. Its far field pattern $\tilde{u}_\infty$ coincides with f at the points $(\hat{x}_j, \hat{\theta}_j) \in S^2 \times S^2$, $j \in \mathcal{Z}_N$. If, in addition, the exact solution m satisfies $m \in X_N$, then the sequence (m_ℓ, u_ℓ) converges to (m, u).

Proof: We define the operator

$$L : C(Q \times S^2) \times C(S^2 \times S^2) \longrightarrow X_N \times C(Q \times S^2)$$

by

$$L(w, \rho) := \left(L_1 \rho, \, w - k^2 V(u^i L_1 \rho) \right).$$

Then L is a left inverse of $T'(0, u^i)$ on $X_N \times C(Q \times S^2)$, i.e.,

$$L T'(0, u^i)(\mu, v) = (\mu, v) \quad \text{for all } (\mu, v) \in X_N \times C(Q \times S^2).$$

Indeed, let $(\mu, v) \in X_N \times C(Q \times S^2)$ and set $(w, \rho) = T'(0, u^i)(\mu, v)$, i.e., $w = v + k^2 V(\mu u^i)$ and $\rho = W(\mu u^i)$. The latter equation implies that

$$\int_Q \mu(y) e^{-ij \cdot y} \, dy = -\frac{4\pi}{k^2} \rho(\hat{x}_j, \hat{\theta}_j), \quad j \in \mathcal{Z}_N.$$

Because $\mu \in X_N$, this yields that $\mu = L_1 \rho$ and thus $L(w, \rho) = (\mu, v)$.

With the abbreviations $z_\ell = (m_\ell, u_\ell)$ and $R = (u^i, f)$, we can write the regularized algorithm in the form

$$z_{\ell+1} = z_\ell - L \left[T(z_\ell) - R \right], \quad \ell = 0, 1, 2, \dots$$

in the space $X_N \times C(Q \times S^2)$. We can now apply a general result about local convergence of the simplified Newton method (see Appendix A, Theorem A.57). This yields the existence of a unique solution $(\tilde{m}, \tilde{u}) \in X_N \times C(Q \times S^2)$ of $L \left[T(\tilde{m}, \tilde{u}) - R \right] = 0$ and linear convergence of the sequence (m_ℓ, u_ℓ) to $(\tilde{m}, \tilde{u})$. The equation $\tilde{u} + k^2 V(\tilde{m}\tilde{u}) = u^i$ is equivalent to the scattering problem by Theorem 5.8. The equation $L_1 W(\tilde{m}\tilde{u}) = L_1 f$ is equivalent to $\tilde{u}_\infty(\hat{x}_j, \hat{\theta}_j) = f(\hat{x}_j, \hat{\theta}_j)$ for all $j \in \mathcal{Z}_N$. Finally, if $m \in X_N$, then (m, u) satisfies $L T(m, u) = L R$ and thus $(\tilde{m}, \tilde{u}) = (m, u)$. This proves the assertion. $\square$

We have formulated the algorithm with respect to the Lippmann–Schwinger integral equation since our analysis on existence and continuous dependence is based on this setting. There is an alternative way to formulate the simplified Newton method in terms of the original scattering problems; see [89]. We note also that our analysis can easily be modified to treat the case where only $n \in L^\infty(K)$. For numerical examples, we refer to [89].

5.5.2 A Modified Gradient Method

The idea of the numerical method proposed and numerically tested by Kleinman and van den Berg (see [126]) is to solve (5.55a), (5.55b) by a gradient-type method. For simplicity, we will describe the method again in the function space setting and refer for discretization aspects to the original literature [126]. Again let $K = K(0,a)$ contain the support of $m = 1 - n$.

(A) Choose $m_0 \in C(\overline{K})$, $u_0 \in C(\overline{K} \times S^2)$ and set $\ell = 0$.

(B) Choose directions $e_\ell \in C(\overline{K} \times S^2)$, $d_\ell \in C(\overline{K})$, and set

$$u_{\ell+1} = u_\ell + \alpha_\ell \, e_\ell, \quad m_{\ell+1} = m_\ell + \beta_\ell \, d_\ell. \qquad (5.65)$$

The stepsizes $\alpha_\ell, \beta_\ell > 0$ are chosen in such a way that they minimize the functional

$$\Psi_\ell(\alpha,\beta) \; := \; \frac{\|r_{\ell+1}\|^2_{L^2(K \times S^2)}}{\|u^i\|^2_{L^2(K \times S^2)}} + \frac{\|s_{\ell+1}\|^2_{L^2(S^2 \times S^2)}}{\|f\|^2_{L^2(S^2 \times S^2)}}, \qquad (5.66a)$$

where the defects $r_{\ell+1}$ and $s_{\ell+1}$ are defined by

$$r_{\ell+1} \; := \; u^i - u_{\ell+1} - k^2 V(m_{\ell+1} \, u_{\ell+1}), \qquad (5.66b)$$

$$s_{\ell+1} \; := \; f - W(m_{\ell+1} \, u_{\ell+1}). \qquad (5.66c)$$

(C) Replace ℓ by $\ell + 1$ and continue with step (B).

There are different choices for the directions d_ℓ and e_ℓ. In [126],

$$d_\ell(x) = -\int_{S^2} \tilde{d}_\ell(x,\hat{\theta}) \, \overline{u_\ell(x,\hat{\theta})} \, ds(\hat{\theta}), \quad x \in K, \quad \text{and} \quad e_\ell := r_\ell \qquad (5.67)$$

have been chosen where

$$\tilde{d}_\ell = -W^*(W(m_\ell u_\ell) - f) \in C(\overline{K} \times S^2).$$

In this case, d_ℓ is the steepest descent direction of the functional $\mu \mapsto \|W(\mu u_\ell) - f\|^2_{L^2(S^2 \times S^2)}$. In [218], for d_ℓ and e_ℓ Polak–Ribière conjugate gradient directions are chosen. A rigorous convergence analysis of either method hasn't been carried out yet.

A severe drawback of the methods discussed in Subsections 5.5.1 and 5.5.2 is that they iterate on functions $m_\ell = m_\ell(x)$ and $u_\ell = u_\ell(x,\hat{\theta})$. To estimate the storage requirements, we choose a grid of order $N \cdot N \cdot N$ gridpoints in K and M directions $\theta_1, \ldots, \theta_M \in S^2$. Then both methods iterate on vectors of dimension $N^6 \cdot M$. From the uniqueness results, M is expected to be large, say, of order N^2. For large values of M, the method described next has proven to be more efficient.

5.5.3   The Dual Space Method

The method described here is due to Colton and Monk [42, 43] based on their earlier work for inverse obstacle scattering problems (see [40, 41]). There exist various modifications of this method, but we will restrict ourselves to the simplest case.

This method consists of two steps. In the first step, one tries to determine a superposition of the incident fields $u^i = u^i(\cdot, \hat{\theta})$ such that the corresponding far field pattern $f(\cdot, \hat{\theta})$ is (close to) the far field pattern of radiating multi-poles. In the second step, the function m is determined from an "interior transmission problem."

We describe both steps separately. Assume for the following that the origin is contained in $K = K(0, a)$.

Step 1: Determine $g \in L^2(S^2)$ with

$$\int_{S^2} f(\hat{x}, \hat{\theta})\, g(\hat{\theta})\, d\hat{\theta} = 1, \quad \hat{\theta} \in S^2. \tag{5.68}$$

This is an integral equation of the first kind for g. Since the kernel is analytic in both variables, this equation represents a severely ill-posed – but linear – equation and can be treated by Tikhonov's regularization method as described in Chapter 2 in detail. (In this connection, see the remarks following Theorems 5.17 and 5.19.)

In Theorem 5.16, we have proven that (5.68) is solvable in $L^2(S^2)$ if and only if the boundary value problem

$$\Delta v + k^2 v = 0 \text{ in } K, \qquad \Delta w + k^2 n w = 0 \text{ in } K, \tag{5.69a}$$

$$w(x) - v(x) = \frac{e^{ik|x|}}{|x|} \quad \text{on } \partial K, \tag{5.69b}$$

$$\frac{\partial w(x)}{\partial \nu} - \frac{\partial v(x)}{\partial \nu} = \frac{\partial}{\partial \nu}\frac{e^{ik|x|}}{|x|} \quad \text{on } \partial K, \tag{5.69c}$$

has a solution (v, w) such that

$$v(x) = \int_{S^2} e^{ikx \cdot \hat{y}}\, g(\hat{y})\, ds(\hat{y}), \quad x \in \mathbb{R}^3. \tag{5.70}$$

We formulate this boundary value problem as an integral equation.

Lemma 5.27
Let v be of the form (5.70). Then v and $w \in C^2(K) \cap C^1(\overline{K})$ solve the

boundary value problem (5.69a)–(5.69c) if and only if v and $w \in C(\overline{K})$ solve the system

$$w(x) \; - \; v(x) \;\; = \;\; -k^2 \int_K m(y)\, w(y)\, \Phi(x,y)\, dy, \quad x \in K, \qquad (5.71a)$$

$$w(x) \; - \; v(x) \;\; = \;\; \frac{e^{ik|x|}}{|x|} \quad \text{on } \partial K. \qquad (5.71b)$$

Proof: First, let v and w solve the boundary value problem (5.69a)–(5.69c). The difference $w - v$ satisfies the differential equation

$$\Delta(w - v) \; + \; k^2(w - v) \;\; = \;\; k^2 m\, w,$$

and its Cauchy data coincide with the Cauchy data of $\exp(ik\,|x|)/\,|x|$ on ∂K. Application of the representation theorem (Theorem 5.11) to $w - v$ in K yields

$$\begin{aligned}
w(x) \; - \; v(x) \;\; &= \;\; -k^2 \int_K m(y)\, w(y)\, \Phi(x,y)\, dy \\
&\quad + \int_{\partial K} \left[\Phi(x,y)\, \frac{\partial}{\partial \nu} \frac{e^{ik|x|}}{|x|} - \frac{e^{ik|x|}}{|x|}\, \frac{\partial}{\partial \nu} \Phi(x,y) \right] ds \\
&= \;\; -k^2 \int_K m(y)\, w(y)\, \Phi(x,y)\, dy \,, \quad x \in K,
\end{aligned}$$

since the boundary term vanishes by Green's second identity applied in the exterior of K. (Note that we assumed that $0 \in K$.)

On the other hand, let v and $w \in C(\overline{K})$ solve the system (5.71a), (5.71b). Then we extend w by

$$w(x) \;\; = \;\; v(x) \; - \; k^2 \int_K m(y)\, w(y)\, \Phi(x,y)\, dy, \quad x \in \mathbb{R}^3.$$

Then $w \in C^2(\mathbb{R}^3 \setminus \partial K) \cap C^1(\mathbb{R}^3)$ by Theorem 5.7. Since $w - v$ radiates and coincides with $\exp(ik\,|x|)/\,|x|$ on ∂K, it has to coincide with $\exp(ik\,|x|)/\,|x|$ in the exterior of K. Thus, the normal derivatives of $w - v$ and $\exp(ik\,|x|)/\,|x|$ also coincide on ∂K. This ends the proof. □

Motivated by this characterization, we describe the second step.

Step 2: With the (approximate) solution $g \in L^2(S^2)$ of (5.68), define the function $v = v_g$ by (5.70). Determine m and w such that m, v_g, and w

solve the interior boundary value problem (5.69a)–(5.69c) or, equivalently, the system

$$w \; - \; v_g \; + \; k^2 V(mw) \; = \; 0 \quad \text{in } K, \tag{5.72a}$$

$$k^2 V(mw) \; + \; 4\pi \, \Phi(\cdot, 0) \; = \; 0 \quad \text{on } \partial K, \tag{5.72b}$$

where V again denotes the volume potential operator (5.51) and Φ the fundamental solution (5.15).

Instead of solving both steps separately, we can combine them and solve the following optimization problem: Given a compact subset $\mathcal{C} \subset C(\overline{K})$, some $\varepsilon > 0$ and $\lambda_1, \lambda_2 > 0$,

$$\text{minimize } J(g, w, m) \text{ on } L^2(S^2) \times L^2(K) \times \mathcal{C}, \tag{5.73a}$$

where

$$J(g, w, m) \quad := \quad \|Fg - 1\|^2_{L^2(S^2)} \; + \; \varepsilon \, \|g\|^2_{L^2(S^2)} \tag{5.73b}$$

$$+ \; \lambda_1 \left\| w - v_g + k^2 V(mw) \right\|^2_{L^2(K)}$$

$$+ \; \lambda_2 \left\| k^2 V(mw) + 4\pi \, \Phi(\cdot, 0) \right\|^2_{L^2(\partial K)},$$

and the far field operator $F : L^2(S^2) \to L^2(S^2)$ is defined by (see (5.32))

$$Fg(\hat{x}) \quad := \quad \int_{S^2} f(\hat{x}, \hat{\theta}) \, g(\hat{\theta}) \, ds\hat{\theta}, \quad \hat{x} \in S^2.$$

Theorem 5.28
This optimization problem (5.73a), (5.73b) has an optimal solution (g, w, m) for every choice of $\varepsilon, \lambda_1, \lambda_2 > 0$ and every compact subset $\mathcal{C} \subset C(\overline{K})$.

Proof: Let $(g_j, w_j, m_j) \in L^2(S^2) \times L^2(K) \times \mathcal{C}$ be a minimizing sequence, i.e., $J(g_j, w_j, m_j) \to J^*$ where the optimal value J^* is defined by

$$J^* := \inf\{J(g, w, m) : (g, w, m) \in L^2(S^2) \times L^2(K) \times \mathcal{C}\}.$$

We can assume that (m_j) converges to some $m \in \mathcal{C}$ since $\mathcal{C}$ is compact. Several tedious applications of the parallelogram equality

$$\|a + b\|^2 \; = \; -\|a - b\|^2 \; + \; 2\,\|a\|^2 \; + \; 2\,\|b\|^2$$

and the binomial formula

$$\|b\|^2 \; = \; \|a\|^2 \; + \; 2\,\text{Re}\,(a, b - a) \; + \; \|a - b\|^2$$

yield

$$J\left(\frac{1}{2}(g_j + g_\ell), \frac{1}{2}(w_j + w_\ell), m_j\right)$$

$$= \frac{1}{2}J(g_j, w_j, m_j) + \frac{1}{2}J(g_\ell, w_\ell, m_j)$$

$$- \frac{1}{4}\|F(g_j - g_\ell)\|_{L^2(S^2)}^2 - \frac{\varepsilon}{4}\|g_j - g_\ell\|_{L^2(S^2)}^2$$

$$- \frac{\lambda_1}{4}\|(w_j - w_\ell) - v_{g_j - g_\ell} + k^2 V(m_j(w_j - w_\ell))\|_{L^2(K)}^2$$

$$- \frac{\lambda_2 k^4}{4}\|V(m_j(w_j - w_\ell))\|_{L^2(\partial K)}^2$$

and

$$J(g_\ell, w_\ell, m_j) = J(g_\ell, w_\ell, m_\ell)$$

$$+ 2k^2 \lambda_1 \mathrm{Re}\left(w_\ell - v_{g_\ell} + k^2 V(m_\ell w_\ell), V((m_j - m_\ell)w_\ell)\right)_{L^2(K)}$$

$$+ 2k^2 \lambda_2 \mathrm{Re}\left(k^2 V(m_\ell w_\ell) + r, V((m_j - m_\ell)w_\ell)\right)_{L^2(\partial K)}$$

$$+ \lambda_1 k^4 \|V((m_j - m_\ell)w_\ell)\|_{L^2(K)}^2$$

$$+ \lambda_2 k^4 \|V((m_j - m_\ell)w_\ell)\|_{L^2(\partial K)}^2$$

$$\leq J(g_\ell, w_\ell, m_\ell)$$

$$+ 2k^2 \lambda_1 \|w_\ell - v_{g_\ell} + k^2 V(m_\ell w_\ell)\|_{L^2(K)} \|V((m_j - m_\ell)w_\ell)\|_{L^2(K)}$$

$$+ 2k^2 \lambda_2 \|k^2 V(m_\ell w_\ell) + r\|_{L^2(\partial K)} \|V((m_j - m_\ell)w_\ell)\|_{L^2(\partial K)}$$

$$+ \lambda_1 k^4 \|V((m_j - m_\ell)w_\ell)\|_{L^2(K)}^2$$

$$+ \lambda_2 k^4 \|V((m_j - m_\ell)w_\ell)\|_{L^2(\partial K)}^2.$$

Substituting this estimate into the first equation and using

$$J^* \leq J((g_j + g_\ell)/2, (w_j + w_\ell)/2, m_j)$$

and $J(g_\ell, w_\ell, m_\ell) \to J^*$, $J(g_j, w_j, m_j) \to J^*$ as $\ell, j \to \infty$ yields that

$$\|g_j - g_\ell\|_{L^2(S^2)} \longrightarrow 0$$

and

$$\|(w_j - w_\ell) - v_{g_j - g_\ell} + k^2 V(m_j(w_j - w_\ell))\|_{L^2(K)} \longrightarrow 0$$

as $\ell, j \to \infty$. Therefore, (g_j) is a Cauchy sequence and thus convergent in $L^2(S^2)$ to some g. Furthermore,

$$\|(w_j - w_\ell) + k^2 V(m_j(w_j - w_\ell))\|_{L^2(K)} \longrightarrow 0$$

as $\ell, j \to \infty$. The operators $I + k^2 V(m_j \cdot)$ converge to the isomorphism $I + k^2 V(m \cdot)$ in the operator norm of $L^2(K)$. Therefore, by Theorem A.35 of Appendix A, we conclude that (w_j) is a Cauchy sequence and thus is convergent in $L^2(K)$ to some w. The continuity of J implies that $J(g_j, w_j, m_j) \to J(g, w, m)$. Therefore, (g, w, m) is optimal. $\qquad\square$

5.6 Problems

5.1 Let $u^b_{1,\infty}(\hat{x}, \hat{\theta}, k)$ and $u^b_{2,\infty}(\hat{x}, \hat{\theta}, k)$ be the far field patterns of the Born approximations corresponding to observation $\hat{x}$, angle of incidence $\hat{\theta}$, wave number k, and indices of refraction n_1 and n_2, respectively. Assume that

$$u^b_{1,\infty}(\hat{x}, \hat{\theta}, k) \;=\; u^b_{2,\infty}(\hat{x}, \hat{\theta}, k)$$

for all $\hat{x} \in S^2$ and $k \in [k_1, k_2] \subset \mathbb{R}^+$ and some $\hat{\theta} \in S^2$. Prove that $n_1 = n_2$.

5.2 Let w_t be the unique solution of (5.49) for $t \geq T$. Extend w_t to a 2π-periodic function into all of $\mathbb{R}^3$. Define

$$u(x) \;:=\; e^{t\hat{e}\cdot x}\left[1 \;+\; \exp(-i/2\,x_1)\,w_t(x)\right], \quad x \in \mathbb{R}^3,$$

where $\hat{e} = (1, i, 0)^\top \in \mathbb{C}^3$.

Prove that u solves (5.46) in the variational sense.

5.3 Prove the following result, sometimes called *Karp's theorem*: Let $u_\infty(\hat{x}; \hat{\theta})$, $\hat{x}, \hat{\theta} \in S^2$, be the far field pattern and assume that there exists a function $f : [-1, 1] \to \mathbb{C}$ with

$$u_\infty(\hat{x}; \hat{\theta}) \;=\; f(\hat{x} \cdot \hat{\theta}) \quad \text{for all } \hat{x}, \hat{\theta} \in S^2.$$

Prove that the index of refraction n has to be radially symmetric, i.e., $n = n(r)$.

Hint: Rotate the geometry and use the uniqueness result.

5.4 Show that for any $a > 0$

$$\max_{|x| \leq a} \int_{|y| < a} \frac{1}{|x - y|}\,dy \;=\; 2\pi\, a^2.$$

Appendix A
Basic Facts from Functional Analysis

In this appendix, we collect some of the basic definitions and theorems from functional analysis. We will prove only those theorems whose proofs are not easily accessible. We recommend the monographs [118, 130, 186, 223] for a comprehensive treatment of linear and nonlinear functional analysis.

A.1 Normed Spaces and Hilbert Spaces

First, we recall two basic definitions.

Definition A.1 *(Scalar Product, Pre-Hilbert Space)*
Let X be a vector space over the field $\mathbb{K} = \mathbb{R}$ or $\mathbb{K} = \mathbb{C}$. A scalar product or inner product *is a mapping*

$$(\cdot, \cdot) : X \times X \longrightarrow \mathbb{K}$$

with the following properties:

(i) $(x + y, z) = (x, z) + (y, z)$ *for all* $x, y, z \in X$,

(ii) $(\alpha x, y) = \alpha (x, y)$ *for all* $x, y \in X$ *and* $\alpha \in \mathbb{K}$,

(iii) $(x, y) = \overline{(y, x)}$ *for all* $x, y \in X$,

(iv) $(x, x) \in \mathbb{R}$ *and* $(x, x) \geq 0$, *for all* $x \in X$,

(v) $(x, x) > 0$ *if* $x \neq 0$.

A vector space X over $\mathbb{K}$ with inner product $(\cdot, \cdot)$ is called a pre-Hilbert space *over $\mathbb{K}$.*

The following properties are easily derived from the definition:

(vi) $(x, y + z) = (x, y) + (x, z)$ *for all* $x, y, z \in X$,

(vii) $(x, \alpha y) = \overline{\alpha}(x, y)$ *for all* $x, y \in X$ *and* $\alpha \in \mathbb{K}$.

Definition A.2 *(Norm)*
Let X be a vector space over the field $\mathbb{K} = \mathbb{R}$ or $\mathbb{K} = \mathbb{C}$. A norm *on X is a mapping*

$$\|\cdot\| : X \longrightarrow \mathbb{R}$$

with the following properties:

(i) $\|x\| > 0$ *for all* $x \in X$ *with* $x \neq 0$,

(ii) $\|\alpha x\| = |\alpha| \, \|x\|$ *for all* $x \in X$ *and* $\alpha \in \mathbb{K}$,

(iii) $\|x + y\| \leq \|x\| + \|y\|$ *for all* $x, y \in X$.

A vector space X over $\mathbb{K}$ with norm $\|\cdot\|$ is called normed space *over $\mathbb{K}$.*

Property (iii) is called *triangle inequality*. Applying it to the identities $x = (x - y) + y$ and $y = (y - x) + x$ yields the second triangle inequality $\|x - y\| \geq |\|x\| - \|y\||$ for all $x, y \in X$.

Theorem A.3
Let X be a pre-Hilbert space. The mapping $\|\cdot\| : X \longrightarrow \mathbb{R}$ defined by

$$\|x\| := \sqrt{(x, x)}, \quad x \in X,$$

is a norm, i.e., it has properties (i), (ii), and (iii) of Definition A.2. Furthermore,

(iv) $|(x, y)| \leq \|x\| \, \|y\|$ *for all* $x, y \in X$ *(Cauchy–Schwarz inequality)*,

(v) $\|x \pm y\|^2 = \|x\|^2 + \|y\|^2 \pm 2\mathrm{Re}\,(x, y)$ *for all* $x, y \in X$

   (binomial formula),

(vi) $\|x + y\|^2 + \|x - y\|^2 = 2\|x\|^2 + 2\|y\|^2$ *for all* $x, y \in X$.

In the following example, we list some of the most important pre-Hilbert and normed spaces.

Example A.4

(a) $\mathbb{C}^n$ is a pre-Hilbert space of dimension n over $\mathbb{C}$ with inner product
$$(x, y) := \sum_{k=1}^{n} x_k \bar{y}_k.$$

(b) $\mathbb{C}^n$ is a pre-Hilbert space of dimension $2n$ over $\mathbb{R}$ with inner product
$$(x, y) := \mathrm{Re} \sum_{k=1}^{n} x_k \bar{y}_k.$$

(c) $\mathbb{R}^n$ is a pre-Hilbert space of dimension n over $\mathbb{R}$ with inner product
$$(x, y) := \sum_{k=1}^{n} x_k y_k.$$

(d) Define the set ℓ^2 of (real-valued) sequences by
$$\ell^2 := \left\{ (x_k) \subset \mathbb{R} : \sum_{k=1}^{\infty} x_k^2 < \infty \right\}. \qquad (\text{A.1})$$

Then ℓ^2 is a linear space since if $(x_k), (y_k) \in \ell^2$, then (λx_k) and $(x_k + y_k)$ are also in ℓ^2. The latter follows from the binomial inequality $(x_k + y_k)^2 \le 2\, x_k^2 + 2\, y_k^2$.

$$(x, y) := \sum_{k=1}^{\infty} x_k\, y_k, \quad x = (x_k),\ y = (y_k) \in \ell^2,$$

defines an inner product on ℓ^2. It is well-defined by the Cauchy–Schwarz inequality.

(e) The space $C[a, b]$ of (real- or complex-valued) continuous functions on $[a, b]$ is a pre-Hilbert space over $\mathbb{R}$ with inner product
$$(x, y)_{L^2} := \int_a^b x(t)\, \overline{y(t)}\, dt, \quad x, y \in C[a, b]. \qquad (\text{A.2a})$$

The corresponding norm is called the *Euclidean norm* and is denoted by
$$\|x\|_{L^2} := \sqrt{(x, x)_{L^2}} = \sqrt{\int_a^b |x(t)|^2\, dt}, \quad x \in C[a, b]. \qquad (\text{A.2b})$$

(f) On the same vector space $C[a, b]$ as in example (e), we introduce a norm by
$$\|x\|_\infty := \max_{a \le t \le b} |x(t)|, \quad x \in C[a, b], \qquad (\text{A.3})$$

which we call the *supremum norm*.

(g) Let $m \in \mathbb{N}$ and $\alpha \in (0, 1]$. We define the spaces $C^m[a, b]$ and $C^{m,\alpha}[a, b]$ by

$$C^m[a, b] \quad := \quad \left\{ x \in C[a, b] : \begin{array}{l} x \text{ is } m \text{ times continuously} \\ \text{differentiable on } [a, b] \end{array} \right\},$$

$$C^{m,\alpha}[a, b] \quad := \quad \left\{ x \in C^m[a, b] : \sup_{t \neq s} \frac{|x^{(m)}(t) - x^{(m)}(s)|}{|t - s|^\alpha} < \infty \right\},$$

and we equip them with norms

$$\|x\|_{C^m} \quad := \quad \max_{0 \leq k \leq m} \|x^{(k)}\|_\infty, \tag{A.4a}$$

$$\|x\|_{C^{m,\alpha}} \quad := \quad \|x\|_{C^m} + \sup_{s \neq t} \frac{|x^{(m)}(t) - x^{(m)}(s)|}{|t - s|^\alpha}. \tag{A.4b}$$

Every normed space carries a topology introduced by the norm, i.e., we can define open, closed, and compact sets; convergent sequences; continuous functions; etc. We introduce balls of radius r and center $x \in X$ by

$$K(x, r) := \{ y \in X : \|y - x\| < r \}, \quad K[x, r] := \{ y \in X : \|y - x\| \leq r \}.$$

Definition A.5
Let X be a normed space over the field $\mathbb{K} = \mathbb{R}$ or $\mathbb{C}$.

(a) *A subset $M \subset X$ is called* bounded *if there exists $r > 0$ with $M \subset K(x, r)$. The set $M \subset X$ is called* open *if for every $x \in M$ there exists $\varepsilon > 0$ such that $K(x, \varepsilon) \subset M$. The set $M \subset X$ is called* closed *if the complement $X \setminus M$ is open.*

(b) *A sequence $(x_k)_k \subset X$ is called* bounded *if there exists $c > 0$ such that $\|x_k\| \leq c$ for all k. The sequence $(x_k)_k \subset X$ is called* convergent *if there exists $x \in X$ such that $\|x - x_k\|$ converges to zero in $\mathbb{R}$. We denote the limit by $x = \lim_{k \to \infty} x_k$, or we write $x_k \to x$ as $k \to \infty$. The sequence $(x_k)_k \subset X$ is called a* Cauchy sequence *if for every $\epsilon > 0$ there exists $N \in \mathbb{N}$ with $\|x_m - x_k\| < \epsilon$ for all $m, k \geq N$.*

(c) *Let $(x_k)_k \subset X$ be a sequence. $x \in X$ is called an* accumulation point *if there exists a subsequence $(a_{k_n})_n$ that converges to x.*

(d) *A set $M \subset X$ is called* compact *if every sequence in M has an accumulation point in M.*

Example A.6

Let $X = C[0,1]$ over $\mathbb{R}$ and $x_k(t) = t^k$, $t \in [0,1]$, $k \in \mathbb{N}$. The sequence (x_k) converges to zero with respect to the Euclidean norm $\|\cdot\|_{L^2}$ introduced in (A.2b). With respect to the supremum norm $\|\cdot\|_\infty$ of (A.3), however, the sequence does not converge to zero.

It is easy to prove that a set M is closed if and only if the limit of every convergent sequence $(x_k)_k \subset M$ also belongs to M. The sets

$$M^o := \{x \in M : \text{there exists } \varepsilon > 0 \text{ with } K(x,\varepsilon) \subset M\}$$

and

$$\overline{M} := \{x \in X : \text{there exists } (x_k)_k \subset M \text{ with } x = \lim_{k \to \infty} x_k\}$$

are called the *interior* and *closure*, respectively, of M. The set $M \subset X$ is called *dense* in X if $\overline{M} = X$.

In general, the topological properties depend on the norm in X. For finite-dimensional spaces, however, these properies are *independent* of the norm. This is seen from the following theorem.

Theorem A.7

Let X be a finite-dimensional space with norms $\|\cdot\|_1$ and $\|\cdot\|_2$. Then both norms are equivalent, i.e., there exist constants $c_2 \geq c_1 > 0$ with

$$c_1 \|x\|_1 \leq \|x\|_2 \leq c_2 \|x\|_1 \quad \text{for all } x \in X.$$

In other words, every ball with respect to $\|\cdot\|_1$ contains a ball with respect to $\|\cdot\|_2$ and vice versa. Further properties are collected in the following theorem.

Theorem A.8

Let X be a normed space over $\mathbb{K}$ and $M \subset X$ be a subset.

(a) *M is closed if and only if $M = \overline{M}$, and M is open if and only if $M = M^o$.*

(b) *If $M \neq X$ is a linear subspace, then $M^o = \emptyset$, and $\overline{M}$ is also a linear subspace.*

(c) *In finite-dimensional spaces, every subspace is closed.*

(d) *Every compact set is closed and bounded. In finite-dimensional spaces, the reverse is also true (Theorem of Bolzano–Weierstrass): In a finite-dimensional normed space, every closed and bounded set is compact.*

A crucial property of the set of real numbers is its *completeness*. It is also a neccessary assumption for many results in functional analysis.

Definition A.9 *(Banach Space, Hilbert Space)*
A normed space X over $\mathbb{K}$ is called complete *or a* Banach space *if every Cauchy sequence converges in X. A complete pre-Hilbert space is called a* Hilbert space.

The spaces $\mathbb{C}^n$ and $\mathbb{R}^n$ are Hilbert spaces with respect to their canonical inner products. The space $C[a, b]$ is not complete with respect to the inner product $(\cdot, \cdot)_{L^2}$ of (A.2a)! As an example, we consider the sequence $x_k(t) = t^k$ for $0 \leq t \leq 1$ and $x_k(t) = 1$ for $1 \leq t \leq 2$. Then (x_k) is a Cauchy sequence in $C[0, 2]$ but does not converge in $C[0, 2]$ with respect to $(\cdot, \cdot)_{L^2}$ since it converges to the function

$$x(t) \;=\; \begin{cases} 0, & t < 1, \\ 1, & t \geq 1, \end{cases}$$

which is not continuous. The space $\big(C[a, b], \|\cdot\|_\infty\big)$, however, is a Banach space.

Every normed space or pre-Hilbert space X can be "completed," i.e., there exists a "smallest" Banach or Hilbert space $\tilde{X}$, respectively, that extends X (i.e., $\|x\|_X = \|x\|_{\tilde{X}}$ or $(x, y)_X = (x, y)_{\tilde{X}}$, respectively, for all $x, y \in X$). More precisely, we have the following (formulated only for normed spaces).

Theorem A.10
Let X be a normed space with norm $\|\cdot\|_X$. There exist a Banach space $\big(\tilde{X}, \|\cdot\|_{\tilde{X}}\big)$ and an injective linear operator $J : X \to \tilde{X}$ such that

(i) The range $J(X) \subset \tilde{X}$ is dense in $\tilde{X}$, and

(ii) $\|Jx\|_{\tilde{X}} = \|x\|_X$ for all $x \in X$, i.e., J preserves the norm.

Furthermore, $\tilde{X}$ is uniquely determined in the sense that if $\hat{X}$ is a second space with properties (i) and (ii) with respect to a linear injective operator $\hat{J}$, then the operator $\hat{J} J^{-1} : J(X) \to \hat{J}(X)$ has an extension to a norm-preserving isomorphism from $\tilde{X}$ onto $\hat{X}$. In other words, $\tilde{X}$ and $\hat{X}$ can be identified.

We denote the completion of the pre-Hilbert space $\big(C[a, b], (\cdot, \cdot)_{L^2}\big)$ by $L^2(a, b)$. Using Lebesgue integration theory, it can be shown that the space $L^2(a, b)$ is characterized as follows. (The notions "measurable," "almost everywhere" (a.e.), and "integrable" are understood with respect to the

Lebesgue measure.) First, we define the vector space

$$\mathcal{L}^2(a,b) := \{x : (a,b) \to \mathbb{C} : x \text{ is measurable and } |x|^2 \text{ integrable}\},$$

where addition and scalar multiplication are defined pointwise almost everywhere. Then $\mathcal{L}^2(a,b)$ is a vector space since, for $x, y \in \mathcal{L}^2(a,b)$ and $\alpha \in \mathbb{C}$, $x + y$ and αx are also measurable and αx, $x + y \in \mathcal{L}^2(a,b)$, the latter by the binomial theorem $|x(t) + y(t)|^2 \leq 2 |x(t)|^2 + 2 |y(t)|^2$. We define a sesquilinear form on $\mathcal{L}^2(a,b)$ by

$$\langle x, y \rangle := \int_a^b x(t) \overline{y(t)} \, dt, \quad x, y \in \mathcal{L}^2(a,b).$$

$\langle \cdot, \cdot \rangle$ is not an inner product on $\mathcal{L}^2(a,b)$ since $\langle x, x \rangle = 0$ only implies that x vanishes almost everywhere, i.e., that $x \in \mathcal{N}$, where $\mathcal{N}$ is defined by

$$\mathcal{N} := \{x \in \mathcal{L}^2(a,b) : x(t) = 0 \text{ a.e. on } (a,b)\}.$$

Now we define $L^2(a,b)$ as the factor space

$$L^2(a,b) := \mathcal{L}^2(a,b)/\mathcal{N}$$

and equip $L^2(a,b)$ with the inner product

$$\big([x], [y]\big)_{L^2} := \int_a^b x(t) \overline{y(t)} \, dt, \quad x \in [x], \; y \in [y].$$

Here, $[x], [y] \in L^2(a,b)$ are equivalence classes of functions in $\mathcal{L}^2(a,b)$. Then it can be shown that this definition is well-defined and yields an inner product on $L^2(a,b)$. From now on, we will write $x \in L^2(a,b)$ instead of $x \in [x] \in L^2(a,b)$. Furthermore, it can be shown by fundamental results of Lebesgue integration theory that $L^2(a,b)$ is complete, i.e., a Hilbert space and contains $C[a,b]$ as a dense subspace.

Definition A.11 *(Separable Space)*
The normed space X is called separable *if there exists a countable dense subset $M \subset X$, i.e., if there exist M and a bijective mappping $j : \mathbb{N} \to M$ with $\overline{M} = X$.*

The spaces $\mathbb{C}^n$, $\mathbb{R}^n$, $L^2(a,b)$, and $C[a,b]$ are all separable. For the first two examples, let M consist of all vectors with rational coefficients; for the latter examples, take polynomials with rational coefficients.

Definition A.12 *(Orthogonal Complement)*
Let X be a pre-Hilbert space (over $\mathbb{K} = \mathbb{R}$ or $\mathbb{C}$).

(a) *Two elements x and y are called* orthogonal *if $(x, y) = 0$.*

(b) *Let $M \subset X$ be a subset. The set*

$$M^\perp := \{x \in X : (x, y) = 0 \text{ for all } y \in M\}$$

is called the orthogonal complement *of M.*

$M^\perp$ is always a closed subspace and $M \subset (M^\perp)^\perp$. Furthermore, $A \subset B$ implies that $B^\perp \subset A^\perp$.

The following theorem is a fundamental result in Hilbert space theory and relies heavily on the completeness property.

Theorem A.13 *(Projection Theorem)*
Let X be a Hilbert space and $V \subset X$ be a closed subspace. Then $V = (V^\perp)^\perp$. Every $x \in X$ possesses a unique decomposition of the form $x = v + w$, where $v \in V$ and $w \in V^\perp$. The operator $P : X \to V$, $x \mapsto v$, is called the orthogonal projection operator *onto V and has the properties*

(a) *$Pv = v$ for $v \in V$, i.e., $P^2 = P$;*

(b) *$\|x - Px\| \leq \|x - v'\|$ for all $v' \in V$.*

This means that $Px \in V$ is the best approximation of $x \in X$ in the closed subspace V.

A.2   Orthonormal Systems

In this section, let X always be a *separable* Hilbert space over the field $\mathbb{K} = \mathbb{R}$ or $\mathbb{C}$.

Definition A.14 *(Orthonormal System)*
A countable set of elements $A = \{x_k : k = 1, 2, 3, \ldots\}$ is called an orthonormal system *(ONS) if*

(i) *$(x_k, x_j) = 0$ for all $k \neq j$ and*

(ii) *$\|x_k\| = 1$ for all $k \in \mathbb{N}$.*

A is called a complete *or a* maximal *orthonormal system if, in addition, there is no ONS B with $A \subset B$ and $A \neq B$.*

One can show using Zorn's Lemma that every separable Hilbert possesses a maximal ONS. Furthermore, it is well-known from linear algebra that every countable set of linearly independent elements of X can be orthonormalized.

For any set $A \subset X$, let

$$\operatorname{span} A := \left\{ \sum_{k=1}^{n} \alpha_k x_k : \alpha_k \in \mathbb{K}, \ x_k \in A, \ n \in \mathbb{N} \right\} \qquad (A.5)$$

be the subspace of X spanned by A.

Theorem A.15

Let $A = \{x_k : k = 1, 2, 3, \ldots\}$ be an orthonormal system. Then

(a) Every finite subset of A is linearly independent.

(b) If A is finite, i.e., $A = \{x_k : k = 1, 2, \ldots, n\}$, then for every $x \in X$ there exist uniquely determined coefficients $\alpha_k \in \mathbb{K}$, $k = 1, \ldots, n$, such that

$$\left\| x - \sum_{k=1}^{n} \alpha_k x_k \right\| \leq \| x - a \| \quad \text{for all } a \in \operatorname{span} A. \qquad (A.6)$$

The coefficients α_k are given by $\alpha_k = (x, x_k)$ for $k = 1, \ldots, n$.

(c) For every $x \in X$, the following Bessel inequality holds:

$$\sum_{k=1}^{\infty} |(x, x_k)|^2 \leq \| x \|^2, \qquad (A.7)$$

and the series converges in X.

(d) A is complete if and only if $\operatorname{span} A$ is dense in X.

(e) A is complete if and only if for all $x \in X$ the following Parseval equation holds:

$$\sum_{k=1}^{\infty} |(x, x_k)|^2 = \| x \|^2. \qquad (A.8)$$

(f) A is complete if and only if every $x \in X$ has a (generalized) Fourier expansion of the form

$$x = \sum_{k=1}^{\infty} (x, x_k) \, x_k, \qquad (A.9)$$

where the convergence is understood in the norm of X. In this case, the Parseval equation holds in the following more general form:

$$(x, y) = \sum_{k=1}^{\infty} (x, x_k) \overline{(y, x_k)}. \tag{A.10}$$

This important theorem includes, as special examples, the classical Fourier expansion of periodic functions and the expansion with respect to orthogonal polynomials. We recall two examples.

Example A.16 *(Fourier Expansion)*
(a) The functions $x_k(t) := \exp(ikt)/\sqrt{2\pi}$, $k \in \mathbb{Z}$, form a complete system of orthonormal functions in $L^2(0, 2\pi)$. By part (f) of the previous theorem, every function $x \in L^2(0, 2\pi)$ has an expansion of the form

$$x(t) = \frac{1}{2\pi} \sum_{k=-\infty}^{\infty} e^{ikt} \int_{0}^{2\pi} x(s) e^{-iks} ds,$$

where the convergence is understood in the sense of L^2, i.e.,

$$\int_{0}^{2\pi} \left| x(t) - \frac{1}{2\pi} \sum_{k=-M}^{N} e^{ikt} \int_{0}^{2\pi} x(s) e^{-iks} ds \right|^2 dt \longrightarrow 0$$

as M, N tend to infinity. For smooth periodic functions, one can even show uniform convergence (see Section A.4).

(b) The *Legendre polynomials* P_k, $k = 0, 1, \ldots$, form a maximal orthonormal system in $L^2(-1, 1)$. They are defined by

$$P_k(t) = \gamma_k \frac{d^k}{dt^k} (1 - t^2)^k, \quad t \in (-1, 1), \ k \in \mathbb{N}_0,$$

with normalizing constants

$$\gamma_k = \sqrt{\frac{2k+1}{2}} \frac{1}{k! \, 2^k}.$$

We refer to [108] for details.

Other important examples will be given later.

A.3   Linear Bounded and Compact Operators

For this section, let X and Y always be normed spaces and $A : X \to Y$ be a linear operator.

Definition A.17 *(Boundedness, Norm of A)*
The linear operator A is called bounded *if there exists $c > 0$ such that*

$$\|Ax\| \leq c\|x\| \quad \textit{for all } x \in X.$$

The smallest of these constants is called the norm of A, *i.e.*,

$$\|A\| := \sup_{x \neq 0} \frac{\|Ax\|}{\|x\|}. \tag{A.11}$$

Theorem A.18
The following assertions are equivalent:

(a) A is bounded.

(b) A is continuous at $x = 0$, i.e., $x_j \to 0$ implies that $Ax_j \to 0$.

(c) A is continuous for every $x \in X$.

The space $\mathcal{L}(X, Y)$ of all linear bounded mappings from X to Y with the operator norm is a normed space, i.e., the operator norm has properties (i), (ii), and (iii) of Definition A.2 and the following: Let $B \in \mathcal{L}(X, Y)$ and $A \in \mathcal{L}(Y, Z)$; then $AB \in \mathcal{L}(X, Z)$ and $\|AB\| \leq \|A\| \|B\|$.

Integral operators are the most important examples for our purposes.

Theorem A.19
(a) Let $k \in L^2\big((c, d) \times (a, b)\big)$. The operator

$$Ax(t) := \int_a^b k(t, s)\, x(s)\, ds, \quad t \in (c, d), \quad x \in L^2(a, b), \tag{A.12}$$

is well-defined, linear, and bounded from $L^2(a, b)$ into $L^2(c, d)$. Furthermore,

$$\|A\|_{L^2} \leq \int_c^d \int_a^b |k(t, s)|\, ds\, dt.$$

(b) Let k be continuous on $[c, d] \times [a, b]$. Then A is also well-defined, linear, and bounded from $C[a, b]$ into $C[c, d]$ and

$$\|A\|_\infty = \max_{t \in [c, d]} \int_a^b |k(t, s)|\, ds.$$

We can extend this theorem to integral operators with weakly singular kernels. We recall that a kernel k is called *weakly singular* on $[a, b] \times [a, b]$

if k is defined and continuous for all $t, s \in [a, b]$, $t \neq s$, and there exist constants $c > 0$ and $\alpha \in [0, 1)$ such that

$$|k(t, s)| \leq c\,|t - s|^{-\alpha} \quad \text{for all } t, s \in [a, b], \ t \neq s.$$

Theorem A.20
Let k be weakly singular on $[a, b]$. Then the integral operator A, defined by (A.12) for $[c, d] = [a, b]$, is well-defined and bounded as an operator in $L^2(a, b)$ as well as in $C[a, b]$.

For the special case $Y = \mathbb{K}$, we denote by $X' := \mathcal{L}(X, \mathbb{K})$ the *dual space* of X. Analogously, the space $(X')'$ is called the *bidual* of X. The canonical embedding $J : X \to (X')'$, defined by

$$(Jx)\ell := \ell(x), \quad x \in X, \ \ell \in X',$$

is linear, bounded, one-to-one, and satisfies $\|Jx\| = \|x\|$ for all $x \in X$.

Definition A.21 *(Reflexive Space)*
The normed space X is called reflexive *if the canonical embedding is surjective, i.e., a norm-preserving isomorphism from X onto the bidual space $(X')'$.*

The following important result gives a characterization of X' in Hilbert spaces.

Theorem A.22 *(Riesz–Fischer)*
Let X be a Hilbert space. For every $x \in X$, the functional $f_x(y) := (y, x)$, $y \in X$, defines a linear bounded mapping from X to $\mathbb{K}$, i.e., $f_x \in X'$. Furthermore, for every $f \in X'$ there exists one and only one $x \in X$ with $f(y) = (y, x)$ for all $y \in X$ and

$$\|f\| := \sup_{y \neq 0} \frac{|f(y)|}{\|y\|} = \|x\|.$$

This theorem implies that every Hilbert space is reflexive. It also yields the existence of a unique adjoint operator for every linear bounded operator $A : X \longrightarrow Y$.

Theorem A.23 *(Adjoint Operator)*
Let $A : X \longrightarrow Y$ be a linear and bounded operator between Hilbert spaces. Then there exists one and only one linear bounded operator $A^* : Y \longrightarrow X$ with the property

$$(Ax, y) = (x, A^*y) \quad \text{for all } x \in X, \ y \in Y.$$

This operator $A^* : Y \longrightarrow X$ is called the adjoint operator *to A. For $X = Y$, the operator A is called* self-adjoint *if $A^* = A$.*

Example A.24

(a) Let $X = L^2(a, b)$, $Y = L^2(c, d)$, and $k \in L^2((c, d) \times (a, b))$. The adjoint A^* of the integral operator

$$Ax(t) = \int_a^b k(t, s) \, x(s) \, ds, \quad t \in (c, d), \quad x \in L^2(a, b),$$

is given by

$$A^*y(t) = \int_c^d \overline{k(s, t)} \, y(s) \, ds, \quad t \in (a, b), \quad y \in L^2(c, d).$$

(b) Let the space $X = C[a, b]$ of continuous function over $\mathbb{C}$ be supplied with the L^2-inner product. Define $f, g : C[a, b] \to \mathbb{R}$ by

$$f(x) := \int_a^b x(t) \, dt \quad \text{and} \quad g(x) := x(a) \quad \text{for } x \in C[a, b].$$

Both f and g are linear. f is bounded but g is unbounded. There is an extension of f to a linear bounded functional (also denoted by f) on $L^2(a, b)$, i.e., $f \in L^2(a, b)'$. By Theorem A.22, we can identify $L^2(a, b)'$ with $L^2(a, b)$ itself. For the given f, the representation function is just the constant function 1 since $f(x) = (x, 1)_{L^2}$ for $x \in L^2(a, b)$. The adjoint of f is calculated by

$$f(x) \cdot \overline{y} = \int_a^b x(t) \, \overline{y} \, dt = (x, y)_{L^2} = (x, f^*(y))_{L^2}$$

for all $x \in L^2(a, b)$ and $y \in \mathbb{C}$. Therefore, $f^*(y) \in L^2(a, b)$ is the constant function with value y.

(c) Let X be the *Sobolev space* $H^1(a, b)$, i.e., the space of L^2-functions that possess generalized L^2-derivatives:

$$H^1(a, b) := \left\{ x \in L^2(a, b) : \begin{array}{l} \text{there exists } \alpha \in \mathbb{K} \text{ and } y \in L^2(a, b) \text{ with} \\ x(t) = \alpha + \int_a^t y(s) \, ds \text{ for } t \in (a, b) \end{array} \right\}.$$

We denote the generalized derivative $y \in L^2(a, b)$ by x'. We observe that $H^1(a, b) \subset C[a, b]$ with bounded embedding. As an inner product in $H^1(a, b)$, we define

$$(x, y)_{H^1} := x(a) \overline{y(a)} + (x', y')_{L^2}, \quad x, y \in H^1(a, b).$$

Now let $Y = L^2(a, b)$ and $A : H^1(a, b) \longrightarrow L^2(a, b)$ be the operator $x \mapsto x'$ for $x \in H^1(a, b)$. Then A is well-defined, linear, and bounded. It is easily seen that the adjoint of A is given by

$$A^* y(t) = \int_a^t y(s) \, ds, \quad t \in (a, b), \quad y \in L^2(a, b).$$

The following theorems are two of the most important results of linear functional analysis.

Theorem A.25 *(Open Mapping Theorem)*
Let X, Y be Banach spaces and $A : X \to Y$ a linear bounded operator from X onto Y. Then A is open, i.e., the images $A(U) \subset Y$ are open in Y for all open sets $U \subset X$. In particular, if A is a bounded isomorphism from X onto Y, then the inverse $A^{-1} : Y \to X$ is bounded. This result is sometimes called the Banach–Schauder theorem.

Theorem A.26 *(Banach–Steinhaus, Principle of Uniform Boundedness)*
Let X be a Banach space, Y be a normed space, I be an index set, and $A_\alpha \in \mathcal{L}(X, Y)$, $\alpha \in I$, be a collection of linear bounded operators such that

$$\sup_{\alpha \in I} \|A_\alpha x\| < \infty \quad \text{for every } x \in X.$$

Then $\sup_{\alpha \in I} \|A_\alpha\| < \infty$.

As an immediate consequence, we have the following.

Theorem A.27
Let X be a Banach space, Y be a normed space, $D \subset X$ be a dense subspace, and $A_n \in \mathcal{L}(X, Y)$ for $n \in \mathbb{N}$. Then the following two assertions are equivalent:

(i) *$A_n x \to 0$ as $n \to \infty$ for all $x \in X$.*

(ii) *$\sup_{n \in \mathbb{N}} \|A_n\| < \infty$ and $A_n x \to 0$ as $n \to \infty$ for all $x \in D$.*

We saw in Theorem A.10 that every normed space X possesses a unique completion $\tilde{X}$. Every linear bounded operator defined on X can also be extended to $\tilde{X}$.

Theorem A.28
Let $\tilde{X}, \tilde{Y}$ be Banach spaces, $X \subset \tilde{X}$ a dense subspace, and $A : X \to \tilde{Y}$ be linear and bounded. Then there exists a linear bounded operator $\tilde{A} : \tilde{X} \to \tilde{Y}$ with

(i) *$\tilde{A} x = A x$ for all $x \in X$, i.e., $\tilde{A}$ is an extension of A, and* *(ii)* $\left\| \tilde{A} \right\| = \|A\|$.

Furthermore, the operator $\tilde{A}$ is uniquely determined.

We will now study equations of the form

$$x \; - \; Kx \; = \; y \, , \qquad (A.13)$$

where the operator norm of the linear bounded operator $K : X \to X$ is small. The following theorem plays an essential role in the study of Volterra integral equations.

Theorem A.29 *(Neumann Series)*
Let X be a Banach space over $\mathbb{R}$ or $\mathbb{C}$ and $K : X \to X$ be a linear bounded operator with

$$\limsup_{n \to \infty} \| K^n \|^{1/n} \; < \; 1. \qquad (A.14)$$

Then $I - K$ is invertible, the Neumann series $\sum_{n=0}^{\infty} K^n$ converges in the operator norm, and

$$\sum_{n=0}^{\infty} K^n \; = \; (I - K)^{-1}.$$

Condition (A.14) is satisfied if, for example, $\| K^m \| < 1$ for some $m \in \mathbb{N}$.

Example A.30
Let $\Delta := \left\{ (t, s) \in \mathbb{R}^2 : a < s < t < b \right\}$.

(a) Let $k \in L^2(\Delta)$. Then the Volterra operator

$$Kx(t) \; := \; \int_a^t k(t, s) \, x(s) \, ds, \quad a < t < b, \; x \in L^2(a, b), \qquad (A.15)$$

is bounded in $L^2(a, b)$. There exists $m \in \mathbb{N}$ with $\left\| K^m \right\|_{L^2} < 1$. The Volterra equation of the second kind

$$x(t) \; - \; \int_a^t k(t, s) \, x(s) \, ds \; = \; y(t), \quad a < t < b, \qquad (A.16)$$

is uniquely solvable in $L^2(a, b)$ for every $y \in L^2(a, b)$, and the solution x depends continuously on y. The solution $x \in L^2(a, b)$ has the form

$$x(t) \; = \; y(t) \; + \; \int_a^t r(t, s) \, y(s) \, ds, \quad t \in (a, b),$$

with some kernel $r \in L^2(\Delta)$.

(b) Let $k \in C(\overline{\Delta})$. Then the operator K defined by (A.15) is bounded in $C[a, b]$, and there exists $m \in \mathbb{N}$ with $\left\| K^m \right\|_\infty < 1$. Equation (A.16) is also uniquely solvable in $C[a, b]$ for every $y \in C[a, b]$, and the solution x depends continuously on y.

For the remaining part of this section, we will assume that X and Y are normed spaces and $K : X \to Y$ a linear and bounded operator.

Definition A.31 *(Compact Operator)*
The operator $K : X \to Y$ is called compact *if it maps every bounded set S into a relatively compact set $K(S)$.*

We recall that a set $M \subset Y$ is called *relatively compact* if every *bounded* sequence $(y_j) \subset M$ has an accumulation point in $\overline{M}$, i.e., if the closure $\overline{M}$ is compact. The set of all compact operators from X into Y is a closed subspace of $\mathcal{L}(X, Y)$ and even a two-sided ideal by the following theorem.

Theorem A.32
(a) If K_1 and K_2 are compact from X into Y, then so are $K_1 + K_2$ and λK_1 for every $\lambda \in \mathbb{K}$.

(b) Let $K_n : X \longrightarrow Y$ be a sequence of compact operators between Banach spaces X and Y. Let $K : X \longrightarrow Y$ be bounded, and let K_n converge to K in the operator norm, i.e.,

$$\| K_n - K \| := \sup_{x \neq 0} \frac{\| K_n x - K x \|}{\| x \|} \longrightarrow 0 \ (n \longrightarrow \infty).$$

Then K is also compact.

(c) If $L \in \mathcal{L}(X, Y)$ and $K \in \mathcal{L}(Y, Z)$, and L or K is compact, then KL is also compact.

(d) Let $A_n \in \mathcal{L}(X, Y)$ be pointwise convergent to some $A \in \mathcal{L}(X, Y)$, i.e., $A_n x \to A x$ for all $x \in X$. If $K : Z \to X$ is compact, then $\| A_n K - A K \| \to 0$, i.e., the operators $A_n K$ converge to AK in the operator norm.

Theorem A.33
(a) Let $k \in L^2\big((c, d) \times (a, b)\big)$. The operator $K : L^2(a, b) \to L^2(c, d)$, defined by

$$Kx(t) := \int_a^b k(t, s)\, x(s)\, ds, \quad t \in (c, d), \quad x \in L^2(a, b), \tag{A.17}$$

is compact from $L^2(a, b)$ into $L^2(c, d)$.

(b) Let k be continuous on $[c, d] \times [a, b]$ or weakly singular on $[a, b] \times [a, b]$ (in this case $[c, d] = [a, b]$). Then K defined by (A.17) is also compact as an operator from $C[a, b]$ into $C[c, d]$.

We will now study equations of the form

$$x - Kx = y, \tag{A.18}$$

where the linear operator $K : X \to X$ is compact. The following theorem extends the well-known existence results for finite linear systems of n equations and n variables to compact perturbations of the identity.

Theorem A.34 *(Riesz)*
Let X be a normed space and $K : X \to X$ be a linear compact operator.

(a) *The nullspace $\mathcal{N}(I - K) = \{x \in X : x = Kx\}$ is finite-dimensional and the range $(I - K)(X)$ is closed in X.*

(b) *If $I - K$ is one-to-one, then $I - K$ is also surjective, and the inverse $(I - K)^{-1}$ is bounded. In other words, if the homogeneous equation $x - Kx = 0$ admits only the trivial solution $x = 0$, then the inhomogeneous equation $x - Kx = y$ is uniquely solvable for every $y \in X$ and the solution x depends continuously on y.*

The next theorem studies approximations of equations of the form $Ax = y$. Again, we have in mind that $A = I - K$.

Theorem A.35
Assume that the operator $A : X \to Y$ between Banach spaces X and Y has a bounded inverse A^{-1}. Let $A_n \in \mathcal{L}(X, Y)$ be a sequence of bounded operators that converge in norm to A, i.e., $\|A_n - A\| \to 0$ as $n \to \infty$. Then, for sufficiently large n, more precisely for all n with

$$\left\| A^{-1}(A_n - A) \right\| < 1, \tag{A.19}$$

the inverse operators $A_n^{-1} : Y \to X$ exist and are uniformly bounded by

$$\left\| A_n^{-1} \right\| \leq \frac{\|A^{-1}\|}{1 - \|A^{-1}(A_n - A)\|} \leq c. \tag{A.20}$$

For the solutions of the equations

$$Ax = y \quad and \quad A_n x_n = y_n,$$

the error estimate

$$\|x_n - x\| \leq c\left\{ \|A_n x - Ax\| + \|y_n - y\| \right\} \tag{A.21}$$

holds with the constant c from (A.20).

A.4 Sobolev Spaces of Periodic Functions

In this section, we will recall definitions and properties of Sobolev (Hilbert) spaces of periodic functions. A complete discussion including proofs can be found in the monograph [130].

From Parseval's identity, we note that $x \in L^2(0, 2\pi)$ if and only if the Fourier coefficients

$$a_k = \frac{1}{2\pi} \int_0^{2\pi} x(s) \, e^{-iks} \, ds, \quad k \in \mathbb{Z}, \tag{A.22}$$

are square summable. In this case

$$\sum_{k \in \mathbb{Z}} |a_k|^2 = \frac{1}{2\pi} \|x\|_{L^2}^2 .$$

If x is periodic and continuously differentiable on $[0, 2\pi]$, partial integration of (A.22) yields the formula

$$a_k = \frac{-i}{2\pi k} \int_0^{2\pi} x'(s) \, e^{-iks} \, ds,$$

i.e., ika_k are the Fourier coefficients of x' and are thus square summable. This motivates the introduction of subspaces $H^r(0, 2\pi)$ of $L^2(0, 2\pi)$ by requiring for their elements a certain decay of the Fourier coefficients a_k.

Definition A.36 *(Sobolev Space)*
For $r \geq 0$, the Sobolev space $H^r(0, 2\pi)$ of order r is defined by

$$H^r(0, 2\pi) := \left\{ \sum_{k \in \mathbb{Z}} a_k \, e^{ikt} : \sum_{k \in \mathbb{Z}} (1 + k^2)^r \, |a_k|^2 < \infty \right\}.$$

We note that $H^0(0, 2\pi)$ coincides with $L^2(0, 2\pi)$.

Theorem A.37
The Sobolev space $H^r(0, 2\pi)$ is a Hilbert space with the inner product defined by

$$(x, y)_{H^r} := \sum_{k \in \mathbb{Z}} (1 + k^2)^r \, a_k \, \overline{b_k} , \tag{A.23}$$

where

$$x(t) = \sum_{k \in \mathbb{Z}} a_k \, e^{ikt} \quad and \quad y(t) = \sum_{k \in \mathbb{Z}} b_k \, e^{ikt}.$$

The norm in $H^r(0, 2\pi)$ is given by

$$\|x\|_{H^r} = \left(\sum_{k \in \mathbb{Z}} (1 + k^2)^r |a_k|^2 \right)^{1/2}.$$

We note that $\|x\|_{L^2} = \sqrt{2\pi} \, \|x\|_{H^0}$, that is, the norms $\|x\|_{L^2}$ and $\|x\|_{H^0}$ are equivalent on $L^2(0, 2\pi)$.

Theorem A.38

(a) For $r \in \mathbb{N}_0 := \mathbb{N} \cup \{0\}$, the space $\{x \in C^r[0, 2\pi] : x \text{ periodic}\}$ is boundedly embedded in $H^r(0, 2\pi)$.

(b) The space $\mathcal{T}$ of all trigonometric polynomials

$$\mathcal{T} := \left\{ \sum_{k=-n}^{n} a_k \, e^{ikt} : a_k \in \mathbb{C}, \ n \in \mathbb{N} \right\}$$

is dense in $H^r(0, 2\pi)$ for every $r \geq 0$.

Definition A.39

For $r \geq 0$, we denote by $H^{-r}(0, 2\pi)$ the dual space of $H^r(0, 2\pi)$, i.e., the space of all linear bounded functionals on $H^r(0, 2\pi)$.

By Theorem A.22, we can identify $H^{-r}(0, 2\pi)$ with $H^r(0, 2\pi)$ by using the inner product $(\cdot, \cdot)_{H^r}$ in $H^r(0, 2\pi)$. The following theorems give characterizations in terms of the Fourier coefficients.

Theorem A.40

Let $F \in H^{-r}(0, 2\pi)$ and define $c_k := F(\exp(ikt))$ for $k \in \mathbb{Z}$. Then

$$\|F\|_{H^{-r}} = \left(\sum_{k \in \mathbb{Z}} (1 + k^2)^{-r} |c_k|^2 \right)^{1/2}.$$

Conversely, let $c_m \in \mathbb{C}$ satisfy

$$\sum_{k \in \mathbb{Z}} (1 + k^2)^{-r} |c_k|^2 < \infty.$$

Then there exists a bounded linear functional F on $H^r(0, 2\pi)$ with $F(\exp(ikt)) = c_k$ for all $k \in \mathbb{Z}$.

Theorem A.41

Let $r \geq 0$. On the space $\mathcal{T}$ of all trigonometric polynomials, define the inner product and norm by

$$(p, q)_{H^{-r}} := \sum_{k=-n}^{n} (1 + k^2)^{-r} a_k \overline{b_k}, \tag{A.24a}$$

$$\|p\|_{-r} \quad := \quad \sum_{k=-n}^{n} (1+k^2)^{-r} \, |a_k|^2, \qquad (\text{A.24b})$$

where

$$p(t) \;=\; \sum_{k=-n}^{n} a_k \, e^{ikt} \quad and \quad q(t) \;=\; \sum_{k=-n}^{n} b_k \, e^{ikt}.$$

Then the completion of $\mathcal{T}$ with respect to $\|\cdot\|_{-r}$ is norm-isomorphic to $H^{-r}(0, 2\pi)$. The isomorphism is given by the extension of

$$J : \mathcal{T} \to H^{-r}(0, 2\pi),$$

where

$$(Jp)x \;:=\; \frac{1}{2\pi} \int_{0}^{2\pi} x(t) \, \overline{(Cp)(t)} \, dt, \quad x \in H^r(0, 2\pi), \; p \in \mathcal{T}, \qquad (\text{A.25})$$

and

$$Cp(t) \;=\; \sum_{k=-n}^{n} \overline{a_k} \, e^{ikt} \quad for \quad p(t) \;=\; \sum_{k=-n}^{n} a_k \, e^{ikt}.$$

Therefore, we identify $\|p\|_{-r}$ with $\|Jp\|_{H^{-r}}$ and simply write $\|p\|_{H^{-r}}$.

Proof: Let

$$p(t) \;=\; \sum_{k=-n}^{n} b_k \, e^{ikt} \in \mathcal{T} \quad and \quad x(t) \;=\; \sum_{k \in \mathbb{Z}} a_k \, e^{ikt} \in H^r(0, 2\pi).$$

Then

$$(Jp)x \;=\; \sum_{k=-n}^{n} a_k \, b_k. \qquad (\text{A.26})$$

Thus, by the Cauchy–Schwarz inequality,

$$
\begin{aligned}
|(Jp)x| \;&\leq\; \sum_{k=-n}^{n} \left\{ (1+k^2)^{r/2} \, |a_k| \right\} \left\{ (1+k^2)^{-r/2} \, |b_k| \right\} \\
&\leq\; \left(\sum_{k=-n}^{n} (1+k^2)^{r} \, |a_k|^2 \right)^{1/2} \left(\sum_{k=-n}^{n} (1+k^2)^{-r} \, |b_k|^2 \right)^{1/2} \\
&=\; \|x\|_{H^r} \, \|p\|_{-r},
\end{aligned}
$$

and thus $\|Jp\|_{H^{-r}} \leq \|p\|_{-r}$. Now we take

$$x(t) \;=\; \sum_{k=-n}^{n} \left\{ (1+k^2)^{-r} \overline{b_k} \, e^{ikt} \right\}.$$

Then $\|x\|_{H^r} = \|p\|_{-r}$ and $(Jp)x = \|p\|_{-r}^2$, i.e.,

$$\|Jp\|_{H^{-r}} \geq \frac{(Jp)x}{\|x\|_{H^r}} = \|p\|_{-r}.$$

This yields $\|Jp\|_{H^{-r}} = \|p\|_{-r}$.

It remains to show that the range of J is dense in $H^{-r}(0, 2\pi)$. Let $F \in H^{-r}(0, 2\pi)$ and $c_k = F(\exp(ikt))$ for $k \in \mathbb{Z}$. Define the polynomial $p_n \in \mathcal{T}$ by $p_n(t) = \sum_{k=-n}^{n} c_k \exp(ikt)$. Then

$$(F - Jp_n)(\exp(ikt)) = \begin{cases} 0, & |k| \leq n, \\ c_k, & |k| > n. \end{cases}$$

Theorem A.40 yields that

$$\|Jp_n - F\|_{H^{-r}}^2 = \sum_{|k| > n} (1 + k^2)^{-r} |c_k|^2 \longrightarrow 0, \quad n \to \infty. \qquad \square$$

Theorem A.42

(a) *For $r > s$, the Sobolev space $H^r(0, 2\pi)$ is a dense subspace of $H^s(0, 2\pi)$. The inclusion operator from $H^r(0, 2\pi)$ into $H^s(0, 2\pi)$ is compact.*

(b) *For all $r \geq 0$ and $x \in H^r(0, 2\pi)$, $y \in L^2(0, 2\pi)$, there holds*

$$\left| \int_0^{2\pi} x(s)\, y(s)\, ds \right| \leq 2\pi \, \|x\|_{H^r} \|y\|_{H^{-r}}.$$

Theorems A.38 and A.41 imply that the space $\mathcal{T}$ of all trigonometric polynomials is dense in $H^r(0, 2\pi)$ for every $r \in \mathbb{R}$. Now we will study the orthogonal projection and the interpolation operators with respect to equidistant knots and the $2n$-dimensional space

$$\mathcal{T}_n := \left\{ \sum_{k=-n}^{n-1} a_k\, e^{ikt} : a_k \in \mathbb{C} \right\}. \tag{A.27}$$

Lemma A.43

Let $P_n : L^2(0, 2\pi) \to \mathcal{T}_n \subset L^2(0, 2\pi)$ be the orthogonal projection operator. Then P_n is given by

$$P_n x(t) = \sum_{k=-n}^{n-1} a_k\, e^{ikt}, \quad x \in L^2(0, 2\pi), \tag{A.28}$$

where

$$a_k = \frac{1}{2\pi} \int_0^{2\pi} x(s) \exp(-iks)\, ds, \quad k \in \mathbb{Z},$$

are the Fourier coefficients of x. Furthermore, the following estimate holds:

$$\|x - P_n x\|_{H^s} \le \frac{1}{n^{r-s}} \|x\|_{H^r} \quad \text{for all } x \in H^r(0, 2\pi), \tag{A.29}$$

where $r \ge s$.

Proof: Let

$$x(t) = \sum_{k \in \mathbb{Z}} a_k e^{ikt} \in L^2(0, 2\pi) \quad \text{and let} \quad z(t) = \sum_{k=-n}^{n-1} a_k e^{ikt} \in \mathcal{T}_n$$

be the right-hand side of (A.28). The orthogonality of $\exp(ikt)$ implies that $x - z$ is orthogonal to $\mathcal{T}_n$. This proves that z coincides with $P_n x$. Now let $x \in H^r(0, 2\pi)$. Then

$$\begin{aligned}
\|x - P_n x\|_{H^s}^2 &\le \sum_{|k| \ge n} (1 + k^2)^s \, |a_k|^2 \\
&= \sum_{|k| \ge n} (1 + k^2)^{-(r-s)} \left[(1 + k^2)^r \, |a_k|^2 \right] \\
&\le (1 + n^2)^{s-r} \|x\|_{H^r}^2 \le n^{2(s-r)} \|x\|_{H^r}^2 . \qquad \square
\end{aligned}$$

Now let $t_j := j \frac{\pi}{n}$, $j = 0, \ldots, 2n - 1$, be equidistantly chosen points in $[0, 2\pi]$. Interpolation of smooth periodic functions by trigonometric polynomials can be found in numerous books, e.g., [52]. Interpolation in Sobolev spaces of integer orders can be found in [30]. We give a different and much simpler proof of the error estimates that are optimal and hold in Sobolev spaces of fractional order.

Theorem A.44
For every $n \in \mathbb{N}$ and every 2π-periodic function $x \in C[0, 2\pi]$, there exists a unique $p_n \in \mathcal{T}_n$ with $x(t_j) = p_n(t_j)$ for all $j = 0, \ldots, 2n - 1$. The trigonometric interpolation operator $Q_n : \{x \in C[0, 2\pi] : x \text{ periodic}\} \to \mathcal{T}_n$ has the form

$$Q_n x = \sum_{k=0}^{2n-1} x(t_k) \, L_k$$

with Lagrange interpolation basis functions

$$L_k(t) = \frac{1}{2n} \sum_{m=-n}^{n-1} e^{im(t - t_k)}, \quad k = 0, \ldots, 2n - 1. \tag{A.30}$$

The interpolation operator Q_n has an extension to a bounded operator from $H^r(0, 2\pi)$ into $\mathcal{T}_n \subset H^r(0, 2\pi)$ for all $r > \frac{1}{2}$. Furthermore, Q_n obeys estimates of the form

$$\|x - Q_n x\|_{H^s} \le \frac{c}{n^{r-s}} \|x\|_{H^r} \quad \text{for all } x \in H^r(0, 2\pi), \tag{A.31}$$

where $0 \le s \le r$ and $r > \frac{1}{2}$. The constant c depends only on s and r.

Proof: The proof of the first part can be found in, e.g., [130]. Let $x(t) = \sum_{m \in \mathbb{Z}} a_m \exp(imt)$. Direct calculation shows that for smooth functions x the interpolation is given by

$$Q_n x(t) = \sum_{j=-n}^{n-1} \hat{a}_j e^{ijt} \quad \text{with}$$

$$\hat{a}_j = \frac{1}{2n} \sum_{k=0}^{2n-1} x(t_k) e^{-ijk\pi/n}, \quad j = -n, \ldots, n-1.$$

The connection between the continuous and discrete Fourier coefficients is simply

$$\hat{a}_j = \frac{1}{2n} \sum_{k=0}^{2n-1} \sum_{m \in \mathbb{Z}} a_m e^{imk\pi/n - ijk\pi/n}$$

$$= \frac{1}{2n} \sum_{m \in \mathbb{Z}} a_m \sum_{k=0}^{2n-1} \left[e^{i(m-j)\pi/n} \right]^k$$

$$= \sum_{\ell \in \mathbb{Z}} a_{j+2n\ell}.$$

It is sufficient to estimate $P_n x - Q_n x$ since the required estimate holds for $x - P_n x$ by formula (A.29). We have

$$\left(P_n x - Q_n x \right)(t) = \sum_{m=-n}^{n-1} \left[a_m - \hat{a}_m \right] e^{imt}$$

and thus by the Cauchy–Schwarz inequality

$$\| P_n x - Q_n x \|_{H^s}^2$$

$$= \sum_{m=-n}^{n-1} |a_m - \hat{a}_m|^2 (1 + m^2)^s \le c n^{2s} \sum_{m=-n}^{n-1} \left| \sum_{\ell \ne 0} a_{m+2n\ell} \right|^2$$

$$\le c n^{2s} \sum_{m=-n}^{n-1} \left| \sum_{\ell \ne 0} [(1 + (m+2n\ell)^2)^{r/2} a_{m+2n\ell}] \frac{1}{(1 + (m+2n\ell)^2)^{r/2}} \right|^2$$

$$\le c n^{2s} \sum_{m=-n}^{n-1} \left[\sum_{\ell \ne 0} (1 + (m+2n\ell)^2)^r |a_{m+2n\ell}|^2 \sum_{\ell \ne 0} \frac{1}{(1 + (m+2n\ell)^2)^r} \right].$$

From the obvious estimate

$$\sum_{\ell \ne 0} (1 + (m+2n\ell)^2)^{-r} \le (2n)^{-2r} \sum_{\ell \ne 0} \left(\frac{m}{2n} + \ell \right)^{-2r} \le c n^{-2r}$$

for all $|m| \leq n$ and $n \in \mathbb{N}$, we conclude that

$$\|P_n x - Q_n x\|_{H^s}^2 \;\leq\; cn^{2(s-r)} \sum_{m=-n}^{n-1} \sum_{\ell \neq 0} \left(1 + (m + 2n\ell)^2\right)^r |a_{m+2n\ell}|^2$$

$$\leq\; cn^{2(s-r)} \|x\|_{H^r}^2 . \qquad \square$$

For real-valued functions, it is more convenient to study the orthogonal projection and interpolation in the $2n$-dimensional space

$$\left\{ \sum_{j=0}^{n} a_j \cos(jt) + \sum_{j=1}^{n-1} b_j \sin(jt) : a_j, b_j \in \mathbb{R} \right\}.$$

In this case, the Lagrange interpolation basis functions L_k are given by (see [130])

$$L_k(t) \;=\; \frac{1}{2n} \left\{ 1 + 2 \sum_{m=1}^{n-1} \cos m(t - t_k) + \cos n(t - t_k) \right\}, \qquad \text{(A.32)}$$

$k = 0, \ldots, 2n - 1$, and the estimates (A.29) and (A.31) are proven by the same arguments.

Theorem A.45
Let $r \in \mathbb{N}$ and $k \in C^r\left([0, 2\pi] \times [0, 2\pi]\right)$ be 2π-periodic with respect to both variables. Then the integral operator K, defined by

$$Kx(t) \;:=\; \int_0^{2\pi} k(t, s)\, x(s)\, ds, \quad t \in (0, 2\pi), \qquad \text{(A.33)}$$

can be extended to a bounded operator from $H^p(0, 2\pi)$ into $H^r(0, 2\pi)$ for every $-r \leq p \leq r$.

Proof: Let $x \in L^2(0, 2\pi)$. From

$$\frac{d^j}{dt^j} Kx(t) \;=\; \int_0^{2\pi} \frac{\partial^j k(t, s)}{\partial t^j}\, x(s)\, ds, \quad j = 0, \ldots, r,$$

we conclude from Theorem A.42 that for $x \in L^2(0, 2\pi)$

$$\left| \frac{d^j}{dt^j} Kx(t) \right| \;\leq\; 2\pi \left\| \frac{\partial^j k(t, \cdot)}{\partial t^j} \right\|_{H^r} \|x\|_{H^{-r}}$$

and thus

$$\|Kx\|_{H^r} \;\leq\; c_1 \|Kx\|_{C^r} \;\leq\; c_2 \|x\|_{H^{-r}}$$

for all $x \in L^2(0, 2\pi)$. Application of Theorem A.28 yields the assertion since $L^2(0, 2\pi)$ is dense in $H^{-r}(0, 2\pi)$. $\qquad \square$

A.5   Spectral Theory for Compact Operators in Hilbert Spaces

Definition A.46 *(Spectrum)*
Let X be a normed space and $A : X \longrightarrow X$ be a linear operator. The spectrum $\sigma(A)$ is defined as the set of (complex) numbers λ such that the operator $A - \lambda I$ does not have a bounded inverse on X. Here, I denotes the identity on X. $\lambda \in \sigma(A)$ is called an eigenvalue *of A if $A - \lambda I$ is not one-to-one. If λ is an eigenvalue, then the nontrivial elements x of the kernel $\mathcal{N}(A - \lambda_n I) = \{ x \in X : Ax - \lambda x = 0 \}$ are called* eigenvectors *of A.*

This definition makes sense for arbitrary linear operators in normed spaces. For noncompact operators A it is possible that the operator $A - \lambda I$ is one-to-one but fails to be bijective. As an example, we consider $X = \ell^2$ and define A by

$$(Ax)_k := \begin{cases} 0, & \text{if} \quad k = 1, \\ x_{k-1}, & \text{if} \quad k \geq 2, \end{cases}$$

for $x = (x_k) \in \ell^2$. Then $\lambda = 1$ belongs to the spectrum of A but is not an eigenvalue of A.

Theorem A.47
Let $A : X \to X$ be a linear operator.

(a) Let $x_j \in X$, $j = 1, \ldots, n$, be a finite set of eigenvectors corresponding to pairwise different eigenvalues $\lambda_j \in \mathbb{C}$. Then $\{x_1, \ldots, x_n\}$ are linearly independent. If X is a Hilbert space and A is self-adjoint, i.e., $A^* = A$, then all eigenvalues λ_j are real-valued and the corresponding eigenvectors $x_1, \ldots, x_n$ are pairwise orthogonal.

(b) Let X be a Hilbert space and $A : X \to X$ be self-adjoint. Then

$$\|A\| = \sup_{\|x\|=1} |(Ax, x)| = r(A),$$

where $r(A) = \sup\{|\lambda| : \lambda \in \sigma(A)\}$ is called the spectral radius *of A.*

The situation is simpler for compact operators. We collect the most important results in the following fundamental theorem.

Theorem A.48 *(Spectral Theorem for Compact Self-Adjoint Operators)*
Let $K : X \to X$ be compact and self-adjoint (and $K \neq 0$). Then the following hold:

(a) The spectrum consists only of eigenvalues and possibly 0. Every eigenvalue of K is real-valued. K has at least one but at most a countable number of eigenvalues with 0 as the only possible accumulation point.

(b) *For every eigenvalue $\lambda \neq 0$, there exist only finitely many linearly independent eigenvectors, i.e., the eigenspaces are finite-dimensional. Eigenvectors corresponding to different eigenvalues are orthogonal.*

(c) *We order the eigenvalues in the form*

$$|\lambda_1| \geq |\lambda_2| \geq |\lambda_3| \geq \ldots$$

and denote by $P_j : X \to \mathcal{N}(K - \lambda_j I)$ the orthogonal projection onto the eigenspace corresponding to λ_j. If there exist only a finite number $\lambda_1, \ldots, \lambda_m$ of eigenvalues, then

$$K = \sum_{j=1}^{m} \lambda_j P_j \, .$$

If there exists an infinite sequence (λ_j) of eigenvalues, then

$$K = \sum_{j=1}^{\infty} \lambda_j P_j \, ,$$

where the series converges in the operator norm. Furthermore,

$$\left\| K - \sum_{j=1}^{m} \lambda_j P_j \right\| = |\lambda_{m+1}| \, .$$

(d) *Let H be the linear span of all of the eigenvectors corresponding to the eigenvalues $\lambda_j \neq 0$ of K. Then*

$$X = \overline{H} \oplus \mathcal{N}(K).$$

Sometimes, part (d) is formulated differently. For a common treatment of the cases of finitely and infinitely many eigenvalues, we introduce the index set $J \subset \mathbb{N}$, where J is finite in the first case and $J = \mathbb{N}$ in the second case. For every eigenvalue λ_j, $j \in J$, we choose an orthonormal basis of the corresponding eigenspace $\mathcal{N}(K - \lambda_j I)$. Again, let the eigenvalues be ordered in the form

$$|\lambda_1| \geq |\lambda_2| \geq |\lambda_3| \geq \ldots > 0.$$

By counting every $\lambda_j \neq 0$ relative to its multiplicity, we can assign an eigenvector x_j to every eigenvalue λ_j. Then every $x \in X$ possesses an abstract Fourier expansion of the form

$$x = x_0 + \sum_{j \in J} (x, x_j) \, x_j$$

for some $x_0 \in \mathcal{N}(K)$ and

$$Kx = \sum_{j \in J} \lambda_j \left(x, x_j \right) x_j.$$

As a corollary, we observe that the set $\{x_j : j \in J\}$ of all eigenvectors forms a complete system in X if K is one-to-one.

The spectral theorem for compact self-adjoint operators has an extension to nonself-adjoint operators $K : X \to Y$. First, we have the following definition.

Definition A.49 *(Singular Values)*
Let X and Y be Hilbert spaces and $K : X \to Y$ be a compact operator with adjoint operator $K^* : Y \to X$. The square roots $\mu_j = \sqrt{\lambda_j}$, $j \in J$, of the eigenvalues λ_j of the self-adjoint operator $K^*K : X \to X$ are called singular values *of K. Here again, $J \subset \mathbb{N}$ could be either finite or $J = \mathbb{N}$.*

Note that every eigenvalue λ of K^*K is nonnegative since $K^*Kx = \lambda x$ implies that $\lambda(x, x) = (K^*Kx, x) = (Kx, Kx) \geq 0$, i.e., $\lambda \geq 0$.

Theorem A.50 *(Singular Value Decomposition)*
Let $K : X \longrightarrow Y$ be a linear compact operator, $K^* : Y \longrightarrow X$ its adjoint operator, and $\mu_1 \geq \mu_2 \geq \mu_3 \ldots > 0$ the ordered sequence of the positive singular values of K, counted relative to its multiplicity. Then there exist orthonormal systems $(x_j) \subset X$ and $(y_j) \subset Y$ with the following properties:

$$Kx_j = \mu_j y_j \quad \text{and} \quad K^*y_j = \mu_j x_j \quad \text{for all } j \in J.$$

The system (μ_j, x_j, y_j) is called a singular system *for K. Every $x \in X$ possesses the* singular value decomposition

$$x = x_0 + \sum_{j \in J} (x, x_j) x_j$$

for some $x_0 \in \mathcal{N}(K)$ and

$$Kx = \sum_{j \in J} \mu_j \left(x, x_j \right) y_j .$$

The following theorem characterizes the range of a compact operator with the help of a singular system.

Theorem A.51 *(Picard)*
Let $K : X \longrightarrow Y$ be a linear compact operator with singular system (μ_j, x_j, y_j). The equation

$$Kx = y \tag{A.34}$$

is solvable if and only if

$$y \in N(K^*)^{\perp} \quad and \quad \sum_{j \in J} \frac{1}{\mu_j^2} |(y, y_j)|^2 < \infty. \tag{A.35}$$

In this case

$$x = \sum_{j \in J} \frac{1}{\mu_j} (y, y_j) x_j$$

is a solution of (A.34).

We note that the solvability conditions (A.35) require a fast decay of the Fourier coefficients of y with respect to the orthonormal system (y_j) in order for the series

$$\sum_{j=1}^{\infty} \frac{1}{\mu_j^2} |(y, y_j)|^2$$

to converge. Of course, this condition is only necessary for the important case where there exist infinitely many singular values. As a simple example, we study the following integral operator.

Example A.52
Let $K : L^2(0, 1) \longrightarrow L^2(0, 1)$ be defined by

$$Kx(t) := \int_0^t x(s) \, ds, \quad t \in (0, 1), \ x \in L^2(0, 1).$$

Then

$$K^*y(t) = \int_t^1 y(s) \, ds \quad and \quad K^*Kx(t) = \int_t^1 \left(\int_0^s x(\tau) \, d\tau \right) ds.$$

The eigenvalue problem $K^*Kx = \lambda x$ is equivalent to

$$\lambda x(t) = \int_t^1 \left(\int_0^s x(\tau) \, d\tau \right) ds, \quad t \in [0, 1].$$

Differentiating twice, we observe that for $\lambda \neq 0$ this is equivalent to the eigenvalue problem

$$\lambda x'' + x = 0 \text{ in } (0, 1), \quad x(1) = x'(0) = 0.$$

Solving this yields

$$x_j(t) = \sqrt{\frac{2}{\pi}} \cos \frac{2j-1}{2} \pi t, \ j \in \mathbb{N}, \quad and \quad \lambda_j = \frac{4}{(2j-1)^2 \pi^2}, \ j \in \mathbb{N}.$$

The singular values μ_j and the ONS $\{y_j : j \in \mathbb{N}\}$ are given by

$$\mu_j = \frac{2}{(2j-1)\pi}, \quad j \in \mathbb{N}, \quad \text{and}$$

$$y_j(t) = \sqrt{\frac{2}{\pi}} \sin \frac{2j-1}{2} \pi t, \quad j \in \mathbb{N}.$$

A.6   The Fréchet Derivative

In this section, we will briefly recall some of the most important results for nonlinear mappings between normed spaces. The notions of continuity and differentiability carry over in a very natural way.

Definition A.53
Let X and Y be normed spaces over the field $\mathbb{K} = \mathbb{R}$ or $\mathbb{C}$, $U \subset X$ an open subset, $\hat{x} \in U$, and $T : X \supset U \to Y$ be a (possibly nonlinear) mapping.

(a) T is called continuous *in $\hat{x}$ if for every $\varepsilon > 0$ there exists $\delta > 0$ such that $\|T(x) - T(\hat{x})\| \leq \varepsilon$ for all $x \in U$ with $\|x - \hat{x}\| \leq \delta$.*

(b) T is called Fréchet differentiable *for $\hat{x} \in U$ if there exists a linear bounded operator $A : X \to Y$ (depending on $\hat{x}$) such that*

$$\lim_{h \to 0} \frac{1}{\|h\|} \|T(\hat{x} + h) - T(\hat{x}) - Ah\| = 0. \tag{A.36}$$

We write $T'(\hat{x}) := A$. In particular, $T'(\hat{x}) \in \mathcal{L}(X, Y)$.

(c) The mapping T is called continuously Fréchet differentiable *for $\hat{x} \in U$ if T is Fréchet differentiable in a neighborhhod V of $\hat{x}$ and the mapping $T' : V \to \mathcal{L}(X, Y)$ is continuous in $\hat{x}$.*

Continuity and differentiability of a mapping depend on the norms in X and Y, in contrast to the finite-dimensional case. If T is differentiable in $\hat{x}$, then the linear bounded mapping A in part (b) of Definition A.53 is unique. Therefore, $T'(\hat{x}) := A$ is well-defined. If T is differentiable in x, then T is also continuous in x. In the finite-dimensional case $X = \mathbb{K}^n$ and $Y = \mathbb{K}^m$, the linear bounded mapping $T'(x)$ is given by the Jacobian (with respect to the Cartesian coordinates).

Example A.54 *(Integral Operator)*
Let $f : [c, d] \times [a, b] \times \mathbb{C} \to \mathbb{C}$ be continuous and continuously differentiable

with respect to the third argument. Let the mapping $T : C[a, b] \to C[c, d]$ be defined by

$$T(x)(t) \; := \; \int_a^b f\big(t, s, x(s)\big) \, ds, \quad t \in [c, d], \; x \in C[a, b].$$

Then T is continuously Fréchet differentiable with derivative

$$\big(T'(x)z\big)(t) \; = \; \int_a^b \frac{\partial}{\partial x} f\big(t, s, x(s)\big) \, z(s) \, ds, \quad t \in [c, d], \; x, z \in C[a, b].$$

The following theorem collects further properties of the Fréchet derivative.

Theorem A.55

(a) *Let $T, S : X \supset U \to Y$ be Fréchet differentiable for $x \in U$. Then $T + S$ and λT are also Fréchet differentiable for all $\lambda \in \mathbb{K}$ and*

$$(T + S)'(x) \; = \; T'(x) + S'(x) \,, \qquad (\lambda T)'(x) \; = \; \lambda T'(x).$$

(b) *Chain rule: Let $T : X \supset U \to V \subset Y$ and $S : Y \supset V \to Z$ be Fréchet differentiable for $x \in U$ and $T(x) \in V$, respectively. Then $S\,T$ is also Fréchet differentiable in x and*

$$(S\,T)'(x) \; = \; \underbrace{S'\big(T(x)\big)}_{\in \mathcal{L}(Y,Z)} \; \underbrace{T'(x)}_{\in \mathcal{L}(X,Y)} \; \in \; \mathcal{L}(X, Z)\,.$$

(c) *Special case: If $T : X \to Y$ is Fréchet differentiable for $\hat{x} \in X$, then so is $\psi : \mathbb{K} \to Y$, defined by $\psi(t) := T(t\hat{x})$, $t \in \mathbb{K}$, for every point $t \in \mathbb{K}$ and $\psi'(t) = T'(t\hat{x})\hat{x} \in Y$. Note that originally $\psi'(t) \in \mathcal{L}(\mathbb{K}, Y)$. In this case, one identifies the linear mapping $\psi'(t) : \mathbb{K} \to Y$ with its generating element $\psi'(t) \in Y$.*

We recall Banach's contraction mapping principle.

Theorem A.56 *(Contraction Mapping Principle)*
Let $K \subset X$ be a closed subset of the Banach space X and $T : X \supset K \to X$ a (nonlinear) mapping with the properties

(a) *T maps K into itself, i.e., $T(x) \in K$ for all $x \in K$, and*

(b) *T is a contraction on K, i.e., there exists $\gamma < 1$ with*

$$\|T(x) - T(y)\| \; \leq \; \gamma \, \|x - y\| \quad \text{for all } x, y \in K. \tag{A.37}$$

Then there exists a unique $\tilde{x} \in K$ with $T(\tilde{x}) = \tilde{x}$. The sequence $(x_\ell) \subset K$, defined by $x_{\ell+1} := T(x_\ell)$, $\ell = 0, 1, \ldots$ converges to $\tilde{x}$ for every $x_0 \in K$. Furthermore, the following error estimates hold:

$$\left\| x_{\ell+1} - \tilde{x} \right\| \leq \gamma \left\| x_\ell - \tilde{x} \right\|, \quad \ell = 0, 1, \ldots, \tag{A.38a}$$

i.e., the sequence converges linearly to $\tilde{x}$,

$$\left\| x_\ell - \tilde{x} \right\| \leq \frac{\gamma^\ell}{1 - \gamma} \left\| x_1 - x_0 \right\|, \quad \text{(a priori estimate)} \tag{A.38b}$$

$$\left\| x_\ell - \tilde{x} \right\| \leq \frac{1}{1 - \gamma} \left\| x_{\ell+1} - x_\ell \right\|, \quad \text{(a posteriori estimate)} \tag{A.38c}$$

for $\ell = 1, 2, \ldots$

The Newton method for systems of nonlinear equations has a direct analogy for equations of the form $T(x) = y$, where $T : X \to Y$ is a continuously Fréchet differentiable mapping between Banach spaces X and Y. We will formulate a simplified Newton method and prove local linear convergence. It differs from the ordinary Newton method not only by replacing the derivative $T'(x_\ell)$ by $T'(\hat{x})$ but also by requiring only the existence of a left inverse.

Theorem A.57 *(Simplified Newton Method)*
Let $T : X \to Y$ be continuously Fréchet differentiable between Banach spaces X and Y. Let $V \subset X$ be a closed subspace, $\hat{x} \in V$ and $\hat{y} := T(\hat{x}) \in Y$. Let $L : Y \to V$ be linear and bounded such that L is a left inverse of $T'(\hat{x}) : X \to Y$ on V, i.e., $L T'(\hat{x})v = v$ for all $v \in V$.

Then there exists $\varepsilon > 0$ such that for any $\bar{y} = T(\bar{x})$ with $\bar{x} \in X$ and $\left\| \bar{x} - \hat{x} \right\| \leq \varepsilon$ the following algorithm converges linearly to some $\tilde{x} \in V$:

$$x_0 = \hat{x}, \quad x_{\ell+1} = x_\ell - L\big[T(x_\ell) - \bar{y}\big], \quad \ell = 0, 1, 2, \ldots. \tag{A.39}$$

The limit $\tilde{x} \in V$ satisfies $L\big[T(\tilde{x}) - \bar{y}\big] = 0$.

Proof: We will apply the contraction mapping principle of the preceding theorem to the mapping

$$S(x) := x - L\big[T(x) - \bar{y}\big] = L\big[T'(\hat{x})x - T(x) + T(\bar{x})\big]$$

on some closed ball $K\big[\hat{x}, \rho\big] \subset V$. We estimate

$$\begin{aligned}
\left\| S(x) - S(z) \right\| &\leq \|L\| \, \left\| T'(\hat{x})(x - z) + T(z) - T(x) \right\| \\
&\leq \|L\| \, \|x - z\| \Big\{ \left\| T'(\hat{x}) - T'(z) \right\| \\
&\quad + \frac{\left\| T(z) - T(x) + T'(z)(x - z) \right\|}{\|x - z\|} \Big\}
\end{aligned}$$

and

$$
\begin{aligned}
\|S(x) - \hat{x}\| &\leq \|L\| \, \|T'(\hat{x})(x - \hat{x}) - T(x) + T(\bar{x})\| \\
&\leq \|L\| \, \|T'(\hat{x})(x - \hat{x}) + T(\hat{x}) - T(x)\| \\
&\quad + \|L\| \, \|T(\hat{x}) - T(\bar{x})\|
\end{aligned}
$$

First, we choose $\rho > 0$ such that

$$
\|L\| \left[\|T'(\hat{x}) - T'(z)\| + \frac{\|T(z) - T(x) + T'(z)(x - z)\|}{\|x - z\|} \right] \leq \frac{1}{2}
$$

for all $x, z \in K[\hat{x}, \rho]$. This is possible since T is continuously differentiable.
Next, we choose $\varepsilon \geq 0$ such that

$$
\|L\| \, \|T(\hat{x}) - T(\bar{x})\| \leq \frac{1}{2} \rho
$$

for $\|\bar{x} - \hat{x}\| \leq \varepsilon$. Then we conclude that

$$
\begin{aligned}
\|S(x) - S(z)\| &\leq \frac{1}{2} \|x - z\| \quad \text{for all } x, z \in K[\hat{x}, \rho], \\
\|S(x) - \hat{x}\| &\leq \frac{1}{2} \|x - \hat{x}\| + \frac{1}{2} \rho \leq \rho \quad \text{for all } x \in K[\hat{x}, \rho].
\end{aligned}
$$

Application of the contraction mapping principle ends the proof. $\square$

The notion of *partial derivatives* of mappings $T : X \times Z \to Y$ is introduced just as for functions of two scalar variables as the Fréchet derivative of the mappings $T(\cdot, z) : X \to Y$ for $z \in Z$ and $T(x, \cdot) : Z \to Y$ for $x \in X$. We denote the partial derivatives in $(x, z) \in X \times Z$ by

$$
\frac{\partial}{\partial x} T(x, z) \in \mathcal{L}(X, Y) \quad \text{and} \quad \frac{\partial}{\partial z} T(x, z) \in \mathcal{L}(Z, Y).
$$

Theorem A.58 *(Implicit Function Theorem)*
Let $T : X \times Z \to Y$ be continuously Fréchet differentiable with partial derivatives $\frac{\partial}{\partial x} T(x, z) \in \mathcal{L}(X, Y)$ and $\frac{\partial}{\partial z} T(x, z) \in \mathcal{L}(Z, Y)$. Furthermore, let $T(\hat{x}, \hat{z}) = 0$ and $\frac{\partial}{\partial z} T(\hat{x}, \hat{z}) : Z \to Y$ be a norm-isomorphism. Then there exists a neighborhood U of $\hat{x}$ and a Fréchet differentiable function $\psi : U \to Z$ such that $\psi(\hat{x}) = \hat{z}$ and $T(x, \psi(x)) = 0$ for all $x \in U$. The Fréchet derivative $\psi' \in \mathcal{L}(X, Z)$ is given by

$$
\psi'(x) = -\left[\frac{\partial}{\partial z} T(x, \psi(x)) \right]^{-1} \frac{\partial}{\partial x} T(x, \psi(x)), \quad x \in U.
$$

The following special case is very important.

Let $Z = Y = \mathbb{K}$; thus $T : X \times \mathbb{K} \to \mathbb{K}$ and $T(\hat{x}, \hat{\lambda}) = 0$ and $\frac{\partial}{\partial \lambda} T(\hat{x}, \hat{\lambda}) \neq$ 0. Then there exists a neighborhood U of $\hat{x}$ and a Fréchet differentiable function $\psi : U \to \mathbb{K}$ such that $\psi(\hat{x}) = \hat{\lambda}$ and $T(x, \psi(x)) = 0$ for all $x \in U$ and

$$\psi'(x) = -\frac{1}{\frac{\partial}{\partial \lambda} T(x, \psi(x))} \frac{\partial}{\partial x} T(x, \psi(x)) \in \mathcal{L}(X, \mathbb{K}) = X', \quad x \in U,$$

where again X' denotes the dual space of X.

Appendix B
Proofs of the Results of Section 2.7

In this appendix, we give the complete proofs of the theorems and lemmas of Chapter 2, Section 2.7. For the convenience of the reader, we formulate the results again.

Theorem 2.20 *(Fletcher–Reeves)*
Let $K : X \to Y$ be a bounded, linear, and injective operator between Hilbert spaces X and Y. The conjugate gradient method is well-defined and either stops or produces sequences (x^m), $(p^m) \subset X$ with the properties

$$\left(\nabla f(x^m), \nabla f(x^j)\right) = 0 \quad \text{for all } j \neq m, \tag{2.36a}$$

and

$$\left(Kp^m, Kp^j\right) = 0 \quad \text{for all } j \neq m, \tag{2.36b}$$

i.e., the gradients are orthogonal and the directions p^m are K-conjugate. Furthermore,

$$\left(\nabla f(x^j), K^* Kp^m\right) = 0 \quad \text{for all } j < m. \tag{2.36c}$$

Proof: First, we note the following identities:

(α) $\nabla f(x^{m+1}) = 2\,K^*(Kx^{m+1} - y) = 2\,K^*(Kx^m - y) - 2t_m K^* Kp^m = \nabla f(x^m) - 2t_m K^* Kp^m$.

(β) $\left(p^m, \nabla f(x^{m+1})\right) = \left(p^m, \nabla f(x^m)\right) - 2t_m \left(Kp^m, Kp^m\right) = 0$ by the definition of t_m.

(γ) $t_m = \frac{1}{2}\big(\nabla f(x^m), p^m\big)\,\|Kp^m\|^{-2} = \frac{1}{4}\,\|\nabla f(x^m)\|^2\,/\,\|Kp^m\|^2$ since $p^m = \frac{1}{2}\nabla f(x^m) + \gamma_{m-1}p^{m-1}$ and (β).

Now we prove the following identities by induction with respect to m:

(i) $\big(\nabla f(x^m), \nabla f(x^j)\big) = 0$ for $j = 0, \ldots, m-1$,

(ii) $\big(Kp^m, Kp^j\big) = 0$ for $j = 0, \ldots, m-1$.

Let $m = 1$. Then, using (α),

(i) $\quad \big(\nabla f(x^1), \nabla f(x^0)\big) = \|\nabla f(x^0)\|^2 - 2t_0\big(Kp^0, K\nabla f(x^0)\big) = 0$,

which vanishes by (γ) since $p^0 = \frac{1}{2}\nabla f(x^0)$.

(ii) By the definition of p^1 and identity (α), we conclude that

$$\big(Kp^1, Kp^0\big) \;=\; \big(p^1, K^*Kp^0\big)$$

$$= \;-\frac{1}{2t_0}\left[\frac{1}{2}\big(\nabla f(x^1) + \gamma_0 p^0, \nabla f(x^1) - \nabla f(x^0)\big)\right]$$

$$= \;-\frac{1}{2t_0}\left[\frac{1}{2}\,\|\nabla f(x^1)\|^2 - \gamma_0\big(p^0, \nabla f(x^0)\big)\right] \;=\; 0,$$

where we have used (β), the definition of p^0, and the choice of γ_0.

Now we assume the validity of (i) and (ii) for m and show it for $m+1$:

(i) For $j = 0, \ldots m-1$ we conclude that (setting $\gamma_{-1} = 0$ in the case $j = 0$)

$$\big(\nabla f(x^{m+1}), \nabla f(x^j)\big) \;=\; \big(\nabla f(x^m) - 2t_m K^*Kp^m, \nabla f(x^j)\big)$$

$$= \;-2t_m\big(\nabla f(x^j), K^*Kp^m\big)$$

$$= \;-4t_m\big(Kp^j - \gamma_{j-1}Kp^{j-1}, Kp^m\big) \;=\; 0,$$

where we have used $\frac{1}{2}\nabla f(x^j) + \gamma_{j-1}p^{j-1} = p^j$ and assertion (ii) for m. For $j = m$, we conclude that

$$\big(\nabla f(x^{m+1}), \nabla f(x^m)\big)$$

$$= \;\|\nabla f(x^m)\|^2 - 2t_m\big(\nabla f(x^m), K^*Kp^m\big)$$

$$= \;\|\nabla f(x^m)\|^2 - \frac{1}{2}\frac{\|\nabla f(x^m)\|^2}{\|Kp^m\|^2}\big(\nabla f(x^m), K^*Kp^m\big)$$

by (γ). Now we write

$$\big(Kp^m, Kp^m\big) \;=\; \left(Kp^m, K\left(\frac{1}{2}\nabla f(x^m) + \gamma_{m-1}p^{m-1}\right)\right)$$

$$= \;\frac{1}{2}\big(Kp^m, K\nabla f(x^m)\big),$$

which implies that $\big(\nabla f(x^{m+1}), \nabla f(x^m)\big)$ vanishes.

(ii) For $j = 0, \ldots, m-1$, we conclude that, using (α),

$$
\begin{aligned}
\big(Kp^{m+1}, Kp^j\big) &= \left(\frac{1}{2} \nabla f(x^{m+1}) + \gamma_m p^m, K^* K p^j \right) \\
&= -\frac{1}{4t_j} \big(\nabla f(x^{m+1}), \nabla f(x^{j+1}) - \nabla f(x^j) \big),
\end{aligned}
$$

which vanishes by (i).

For $j = m$ by (α) and the definition of p^{m+1}, we have

$$
\begin{aligned}
\big(Kp^{m+1}, Kp^m\big) &= \frac{1}{2t_m} \left(\frac{1}{2} \nabla f(x^{m+1}) + \gamma_m p^m, \nabla f(x^m) - \nabla f(x^{m+1}) \right) \\
&= \frac{1}{2t_m} \bigg\{ \frac{1}{2} \big(\nabla f(x^{m+1}), \nabla f(x^m) \big) - \frac{1}{2} \left\| \nabla f(x^{m+1}) \right\|^2 \\
&\quad + \gamma_m \underbrace{\big(p^m, \nabla f(x^m)\big)}_{= \frac{1}{2} \| \nabla f(x^m) \|^2} - \gamma_m \big(p^m, \nabla f(x^{m+1})\big) \bigg\} \\
&= \frac{1}{4t_m} \left\{ \gamma_m \left\| \nabla f(x^m) \right\|^2 - \left\| \nabla f(x^{m+1}) \right\|^2 \right\}
\end{aligned}
$$

by (i) and (β). This term vanishes by the definition of γ_m. Thus we have proven (i) and (ii) for $m+1$ and thus for all $m = 1, 2, 3 \ldots$ To prove (2.36c) we write

$$
\big(\nabla f(x^j), K^* K p^m\big) = 2 \big(p^j - \gamma_{j-1} p^{j-1}, K^* K p^m \big) = 0 \quad \text{for } j < m,
$$

and note that we have already shown this in the proof. $\qquad\square$

Theorem 2.21

Let (x^m) and (p^m) be the sequences of the conjugate gradient method. Define the space $V_m := \operatorname{span} \{p^0, \ldots, p^m\}$. Then we have the following equivalent characterizations of V_m:

$$
\begin{aligned}
V_m &= \operatorname{span} \{ \nabla f(x^0), \ldots, \nabla f(x^m) \} & (2.37a) \\
&= \operatorname{span} \{ p^0, K^* K p^0, \ldots, (K^* K)^m p^0 \} & (2.37b)
\end{aligned}
$$

for $m = 0, 1, \ldots$. Furthermore, x^m is the minimum of f on V_{m-1} for every $m \geq 1$.

Proof: Let $\tilde{V}_m = \operatorname{span} \{ \nabla f(x^0), \ldots, \nabla f(x^m) \}$. Then $V_0 = \tilde{V}_0$. Assume that we have already shown that $V_m = \tilde{V}_m$. Since $p^{m+1} = \frac{1}{2} \nabla f(x^{m+1}) + \gamma_m p^m$ we also have that $V_{m+1} = \tilde{V}_{m+1}$. Similarly, define the space $\hat{V}_m :=$

span $\{p^0, \ldots, (K^*K)^m p^0\}$. Then $V_1 = \hat{V}_1$. Assume that we have already shown that $V_m = \hat{V}_m$. Then we conclude that

$$
\begin{aligned}
p^{m+1} &= K^*(Kx^{m+1} - y) + \gamma_m p^m \\
&= K^*(Kx^m - y) - t_m K^* K p^m + \gamma_m p^m \\
&= p^m - \gamma_{m-1} p^{m-1} - t_m K^* K p^m + \gamma_m p^m \in \hat{V}_{m+1}. \qquad (*)
\end{aligned}
$$

On the other hand, from

$$
(K^*K)^{m+1} p^0 = (K^*K)\left[(K^*K)^m p^0\right] \in (K^*K)(V_m)
$$

and $K^* K p^j \in V_{j+1} \subset V_{m+1}$ by $(*)$ for $j = 0, \ldots, m$, we conclude also that $V_{m+1} = \hat{V}_{m+1}$.

Now every x^m lies in V_{m-1}. This is certainly true for $m = 1$, and if it holds for m then it holds also for $m+1$ since $x^{m+1} = x^m - t_m p^m \in V_m$. x^m is the minimum of f on V_{m-1} if and only if $(Kx^m - y, Kz) = 0$ for all $z \in V_{m-1}$. By (2.37a), this is the case if and only if $(\nabla f(x^m), \nabla f(x^j)) = 0$ for all $j = 0, \ldots, m-1$. This holds by the preceding theorem.   □

Lemma 2.22

(a) The polynomial $\mathbb{Q}_m$, defined by $\mathbb{Q}_m(t) = 1 - t\,\mathbb{P}_{m-1}(t)$ with $\mathbb{P}_{m-1}$ from (2.38), minimizes the functional

$$
H(\mathbb{Q}) := \|\mathbb{Q}(KK^*)y\|^2 \quad on \quad \{\mathbb{Q} \in \mathcal{P}_m : \mathbb{Q}(0) = 1\}
$$

and satisfies

$$
H(\mathbb{Q}_m) = \|Kx^m - y\|^2.
$$

(b) For $k \neq \ell$, the following orthogonality relation holds:

$$
\langle \mathbb{Q}_k, \mathbb{Q}_\ell \rangle := \sum_{j=1}^{\infty} \mu_j^2 \, \mathbb{Q}_k(\mu_j^2) \, \mathbb{Q}_\ell(\mu_j^2) \, |(y, y_j)|^2 = 0. \qquad (2.39)
$$

If $y \notin \operatorname{span}\{y_1, \ldots, y_N\}$ for any $N \in \mathbb{N}$, then $\langle \cdot, \cdot \rangle$ defines an inner product on the space $\mathcal{P}$ of all polynomials.

Proof: (a) Let $\mathbb{Q} \in \mathcal{P}_m$ be an arbitrary polynomial with $\mathbb{Q}(0) = 1$. Set $\mathbb{P}(t) := (1 - \mathbb{Q}(t))/t$ and $x := \mathbb{P}(K^*K)K^*y = -\mathbb{P}(K^*K)p^0 \in V_{m-1}$. Then

$$
y - Kx = y - K\mathbb{P}(K^*K)K^*y = \mathbb{Q}(KK^*)y.
$$

Thus

$$
H(\mathbb{Q}) = \|Kx - y\|^2 \geq \|Kx^m - y\|^2 = H(\mathbb{Q}_m).
$$

(b) Let $k \neq \ell$. From the identity

$$\frac{1}{2}\nabla f(x^k) = K^*(Kx^k - y) = -\sum_{j=1}^{\infty} \mu_j \mathbb{Q}_k(\mu_j^2)(y, y_j)\, x_j\,,$$

we conclude that

$$0 = \frac{1}{4}\big(\nabla f(x^k), \nabla f(x^\ell)\big) = \langle \mathbb{Q}_k, \mathbb{Q}_\ell \rangle.$$

The properties of the inner product are obvious, except perhaps the definiteness. If $\langle \mathbb{Q}_k, \mathbb{Q}_k \rangle = 0$, then $\mathbb{Q}_k(\mu_j^2)\,(y, y_j)$ vanishes for all $j \in \mathbb{N}$. The assumption on y implies that $(y, y_j) \neq 0$ for infinitely many j. But then the polynomial $\mathbb{Q}_k$ has infinitely many zeros μ_j^2. This implies $\mathbb{Q}_k = 0$, which ends the proof. $\qquad\square$

The following lemma is needed for the proof of Theorem 2.24.

Lemma
Let $0 < m \leq m(\delta)$, $x \in X^\sigma$ for some $\sigma > 0$, and $\|x\|_\sigma \leq E$. Then

$$\left\|Kx^{m,\delta} - y^\delta\right\| \leq \delta + (1+\sigma)^{(\sigma+1)/2}\, \frac{E}{\left|\frac{d}{dt}\mathbb{Q}_m^\delta(0)\right|^{(\sigma+1)/2}}.$$

Before we prove this lemma, we recall some properties of orthogonal functions (see [208]).

As we saw in Lemma 2.22, the polynomials $\mathbb{Q}_m$ are orthogonal with respect to the inner product $\langle \cdot, \cdot \rangle$. Therefore, the zeros $\lambda_{j,m}$, $j = 1, \ldots, m$, of $\mathbb{Q}_m$ are all real and positive and lie in the interval $(0, \|K\|^2)$. By their normalization, $\mathbb{Q}_m$ must have the form

$$\mathbb{Q}_m(t) = \prod_{j=1}^{m}\left(1 - \frac{t}{\lambda_{j,m}}\right).$$

Furthermore, the zeros of two subsequent polynomials interlace, i.e.,

$$0 < \lambda_{1,m} < \lambda_{1,m-1} < \lambda_{2,m} < \lambda_{2,m-1} < \cdots < \lambda_{m-1,m-1} < \lambda_{m,m} < \|K\|^2.$$

Finally, from the factorization of $\mathbb{Q}_m$, we see that

$$\frac{d}{dt}\mathbb{Q}_m(t) = -\mathbb{Q}_m(t)\sum_{j=1}^{m}\frac{1}{\lambda_{j,m} - t} \quad \text{and}$$

$$\frac{d^2}{dt^2}\mathbb{Q}_m(t) = \mathbb{Q}_m(t)\left[\left(\sum_{j=1}^{m}\frac{1}{\lambda_{j,m} - t}\right)^2 - \sum_{j=1}^{m}\frac{1}{(\lambda_{j,m} - t)^2}\right].$$

For $0 \le t \le \lambda_{1,m}$, we conclude that $\frac{d}{dt}\mathbb{Q}_m(t) \le 0$, $\frac{d^2}{dt^2}\mathbb{Q}_m(t) \ge 0$, and $0 \le \mathbb{Q}_m(t) \le 1$.

For the proof of the lemma and the following theorem, it is convenient to introduce two orthogonal projections. For any $\varepsilon > 0$, we denote by $L_\varepsilon : X \to X$ and $M_\varepsilon : Y \to Y$ the orthogonal projections

$$L_\varepsilon z := \sum_{\mu_n^2 \le \varepsilon} (z, x_n)\, x_n, \quad z \in X,$$

$$M_\varepsilon z := \sum_{\mu_n^2 \le \varepsilon} (z, y_n)\, y_n, \quad z \in Y.$$

The following estimates are easily checked:

$$\|M_\varepsilon K x\| \le \sqrt{\varepsilon}\, \|L_\varepsilon x\| \quad \text{and} \quad \|(I - L_\varepsilon) x\| \le \frac{1}{\sqrt{\varepsilon}}\, \|(I - M_\varepsilon) K x\|$$

for all $x \in X$.

Proof of the lemma: Let $\lambda_{j,m}$ be the zeros of $\mathbb{Q}_m^\delta$. We suppress the dependence on δ. The orthogonality relation (2.39) implies that $\mathbb{Q}_m^\delta$ is orthogonal to the polynomial $t \mapsto \mathbb{Q}_m^\delta(t)/(\lambda_{1,m} - t)$ of degree $m - 1$, i.e.,

$$\sum_{n=1}^\infty \mu_n^2\, \mathbb{Q}_m^\delta(\mu_n^2)\, \frac{\mathbb{Q}_m^\delta(\mu_n^2)}{\lambda_{1,m} - \mu_n^2}\, \left|(y^\delta, y_n)\right|^2 = 0.$$

This implies that

$$\sum_{\mu_n^2 \le \lambda_{1,m}} \mathbb{Q}_m^\delta(\mu_n^2)^2\, \frac{\mu_n^2}{\lambda_{1,m} - \mu_n^2}\, \left|(y^\delta, y_n)\right|^2$$

$$= \sum_{\mu_n^2 > \lambda_{1,m}} \mathbb{Q}_m^\delta(\mu_n^2)^2\, \frac{\mu_n^2}{\mu_n^2 - \lambda_{1,m}}\, \left|(y^\delta, y_n)\right|^2$$

$$\ge \sum_{\mu_n^2 > \lambda_{1,m}} \mathbb{Q}_m^\delta(\mu_n^2)^2 \left|(y^\delta, y_n)\right|^2.$$

From this, we see that

$$\|y^\delta - K x^{m,\delta}\|^2 = \left(\sum_{\mu_n^2 \le \lambda_{1,m}} + \sum_{\mu_n^2 > \lambda_{1,m}}\right) \mathbb{Q}_m^\delta(\mu_n^2)^2 \left|(y^\delta, y_n)\right|^2$$

$$\le \sum_{\mu_n^2 \le \lambda_{1,m}} \underbrace{\mathbb{Q}_m^\delta(\mu_n^2)^2 \left\{1 + \frac{\mu_n^2}{\lambda_{1,m} - \mu_n^2}\right\}}_{=:\Phi_m(\mu_n^2)^2} \left|(y^\delta, y_n)\right|^2$$

$$= \left\|M_{\lambda_{1,m}} \Phi_m(KK^*) y^\delta\right\|^2,$$

where we have set

$$\Phi_m(t) := \mathbb{Q}_m^\delta(t) \sqrt{1 + \frac{t}{\lambda_{1,m} - t}} = \mathbb{Q}_m^\delta(t) \sqrt{\frac{\lambda_{1,m}}{\lambda_{1,m} - t}}.$$

Therefore,

$$\left\| y^\delta - K x^{m,\delta} \right\| \leq \left\| M_{\lambda_{1,m}} \Phi_m(KK^*)(y^\delta - y) \right\| + \left\| M_{\lambda_{1,m}} \Phi_m(KK^*)Kx \right\|.$$

We estimate both terms on the right-hand side separately:

$$\left\| M_{\lambda_{1,m}} \Phi_m(KK^*)(y^\delta - y) \right\|^2 = \sum_{\mu_n^2 \leq \lambda_{1,m}} \Phi_m(\mu_n^2)^2 \left| (y^\delta - y, y_n) \right|^2$$

$$\leq \max_{0 \leq t \leq \lambda_{1,m}} \Phi_m(t)^2 \left\| y^\delta - y \right\|^2,$$

$$\left\| M_{\lambda_{1,m}} \Phi_m(KK^*)Kx \right\|^2 = \sum_{\mu_n^2 \leq \lambda_{1,m}} \left[\Phi_m(\mu_n^2)^2 \mu_n^{2+2\sigma} \right] \mu_n^{-2\sigma} \left| (x, x_n) \right|^2$$

$$\leq \max_{0 \leq t \leq \lambda_{1,m}} \left[t^{1+\sigma} \Phi_m(t)^2 \right] \left\| x \right\|_\sigma^2.$$

The proof is finished provided we can show that $0 \leq \Phi_m(t) \leq 1$ and

$$t^{1+\sigma} \Phi_m^2(t) \leq \left(\frac{1+\sigma}{\left| \frac{d}{dt} \mathbb{Q}_m^\delta(0) \right|} \right)^{\sigma+1} \qquad \text{for all } 0 \leq t \leq \lambda_{1,m}.$$

The first assertion follows from $\Phi_m(0) = 1$, $\Phi_m(\lambda_{1,m}) = 0$, and

$$\frac{d}{dt} \left[\Phi_m(t)^2 \right] = 2 \mathbb{Q}_m(t) \frac{d}{dt} \mathbb{Q}_m(t) \frac{\lambda_{1,m}}{\lambda_{1,m} - t} + \mathbb{Q}_m(t)^2 \frac{\lambda_{1,m}}{(\lambda_{1,m} - t)^2}$$

$$= \Phi_m(t)^2 \left[\frac{1}{\lambda_{1,m} - t} - 2 \sum_{j=1}^m \frac{1}{\lambda_{j,m} - t} \right] \leq 0.$$

Now we set $\psi(t) := t^{1+\sigma} \Phi_m(t)^2$. Then $\psi(0) = \psi(\lambda_{1,m}) = 0$. Let $\hat{t} \in (0, \lambda_{1,m})$ be the maximum of ψ in this interval. Then $\psi'(\hat{t}) = 0$, and thus by differentiation

$$(\sigma + 1) \hat{t}^\sigma \Phi_m(\hat{t})^2 + \hat{t}^{\sigma+1} \frac{d}{dt} \left[\Phi_m(\hat{t})^2 \right] = 0,$$

i.e.,

$$\sigma + 1 = \hat{t} \left[2 \sum_{j=1}^m \frac{1}{\lambda_{j,m} - \hat{t}} - \frac{1}{\lambda_{1,m} - \hat{t}} \right] \geq \hat{t} \sum_{j=1}^m \frac{1}{\lambda_{j,m} - \hat{t}}$$

$$\geq \hat{t} \sum_{j=1}^m \frac{1}{\lambda_{j,m}} = \hat{t} \left| \frac{d}{dt} \mathbb{Q}_m^\delta(0) \right|.$$

This implies that $\hat{t} \leq (\sigma + 1)/\left|\frac{d}{dt}\mathbb{Q}_m^\delta(0)\right|$. With $\psi(t) \leq \hat{t}^{\sigma+1}$ for all $t \in [0, \lambda_{1,m}]$, the assertion follows. $\qquad\qquad\qquad\qquad\qquad\qquad\square$

Theorem 2.24
Assume that y and y^δ do not belong to the linear span of finitely many y_j. Let the sequence $x^{m(\delta),\delta}$ be constructed by the conjugate gradient method with stopping rule (2.41) for fixed parameter $\tau > 1$. Let $x \in X^\sigma$ for some $\sigma > 0$ and $\|x\|_\sigma \leq E$. Then there exists $c > 0$ with

$$\left\|x - x^{m(\delta),\delta}\right\| \leq c\,\delta^{\sigma/(\sigma+1)}\,E^{1/(\sigma+1)}. \qquad (2.43)$$

Proof: Similar to the analysis of Landweber's method, we estimate the error by the sum of two terms: the first converges to zero as $\delta \to 0$ independently of m, and the second term tends to infinity as $m \to \infty$. The role of the norm $\|R_\alpha\|$ here is played by $\left|\frac{d}{dt}\mathbb{Q}_m^\delta(0)\right|^{1/2}$.

First, let δ and $m := m(\delta)$ be fixed. Set for abbreviation

$$q := \left|\frac{d}{dt}\mathbb{Q}_m^\delta(0)\right|.$$

Choose $0 < \varepsilon \leq 1/q \leq \lambda_{1,m}$. With

$$\bar{x} := x - \mathbb{Q}_m^\delta(K^*K)x = \mathbb{P}_{m-1}^\delta(K^*K)K^*y,$$

we conclude that

$$
\begin{aligned}
\left\|x - x^{m,\delta}\right\| &\leq \left\|L_\varepsilon(x - x^{m,\delta})\right\| + \left\|(I - L_\varepsilon)(x - x^{m,\delta})\right\| \\
&\leq \left\|L_\varepsilon(x - \bar{x})\right\| + \left\|L_\varepsilon(\bar{x} - x^{m,\delta})\right\| \\
&\quad + \frac{1}{\sqrt{\varepsilon}}\left\|(I - M_\varepsilon)(y - Kx^{m,\delta})\right\| \\
&\leq \left\|L_\varepsilon\mathbb{Q}_m^\delta(K^*K)x\right\| + \left\|L_\varepsilon\mathbb{P}_{m-1}^\delta(K^*K)K^*(y - y^\delta)\right\| \\
&\quad + \frac{1}{\sqrt{\varepsilon}}\left\|y - y^\delta\right\| + \frac{1}{\sqrt{\varepsilon}}\left\|y^\delta - Kx^{m,\delta}\right\| \\
&\leq E \max_{0\leq t\leq\varepsilon}\left|t^{\sigma/2}\mathbb{Q}_m^\delta(t)\right| + \delta \max_{0\leq t\leq\varepsilon}\left|\sqrt{t}\,\mathbb{P}_{m-1}^\delta(t)\right| + \frac{1+\tau}{\sqrt{\varepsilon}}\delta.
\end{aligned}
$$

From $\varepsilon \leq \lambda_{1,m}$ and $0 \leq \mathbb{Q}_m^\delta(t) \leq 1$ for $0 \leq t \leq \lambda_{1,m}$, we conclude that

$$0 \leq t^{\sigma/2}\mathbb{Q}_m^\delta(t) \leq \varepsilon^{\sigma/2} \quad \text{for } 0 \leq t \leq \varepsilon.$$

Furthermore,

$$0 \leq t\mathbb{P}_{m-1}^\delta(t)^2 = \underbrace{[1 - \mathbb{Q}_m^\delta(t)]}_{\leq 1}\underbrace{\frac{1 - \mathbb{Q}_m^\delta(t)}{t}}_{= -\frac{d}{dt}\mathbb{Q}_m^\delta(s)} \leq \left|\frac{d}{dt}\mathbb{Q}_m^\delta(0)\right|$$

for some $s \in [0, \varepsilon]$. Thus we have proven the basic estimate

$$\|x - x^{m,\delta}\| \leq E\varepsilon^{\sigma/2} + (1+\tau)\frac{\delta}{\sqrt{\varepsilon}} + \sqrt{q}\,\delta \quad \text{for} \quad 0 < \varepsilon \leq \frac{1}{q}. \quad (2.44)$$

$\varepsilon \in (0, 1/q)$ is a free parameter in this expression. We minimize the right-hand side with respect to ε. This gives

$$\varepsilon_*^{(\sigma+1)/2} = \frac{\tau+1}{\sigma}\frac{\delta}{E}.$$

Since we do not know if ε_* lies in the interval $(0, 1/q)$, we have to distinguish between two cases.

Case I: $\varepsilon_* \leq 1/q$. Then

$$\sqrt{q} \leq \frac{1}{\sqrt{\varepsilon_*}} = \left(\frac{\sigma}{\tau+1}\right)^{1/(\sigma+1)} \left(\frac{E}{\delta}\right)^{1/(\sigma+1)}$$

and thus

$$\|x - x^{m,\delta}\| \leq c\,\delta^{\sigma/(\sigma+1)}\,E^{1/(\sigma+1)}$$

with constant $c > 0$, which depends only on σ and τ. This case is finished.

Case II: $\varepsilon_* > 1/q$. In this case, we substitute $\varepsilon = 1/q$ in (2.44) and conclude that

$$\begin{aligned}
\|x - x^{m,\delta}\| &\leq E q^{-\sigma/2} + (\tau+2)\sqrt{q}\,\delta \leq E\varepsilon_*^{\sigma/2} + (\tau+2)\sqrt{q}\,\delta \\
&\leq \left(\frac{\tau+1}{\sigma}\right)^{\sigma/(\sigma+1)} \delta^{\sigma/(\sigma+1)}\,E^{1/(\sigma+1)} + (\tau+2)\sqrt{q}\,\delta.
\end{aligned}$$

It remains to estimate the quantity $q = q_m = \left|\frac{d}{dt}\mathbb{Q}_{m(\delta)}^{\delta}(0)\right|$. Until now, we have not used the stopping rule. We will now use this rule to prove the estimate

$$q_m \leq c\left(\frac{E}{\delta}\right)^{2/(\sigma+1)} \quad (2.45)$$

for some $c > 0$, which depends only on σ and τ. Analogously to q_m, we define $q_{m-1} := \left|\frac{d}{dt}\mathbb{Q}_{m(\delta)-1}^{\delta}(0)\right|$. By the previous lemma, we already know that

$$\tau\delta < \|y^{\delta} - Kx^{m(\delta)-1,\delta}\| \leq \delta + (1+\sigma)^{(\sigma+1)/2}\frac{E}{q_{m-1}^{(\sigma+1)/2}},$$

i.e.,

$$q_{m-1}^{(\sigma+1)/2} \leq \frac{(1+\sigma)^{(1+\sigma)/2}}{\tau-1}\frac{E}{\delta}. \quad (2.46)$$

We have to prove such an estimate for m instead of $m - 1$.

Choose $T > 1$ and $\rho^* \in (0, 1)$ with

$$\frac{T}{T-1} < \tau \quad \text{and} \quad T\,\frac{\rho^*}{1-\rho^*} \le 2.$$

If $q_m \le q_{m-1}/\rho^*$, then we are finished by (2.46). Therefore, we assume that $q_m > q_{m-1}/\rho^*$. From $\lambda_{j,m} \ge \lambda_{j-1,m-1}$ for all $j = 2, \dots, m$, we conclude that

$$q_m = \left|\frac{d}{dt}\mathbb{Q}_m^\delta(0)\right| = \sum_{j=1}^m \frac{1}{\lambda_{j,m}} \le \frac{1}{\lambda_{1,m}} + \sum_{j=1}^{m-1} \frac{1}{\lambda_{j,m-1}} = \frac{1}{\lambda_{1,m}} + q_{m-1}.$$

This implies that

$$q_{m-1} \le \rho^* q_m \le \frac{\rho^*}{\lambda_{1,m}} + \rho^* q_{m-1},$$

i.e.,

$$q_{m-1} \le \frac{\rho^*}{(1-\rho^*)\,\lambda_{1,m}}.$$

Finally, we need

$$\frac{1}{\lambda_{2,m}} \le \frac{1}{\lambda_{1,m-1}} \le \sum_{j=1}^{m-1} \frac{1}{\lambda_{j,m-1}} = q_{m-1} \le \frac{\rho^*}{1-\rho^*}\,\frac{1}{\lambda_{1,m}}.$$

Now we set $\varepsilon := T\,\lambda_{1,m}$. Then

$$\varepsilon \le T\,\frac{\rho^*}{1-\rho^*}\,\lambda_{2,m} \le 2\,\lambda_{2,m}.$$

Define the polynomial $\varphi \in \mathcal{P}_{m-1}$ by

$$\varphi(t) := \mathbb{Q}_m^\delta(t)\left(1 - \frac{t}{\lambda_{1,m}}\right)^{-1} = \prod_{j=2}^m \left(1 - \frac{t}{\lambda_{j,m}}\right).$$

For $t \le \varepsilon$ and $j \ge 2$, we note that

$$1 \ge 1 - \frac{t}{\lambda_{j,m}} \ge 1 - \frac{\varepsilon}{\lambda_{2,m}} \ge -1,$$

i.e.,

$$|\varphi(t)| \le 1 \quad \text{for all } 0 \le t \le \varepsilon.$$

For $t \ge \varepsilon$, we conclude that

$$\left|1 - \frac{t}{\lambda_{1,m}}\right| = \frac{t - \lambda_{1,m}}{\lambda_{1,m}} \ge \frac{\varepsilon}{\lambda_{1,m}} - 1 = T - 1,$$

i.e.,

$$|\varphi(t)| \leq \frac{1}{T-1} |Q_m^\delta(t)| \quad \text{for all } t \geq \varepsilon.$$

Since $\varphi(0) = 1$, we can apply Lemma 2.22. Using the projector M_ε, we conclude that

$$
\begin{aligned}
\tau\delta \;<\; & \|y^\delta - Kx_{m-1}^\delta\| \;\leq\; \|\varphi(KK^*)y^\delta\| \\
\leq\; & \|M_\varepsilon\varphi(KK^*)y^\delta\| + \|(I - M_\varepsilon)\varphi(KK^*)y^\delta\| \\
\leq\; & \|M_\varepsilon(y^\delta - y)\| + \|M_\varepsilon y\| + \frac{1}{T-1} \underbrace{\|Q_m^\delta(KK^*)y^\delta\|}_{=\|y^\delta - Kx^{m,\delta}\|} \\
\leq\; & \delta + \varepsilon^{(\sigma+1)/2}E + \frac{1}{T-1}\delta \;=\; \frac{T}{T-1}\delta + \left(T\lambda_{1,m}\right)^{(\sigma+1)/2}E,
\end{aligned}
$$

since $\|M_\varepsilon y\| = \|M_\varepsilon Kx\| \leq \varepsilon^{(\sigma+1)/2}\|x\|_\sigma$. Defining $c := \tau - \frac{T}{T-1}$, we conclude that $c\frac{\delta}{E} \leq \left(T\lambda_{1,m}\right)^{(\sigma+1)/2}$ and thus finally

$$q_m \leq \frac{1}{\lambda_{1,m}} + q_{m-1} \leq T\left(\frac{E}{c\delta}\right)^{2/(\sigma+1)} + q_{m-1}.$$

Combining this with (2.46) proves (2.45) and ends the proof. □

References

[1] L.V. Ahlfors. *Complex Analysis*. McGraw–Hill, New York, 1966.

[2] R.S. Anderssen and P. Bloomfield. Numerical differentiation procedures for non-exact data. *Numer. Math.*, 22:157–182, 1973.

[3] L.E. Andersson. Algorithms for solving inverse eigenvalue problems for Sturm–Liouville equations. In: *Inverse Methods in Action* (P.C. Sabatier, editor). Springer–Verlag, Berlin, 1990.

[4] G. Anger. Uniquely determined mass distributions in inverse problems. Veröffentlichungen des Zentralinstituts für Physik der Erde, 52,2, Potsdam, 1977.

[5] G. Anger, editor. *Inverse and Improperly Posed Problems in Differential Equations*. Akademie Verlag, Berlin, 1979.

[6] G. Anger. Einige Betrachtungen über inverse Probleme, Identifikationsprobleme und inkorrekt gestellte Probleme. In: *Jahrbuch Überblicke Mathematik*. Springer–Verlag, New York, pages 55–71, 1982.

[7] D.N. Arnold and W.L. Wendland. On the asymptotic convergence of collocation methods. *Math. Comput.*, 41:349–381, 1983.

[8] D.N. Arnold and W.L. Wendland. The convergence of spline collocation for strongly elliptic equations on curves. *Numer. Math.*, 47:310–
341, 1985.

[9] K.E. Atkinson. A discrete Galerkin method for first kind integral
equations with a logarithmic kernel. *J. Integral Equations Appl.*,
1:343–363, 1988.

[10] K.E. Atkinson and I.H. Sloan. The numerical solution of first-kind
logarithmic-kernel integral equations on smooth open arcs. *Math.
Comput.*, 56:119–139, 1991.

[11] G. Backus and F. Gilbert. The resolving power of gross earth data.
Geophys. J. R. Astron. Soc., 16:169–205, 1968.

[12] G. Backus and F. Gilbert. Uniqueness in the inversion of inaccurate
gross earth data. *Philos. Trans. R. Soc. London*, 266:123–197, 1970.

[13] H.T. Banks and K. Kunisch. *Estimation Techniques for Distributed
Parameter Systems*. Birkhäuser–Verlag, Basel, 1989.

[14] V. Barcilon. Iterative solution of the inverse Sturm–Liouville problem.
J. Math. Phys., 15:429–436, 1974.

[15] J. Baumeister. *Stable Solutions of Inverse Problems*. Vieweg–Verlag,
Braunschweig, 1987.

[16] H. Bialy. Iterative Behandlung linearer Funktionalgleichungen. *Arch.
Rat. Mech. Anal.*, 4:166, 1959.

[17] H. Brakhage. On ill-posed problems and the method of conjugate
gradients. In: *Inverse and Ill-Posed Problems* (H.W. Engl and C.W.
Groetsch, editors). Academic Press, New York, pages 165–175, 1987.

[18] G. Bruckner. On the regularization of the ill-posed logarithmic kernel
integral equation of the first kind. *Inverse Problems*, 11:65–78, 1995.

[19] H. Brunner. Discretization of Volterra integral equations of the first
kind (II). *Numer. Math.*, 30:117–136, 1978.

[20] H. Brunner. A survey of recent advances in the numerical treatment of
Volterra integral and integro–differential equations. *J. Comp. Appl.
Math.*, 8:213–229, 1982.

[21] L. Brynielson. On Fredholm integral equations of the first kind with
convex constraints. *SIAM J. Math. Anal.*, 5:955–962, 1974.

[22] K.E. Bullen and B. Bolt. *An Introduction to the Theory of Seismology.* Cambridge University Press, Cambridge, 1985.

[23] J.P. Butler, J.A. Reeds, and S.V. Dawson. Estimating solutions of first kind integral equations with nonnegative constraints and optimal smoothing. *SIAM J. Numer. Anal.*, 18:381–397, 1981.

[24] B.L. Buzbee and A. Carasso. On the numerical computation of parabolic problems for preceding times. *Math. Comput.*, 27:237–266, 1973.

[25] B. Caccin, C. Roberti, P. Russo, and L.A. Smaldone. The Backus–Gilbert inversion method and the processing of sampled data. *IEEE Trans. on Signal Processing*, 40:2823–2825, 1992.

[26] A.P. Calderón. On an inverse boundary value problem. In *Seminar on Numerical Analysis and its Applications to Continuum Mechanics.* Soc. Brasileira de Matemática, Rio de Janerio, pages 65–73, 1980.

[27] J.R. Cannon. *The One-Dimensional Heat Equation.* Addison–Wesley, Reading, 1984.

[28] J.R. Cannon and C.D. Hill. Existence, uniqueness, stability and monotone dependence in a Stefan problem for the heat equation. *J. Math. Mech.*, 17:1–19, 1967.

[29] J.R. Cannon and U. Hornung, editors. *Inverse Problems.* ISNM 77, Birkhäuser–Verlag, Basel, 1986.

[30] C. Canuto, M. Hussani, A. Quarteroni, and T. Zang. *Spectral Methods in Fluid Dynamics.* Springer–Verlag, Berlin, 1987.

[31] A. Carasso. Error bounds in the final value problem for the heat equation. *SIAM J. Math. Anal.*, 7:195–199, 1976.

[32] K. Chandrasekharan. *Classical Fourier Transforms.* Springer–Verlag, Berlin, 1989.

[33] D. Colton. The approximation of solutions to the backwards heat equation in a nonhomogeneous medium. *J. Math. Anal. Appl.*, 72:418–429, 1979.

[34] D. Colton. The inverse scattering problem for time harmonic acoustic waves. *SIAM Rev.*, 26:323–350, 1984.

[35] D. Colton and A. Kirsch. Dense sets and far field patterns in acoustic wave propagation. *SIAM J. Math. Anal.*, 15:996–1006, 1984.

[36] D. Colton, A. Kirsch, and L. Päivärinta. Far field patterns for acoustic waves in an inhomogeneous medium. *SIAM J. Math. Anal.*, 20:1472–1483, 1989.

[37] D. Colton and R. Kress. *Integral Equation Methods in Scattering Theory*. John Wiley, New York, 1983.

[38] D. Colton and R. Kress. *Inverse Acoustic and Electromagnetic Scattering Theory*. Springer–Verlag, New York, 1992.

[39] D. Colton and R. Kress. Eigenvalues of the far field operator and inverse scattering theory. *SIAM J. Math. Anal.*, 26:601–615, 1995.

[40] D. Colton and P. Monk. A novel method for solving the inverse scattering problem for time-harmonic acoustic waves in the resonance region. *SIAM J. Appl. Math.*, 45:1039–1053, 1985.

[41] D. Colton and P. Monk. A novel method for solving the inverse scattering problem for time-harmonic acoustic waves in the resonance region II. *SIAM J. Appl. Math.*, 46:506–523, 1986.

[42] D. Colton and P. Monk. The inverse scattering problem for time-harmonic acoustic waves in a penetrable medium. *Q. J. Mech. Appl. Math.*, 40:189–212, 1987.

[43] D. Colton and P. Monk. The inverse scattering problem for time-harmonic acoustic waves in an inhomogeneous medium. *Q. J. Mech. Appl. Math.*, 41:97–125, 1988.

[44] D. Colton and P. Monk. A new method for solving the inverse scattering problem for acoustic waves in an inhomogeneous medium. *Inverse Problems*, 5:1013–1026, 1989.

[45] M. Costabel. Boundary integral operators on Lipschitz domains: elementary results. *SIAM J. Math. Anal.*, 19:613–626, 1988.

[46] M. Costabel, V.J. Ervin, and E.P. Stephan. On the convergence of collocation methods for Symm's integral equations on smooth open arcs. *Math. Comput.*, 51:167–179, 1988.

[47] M. Costabel and E.P. Stephan. On the convergence of collocation methods for boundary integral equations on polygons. *Math. Comput.*, 49:461–478, 1987.

[48] M. Costabel and W. Wendland. Strong ellipticity of boundary integral operators. *J. Reine Angew. Math.*, 372:39–63, 1986.

[49] J. Cullum. Numerical differentiation and regularization. *SIAM J. Numer. Anal.*, 8:254–265, 1971.

[50] J. Cullum. The effective choice of the smoothing norm in regularization. *Math. Comput.*, 33:149–170, 1979.

[51] J.W. Daniel. The conjugate gradient method for linear and nonlinear operator equations. *SIAM J. Numer. Anal.*, 4:10–26, 1967.

[52] P.J. Davis. *Interpolation and Approximation.* Blaisdell Publishing Company, New York, 1963.

[53] E. Deuflhard and E. Hairer, editors. *Numerical treatment of inverse problems in differential and integral equations.* Springer–Verlag, Berlin, 1983.

[54] T.F. Dolgopolova and V.K. Ivanov. Numerical differentiation. *Comput. Math. Math. Phys.*, 6:570–576, 1966.

[55] P.P.B. Eggermont. Approximation properties of quadrature methods for Volterra equations of the first kind. *Math. Comput.*, 43:455–471, 1984.

[56] P.P.B. Eggermont. Beyond superconvergence of collocation methods for Volterra equations of the first kind. In: *Constructive Methods for the Practical Treatment of Integral Equations* (G. Hämmerlin and K.H. Hoffmann, editors). ISNM 73, Birkhäuser–Verlag, Basel, pages 110–119, 1985.

[57] B. Eicke, A.K. Louis, and R. Plato. The instability of some gradient methods for ill-posed problems. *Numer. Math.*, 58:129–134, 1990.

[58] L. Eldén. Algorithms for the regularization of ill-conditioned least squares problems. *BIT*, 17:134–145, 1977.

[59] L. Eldén. Regularization of the backwards solution of parabolic problems. In: *Inverse and Improperly Posed Problems in Differential Equations* (G. Anger, editor). Akademie Verlag, Berlin, 1979.

[60] L. Eldén. Time discretization in the backward solution of parabolic equations. *Math. Comput.*, 39:53–84, 1982.

[61] L. Eldén. An algorithm for the regularization of ill-conditioned banded least squares problems. *SIAM J. Sci. Stat. Comput.*, 5:237–254, 1984.

[62] J. Elschner. On spline approximation for a class of integral equations. I: Galerkin and collocation methods with piecewise polynomials. *Math. Meth. Appl. Sci.*, 10:543–559, 1988.

[63] H. Engl. Necessary and sufficient conditions for convergence of regularization methods for solving linear operator equations of the first kind. *Numer. Funct. Anal. Optim.*, 3:201–222, 1981.

[64] H. Engl. On least-squares collocation for solving linear integral equations of the first kind with noisy right-hand-side. *Boll. Geodesia Sc. Aff.*, 41:291–313, 1982.

[65] H. Engl. Discrepancy principles for Tikhonov regularization of ill-posed problems leading to optimal convergence rates. *J. Optim. Theory Appl.*, 52:209–215, 1987.

[66] H. Engl. On the choice of the regularization parameter for iterated Tikhonov-regularization of ill-posed problems. *J. Approx. Theory*, 49:55–63, 1987.

[67] H. Engl. Regularization methods for the stable solution of inverse problems. *Surv. Math. Ind.*, 3:71–143, 1993.

[68] H. Engl and H. Gfrerer. A posteriori parameter choice for general regularization methods for solving linear ill-posed problems. *Appl. Numer. Math.*, 4:395–417, 1988.

[69] H. Engl and W. Grever. Using the L-curve for determining optimal regularization parameters. *Numer. Math.*, 69:25–31, 1994.

[70] H. Engl and C.W. Groetsch, editors. *Inverse and Ill-Posed Problems*. Academic Press, Boston, 1987.

[71] H. Engl and A. Neubauer. Optimal discrepancy principles for the Tikhonov regularization of integral equations of the first kind. In: *Constructive Methods for the Practical Treatment of Integral Equations* (G. Hämmerlin and K.H. Hoffmann, editors). ISNM 73, Birkhäuser Verlag, Basel, pages 120–141, 1985.

[72] R.E. Ewing. The approximation of certain parabolic equations backwards in time by Sobolev equations. *SIAM J. Math. Anal.*, 6:283–294, 1975.

[73] A. Fasano and M. Primicerio. General free-boundary problems for the heat equation. Parts I, II. *J. Math. Anal. Appl.*, 57:694–723 and 58:202–231, 1977.

References

267

[74] J.N. Franklin. On Tikhonov's method for ill-posed problems. *Math. Comput.*, 28:889–907, 1974.

[75] V. Fridman. A method of successive approximations for Fredholm integral equations of the first kind (in Russian). *Uspeki Mat. Nauk.*, 11:233–234, 1956.

[76] B.G. Galerkin. Expansions in stability problems for elastic rods and plates (in Russian). *Vestnik Inzkenorov*, 19:897–908, 1915.

[77] I.M. Gel'fand and B.M. Levitan. On the determination of a differential operator from its spectral function. *Am. Math. Soc. Trans.*, 1:253–304, 1951.

[78] H. Gfrerer. An a posteriori parameter choice for ordinary and iterated Tikhonov regularization of ill-posed problems leading to optimal convergence rates. *Math. Comput.*, 49:507–522, 1987.

[79] D. Gilbarg and N.S. Trudinger. *Elliptic Partial Differential Equations of Second Order*. Springer–Verlag, Berlin, 1977.

[80] S.F. Gilyazov. Regularizing algorithms based on the conjugate gradient method. *USSR Comput. Math. Math. Phys.*, 26:9–13, 1986.

[81] V.B. Glasko. *Inverse Problems of Mathematical Physics*. American Institute of Physics, New York, 1984.

[82] G.H. Golub and D.P. O'Leary. Some history of the conjugate gradient method and Lanczos algorithms: 1948–1976. *SIAM Rev.*, 31:50–102, 1989.

[83] G.H. Golub and C. Reinsch. Singular value decomposition and least squares solutions. *Numer. Math.*, 14:403–420, 1970.

[84] R. Gorenflo and S. Vessella. *Abel Integral Equations, Analysis and Applications*. Lecture Notes in Mathematics, volume 1461, Springer–Verlag, Berlin, 1991.

[85] J. Graves and P.M. Prenter. Numerical iterative filters applied to first kind Fredholm integral equations. *Numer. Math.*, 30:281–299, 1978.

[86] C.W. Groetsch. *The Theory of Tikhonov Regularization for Fredholm Equations of the First Kind*. Pitman, Boston, 1984.

[87] C.W. Groetsch. *Inverse Problems in the Mathematical Sciences*. Vieweg–Verlag, Braunschweig, Wiesbaden, 1993.

[88] S. Gutman and M. Klibanov. Regularized quasi-Newton method for inverse scattering problems. *Math. Comput. Modelling*, 18:5–31, 1993.

[89] S. Gutman and M. Klibanov. Iterative method for multi-dimensional inverse scattering problems at fixed frequencies. *Inverse Problems*, 10:573–599, 1994.

[90] H. Haario and E. Somersalo. The Backus–Gilbert method revisited: Background, implementation and examples. *Numer. Funct. Anal. Optim.*, 9:917–943, 1985.

[91] J. Hadamard. *Lectures on the Cauchy Problem in Linear Partial Differential Equations*. Yale University Press, New Haven, 1923.

[92] P. Hähner. A periodic Faddeev-type solution operator. To appear in J. Diff. Equations, 1996.

[93] G. Hämmerlin and K.H. Hoffmann, editors. *Improperly Posed Problems and their Numerical Treatment*. ISNM 63, Birkhäuser–Verlag, Basel, 1983.

[94] M. Hanke. Accelerated Landweber iterations for the solution of ill-posed equations. *Numer. Math.*, 60:341–373, 1991.

[95] M. Hanke. Regularization with differential operators. An iterative approach. *Numer. Funct. Anal. Optim.*, 13:523–540, 1992.

[96] M. Hanke. An ϵ-free a posteriori stopping rule for certain iterative regularization methods. *SIAM J. Numer. Anal.*, 30:1208–1228, 1993.

[97] M. Hanke. *Regularisierung schlecht gestellter Gleichungen*. Lecture Notes, University of Karlsruhe, Karlsruhe, 1993.

[98] M. Hanke. *Conjugate Gradient Type Methods for Ill-Posed Problems*. Pitman Research Notes in Mathematics. Longman Scientific & Technical, Harlow, 1995.

[99] M. Hanke and H. Engl. An optimal stopping rule for the ν-method for solving ill-posed problems using Christoffel functions. *J. Approx. Theory*, 79:89–108, 1994.

[100] M. Hanke and C. Hansen. Regularization methods for large-scale problems. *Surv. Math. Ind.*, 3:253–315, 1993.

[101] C. Hansen. Analysis of discrete ill-posed problems by means of the L-curve. *SIAM Rev.*, 34:561–580, 1992.

[102] S. Helgason. *The Radon Transform*. Birkhäuser–Verlag, Boston, 1980.

[103] G. Hellwig. *Partielle Differentialgleichungen*. Teubner–Verlag, Stuttgart, 1960.

[104] G.T. Herman, editor. *Image Reconstruction from Projections: The Fundamentals of Computerized Tomography*. Academic Press, New York, 1980.

[105] G.T. Herman and F. Natterer, editors. *Mathematical Aspects of Computerized Tomography*. Lecture Notes in Medical Informatics, volume 8, Springer–Verlag, New York, 1981.

[106] M.R. Hestenes and E. Stiefel. Methods of conjugate gradients for solving linear systems. *J. Research Nat. Bur. Standards*, 49:409–436, 1952.

[107] J.W. Hilgers. On the equivalence of regularization and certain reproducing kernel Hilbert space approaches for solving first kind problems. *SIAM J. Numer. Anal.*, 13:172–184, 1976.

[108] H. Hochstadt. *The Functions of Mathematical Physics*. John Wiley, New York, 1971.

[109] B. Hofmann. *Regularization for Applied Inverse and Ill-Posed Problems*. Teubner–Verlag, Leipzig, 1986.

[110] G.C. Hsiao, P. Kopp, and W.L. Wendland. A Galerkin collocation method for some integral equations of the first kind. *Computing*, 25:89–113, 1980.

[111] G.C. Hsiao and R.C. MacCamy. Solution of boundary value problems by integral equations of the first kind. *SIAM Rev.*, 15:687–705, 1973.

[112] G.C. Hsiao and W.L. Wendland. A finite element method for some integral equations of the first kind. *J. Math. Anal. Appl.*, 58:449–481, 1977.

[113] G.C. Hsiao and W.L. Wendland. The Aubin–Nitsche lemma for integral equations. *J. Integral Equations*, 3:299–315, 1981.

[114] S.P. Huestis. The Backus–Gilbert problem for sampled band-limited functions. *Inverse Problems*, 8:873–887, 1992.

[115] S. Joe and Y. Yan. A piecewise constant collocation method using cosine mesh grading for Symm's equation. *Numer. Math.*, 65:423–433, 1993.

[116] M. Kac. Can one hear the shape of the drum? *Am. Math. Mon.*, 73:1–23, 1966.

[117] W.J. Kammerer and M.Z. Nashed. Iterative methods for best approximate solutions of integral equations of the first and second kinds. *J. Math. Anal. Appl.*, 40:547–573, 1972.

[118] L.V. Kantorovic and G.P. Akilov. *Functional Analysis in Normed Spaces*. Pergamon Press, Oxford, 1964.

[119] J.B. Keller. Inverse problems. *Am. Math. Mon.*, 83:107–118, 1976.

[120] J.T. King and D. Chillingworth. Approximation of generalized inverses by iterated regularization. *Numer. Funct. Anal. Optim.*, 1:499–513, 1979.

[121] A. Kirsch. The denseness of the far field patterns for the transmission problem. *IMA J. Appl. Math.*, 37:213–225, 1986.

[122] A. Kirsch. Inverse problems. In: *Trends in Mathematical Optimization* (K.-H. Hoffmann, J.-B. Hiriart-Urruty, C. Lemarechal, and J. Zowe, editors). ISNM 84, Birkhäuser–Verlag, Basel, pages 117–137, 1988.

[123] A. Kirsch. An inverse scattering problem for periodic structures. In: *Inverse Scattering and Potential Problems in Mathematical Physics* (R.E. Kleinman, R. Kress, and E. Martensen, editors). Peter Lang, Frankfurt, pages 75–93, 1995.

[124] A. Kirsch, B. Schomburg, and G. Berendt. The Backus–Gilbert method. *Inverse Problems*, 4:771–783, 1988.

[125] A. Kirsch, B. Schomburg, and G. Berendt. Mathematical aspects of the Backus–Gilbert method. In: *Inverse Modeling in Exploration Geophysics* (B. Kummer, A. Vogel, R. Gorenflo, and C.O. Ofoegbu, editors). Vieweg–Verlag, Braunschweig, Wiesbaden, 1989.

[126] R.E. Kleinman and P.M. van den Berg. A modified gradient method for two-dimensional problems in tomography. *J. Comput. Appl. Math.*, 42:17–36, 1992.

[127] I. Knowles and R. Wallace. A variational method for numerical differentiation. *Numer. Math.*, 70:91–110, 1995.

[128] C. Kravaris and J.H. Seinfeld. Identification of parameters in distributed parameter systems by regularization. *SIAM J. Control Optim.*, 23:217–241, 1985.

[129] C. Kravaris and J.H. Seinfeld. Identifiabilty of spatially-varying conductivity from point obervation as an inverse Sturm–Liouville problem. *SIAM J. Control Optim.*, 24:522–542, 1986.

[130] R. Kress. *Linear Integral Equations.* Springer–Verlag, Berlin, 1989.

[131] R. Kress. Personal communication, 1994.

[132] R. Kress and I.H. Sloan. On the numerical solution of a logarithmic integral equation of the first kind for the Helmholtz equation. *Numer. Math.*, 66:199–214, 1993.

[133] O.A. Ladyzenskaja, V.A. Solonnikov, and N.N. Uralceva. *Linear and Quasi-linear Equations of Parabolic Type.* American Mathematical Society, Providence, Rhode Island, 1986.

[134] C. Lanczos. *Linear Differential Operators.* Van Nostrand, New York, 1961.

[135] L. Landweber. An iteration formula for Fredholm integral equations of the first kind. *Am. J. Math.*, 73:615–624, 1951.

[136] M.M. Lavrentiev. *Some Improperly Posed Problems of Mathematical Physics.* Springer–Verlag, Berlin, 1967.

[137] M.M. Lavrentiev, K.G. Reznitskaya, and V.G. Yakhov. *One-Dimensional Inverse Problems of Mathematical Physics.* American Mathematical Society Translations, Providence, Rhode Island, 1986.

[138] M.M. Lavrentiev, V.G. Romanov, and Vasiliev. *Multidimensional Inverse Problems for Differential Equations.* Lecture Notes in Mathematics, volume 167, Springer–Verlag, Berlin, 1970.

[139] P.D. Lax and R.S. Phillips. *Scattering Theory.* Academic Press, New York, 1967.

[140] P. Linz. Numerical methods for Volterra equations of the first kind. *Comput. J.*, 12:393–397, 1969.

[141] P. Linz. *Analytical and Numerical Methods for Volterra Equations.* SIAM Publications, Philadelphia, 1985.

[142] J. Locker and P.M. Prenter. Regularization with differential operators. I: General theory. *J. Math. Anal. Appl.*, 74:504–529, 1980.

[143] A.K. Louis. Convergence of the conjugate gradient method for compact operators. In: *Inverse and Ill-posed Problems* (H.W. Engl and C.W. Groetsch, editors). Academic Press, Boston, pages 177–183, 1987.

[144] A.K. Louis. *Inverse und schlecht gestellte Probleme*. Teubner–Verlag, Stuttgart, 1989.

[145] A.K. Louis. Medical imaging, state of art and future developments. *Inverse Problems*, 8:709–738, 1992.

[146] A.K. Louis and P. Maass. Smoothed projection methods for the moment problem. *Numer. Math.*, 59:277–294, 1991.

[147] A.K. Louis and F. Natterer. Mathematical problems in computerized tomography. *Proc. IEEE*, 71:379–389, 1983.

[148] B.D. Lowe, M. Pilant, and W. Rundell. The recovery of potentials from finite spectral data. *SIAM J. Math. Anal.*, 23:482–504, 1992.

[149] B.D. Lowe and W. Rundell. The determination of a coefficient in a parabolic equation from input sources. *IMA J. of Appl. Math.*, 52:31–50, 1994.

[150] J.T. Marti. An algorithm for computing minimum norm solutions of Fredholm integral equations of the first kind. *SIAM J. Numer. Anal.*, 15:1071–1076, 1978.

[151] J.T. Marti. On the convergence of an algorithm computing minimum-norm solutions to ill-posed problems. *Math. Comput.*, 34:521–527, 1980.

[152] C.A. Miccelli and T.J. Rivlin. A survey of optimal recovery. In: *Optimal Estimation in Approximation Theory* (C.A. Miccelli and T.J. Rivlin, editors). Plenum Press, New York, 1977.

[153] K. Miller. Efficient numerical methods for backward solution of parabolic problems with variable coefficients. In: *Improperly Posed Problems* (A. Carasso, editor). Pitman, Boston, 1975.

[154] V.A. Morozov. Choice of parameter for the solution of functional equations by the regularization method. *Sov. Math. Doklady*, 8:1000–1003, 1967.

[155] V.A. Morozov. The error principle in the solution of operational equations by the regularization method. *USSR Comput. Math. Math. Phys.*, 8:63–87, 1968.

[156] V.A. Morozov. *Methods for Solving Incorrectly Posed Problems.* Springer–Verlag, Berlin, 1984.

[157] V.A. Morozov. *Regularization Methods for Ill-Posed Problems.* CRC Press, Boca Raton, 1993.

[158] D.A. Murio. *The Mollification Method and the Numerical Solution of Ill-Posed Problems.* John Wiley, New York, 1993.

[159] A. Nachman. Reconstructions from boundary measurements. *Ann. Math.*, 128:531–576, 1988.

[160] S.I. Nakagiri. Review of Japanese work of the last ten years on identifiability in distributed parameter systems. *Inverse Problems*, 9:143–191, 1993.

[161] M.Z. Nashed. On moment discretization and least-squares solution of linear integral equations of the first kind. *J. Math. Anal. Appl.*, 53:359–366, 1976.

[162] M.Z. Nashed and G. Wahba. Convergence rates of approximate least squares solution of linear integral and operator equations of the first kind. *Math. Comput.*, 28:69–80, 1974.

[163] I.P. Natanson. *Constructive Function Theory.* Frederick Ungar Publ. Co., New York, 1965.

[164] F. Natterer. Regularisierung schlecht gestellter Probleme durch Projektionsverfahren. *Numer. Math.*, 28:329–341, 1977.

[165] F. Natterer. Error bounds for Tikhonov regularization in Hilbert scales. *Appl. Anal.*, 18:29–37, 1984.

[166] F. Natterer. *The Mathematics of Computerized Tomography.* Teubner–Verlag, Stuttgart, 1986.

[167] A.S. Nemirov'ski and B.T. Polyak. Iterative methods for solving linear ill-posed problems and precise information I. *Eng. Cybernetics*, 22:1–11, 1984.

[168] A.S. Nemirov'ski and B.T. Polyak. Iterative methods for solving linear ill-posed problems and precise information II. *Eng. Cybernetics*, 22:50–56, 1984.

[169] A. Neubauer. An a posteriori parameter choice for Tikhonov regularization in Hilbert scales leading to optimal convergence rates. *SIAM J. Numer. Anal.*, 25:1313–1326, 1988.

[170] Y. Notay. On the convergence rate of the conjugate gradients in the presence of rounding errors. *Numer. Math.*, 65:301–318, 1993.

[171] R. Novikov. Multidimensional inverse spectral problems for the equation $-\triangle\psi + \big(v(x) - Eu(x)\big)\psi = 0$. *Trans. Func. Anal. Appl.*, 22:263–272, 1988.

[172] L. Päivärinta and E. Somersalo, editors. *Inverse Problems in Mathematical Physics*. Lecture Notes in Physics, volume 422, Springer–Verlag, Berlin, 1993.

[173] R.L. Parker. Understanding inverse theory. *Annual Rev. Earth Planet. Sci.*, 5:35–64, 1977.

[174] L.E. Payne. *Improperly Posed Problems in Partial Differential Equations*. SIAM Publications, Philadelphia, 1975.

[175] G.I. Petrov. Application of Galerkin's method to a problem of the stability of the flow of a viscous fluid (in Russian). *Priklad. Matem. Mekh.*, 4:3–12, 1940.

[176] D.L. Phillips. A technique for the numerical solution of certain integral equations of the first kind. *J. Assoc. Comput. Mach.*, 9:84–97, 1962.

[177] J. Pöschel and E. Trubowitz. *Inverse Spectral Theory*. Academic Press, London, 1987.

[178] A.G. Ramm. Recovery of the potential from fixed energy scattering data. *Inverse Problems*, 4:877–886, 1988.

[179] Lord Rayleigh. *The Theory of Sound*. London, 1896.

[180] F. Rellich. Über das asymptotische Verhalten von Lösungen von $\Delta u + \lambda u = 0$ in unendlichen Gebieten. *Jber. Deutsch. Math. Verein.*, 53:57–65, 1943.

[181] G.R. Richter. Numerical solution of integral equations of the first kind with nonsmooth kernels. *SIAM J. Numer. Anal.*, 15:511–522, 1978.

[182] G.R. Richter. An inverse problem for the steady state diffusion equation. *SIAM J. Appl. Math.*, 41:210–221, 1981.

[183] G.R. Richter. Numerical identification of a spacially varying diffusion coefficient. *Math. Comput.*, 36:375–386, 1981.

[184] W. Ritz. Über lineare Funktionalgleichungen. *Acta Math.*, 41:71–98, 1918.

[185] G. Rodriguez and S. Seatzu. Numerical solution of the finite moment problem in a reproducing kernel Hilbert space. *J. Comput. Appl. Math.*, 33:233–244, 1990.

[186] W. Rudin. *Functional Analysis*. McGraw–Hill, New York, 1973.

[187] W. Rundell. Inverse Sturm–Liouville Problems. Lecture Notes, University of Oulu, Oulu, 1996.

[188] W. Rundell and P.E. Sacks. The reconstruction of Sturm–Liouville operators. *Inverse Problems*, 8:457–482, 1992.

[189] W. Rundell and P.E. Sacks. Reconstruction techniques for classical Sturm–Liouville problems. *Math. Comput.*, 58:161–183, 1992.

[190] P.C. Sabatier. Positivity constraints in linear inverse problem - I. General theory. *Geophys. J.R. Astron. Soc.*, 48:415–441, 1977.

[191] P.C. Sabatier. Positivity constraints in linear inverse problem - II. Applications. *Geophys. J.R. Astron. Soc.*, 48:443–459, 1977.

[192] P.C. Sabatier, editor. *Applied Inverse Problems*. Lecture Notes in Physics, volume 85, Springer–Verlag, New York, 1978.

[193] G. Santhosh and M. Thamban Nair. A class of discrepancy principles for the simplified regularization of ill-posed problems. *J. Austr. Math. Soc. Ser. B*, 36:242–248, 1995.

[194] J. Saranen. The modified quadrature method for logarithmic-kernel integral equations on closed curves. *J. Integral Equations Appl.*, 3:575–600, 1991.

[195] J. Saranen and I.H. Sloan. Quadrature methods for logarithmic-kernel integral equations on closed curves. *IMA J. Numer. Anal.*, 12:167–187, 1992.

[196] J. Saranen and W.L. Wendland. On the asymptotic convergence of collocation methods with spline functions of even degree. *Math. Comput.*, 45:91–108, 1985.

[197] G. Schmidt. On spline collocation methods for boundary integral equations in the plane. *Math. Meth. Appl. Sci.*, 7:74–89, 1985.

[198] E. Schock. On the asymptotic order of accuracy of Tikhonov regularization. *J. Optim. Theory Appl.*, 44:95–104, 1984.

[199] E. Schock. What are the proper condition numbers of discretized ill-posed problems? *Lin. Alg. Appl.*, 81:129–136, 1986.

[200] B. Schomburg and G. Berendt. On the convergence of the Backus–Gilbert algorithm. *Inverse Problems*, 3:341–346, 1987.

[201] T.I. Seidman. Nonconvergence results for the application of least squares estimation to ill-posed problems. *J. Optim. Theory Appl.*, 30:535–547, 1980.

[202] R.E. Showalter. The final value problem for evolution equations. *J. Math. Anal. Appl.*, 47:563–572, 1974.

[203] B.D. Sleeman. The inverse problem of acoustic scattering. *IMA J. Appl. Math.*, 29:113–142, 1982.

[204] I.H. Sloan. Error analysis of boundary integral methods. *Acta Numerica*, 1:287–339, 1992.

[205] I.H. Sloan and B.J. Burn. An unconventional quadrature method for logarithmic-kernel integral equations on closed curves. *J. Integral Equations Appl.*, 4:117–151, 1992.

[206] I.H. Sloan and W.L. Wendland. A quadrature-based approach to improving the collocation method for splines of even degree. *Z. Anal. Anw.*, 8:362–376, 1989.

[207] J. Stefan. über einige Probleme der Theorie der Wärmeleitung. *S.-Ber. Wien Akad. Mat. Nat.*, 98:173, 616, 956, 1418, 1889.

[208] J. Stoer and R. Bulirsch. *Introduction to Numerical Analysis.* Springer–Verlag, Heidelberg, 1980.

[209] T. Suzuki and R. Muayama. A uniqueness theorem in an identification problem for coefficients of parabolic equations. *Proc. Jpn. Acad. Ser. A*, 56:259–263, 1980.

[210] J. Sylvester and G. Uhlmann. A global uniqueness theorem for an inverse boundary value problem. *Ann. Math.*, 125:153–169, 1987.

[211] J. Sylvester and G. Uhlmann. The Dirichlet to Neumann map and applications. In: *Inverse Problems in Partial Differential Equations* (D. Colton, R. Ewing, and W. Rundell, editors). SIAM Publications, Philadelphia, pages 101–139, 1990.

[212] G. Talenti, editor. *Inverse Problems.* Springer–Verlag, Berlin, 1986.

[213] A.N. Tikhonov. Regularization of incorrectly posed problems. *Sov. Doklady*, 4:1624–1627, 1963.

[214] A.N. Tikhonov. Solution of incorrectly formulated problems and the regularization method. *Sov. Doklady*, 4:1035–1038, 1963.

[215] A.N. Tikhonov and V.Y. Arsenin. *Solutions of Ill-Posed Problems.* V.H. Winston & Sons, Washington, D.C., 1977.

[216] A.N. Tikhonov, A.V. Goncharsky, V.V. Stepanov, and A.G. Yagola. *Numerical Methods for the Solution of Ill-Posed Problems.* Kluwer, Dordrecht, 1995.

[217] S. Twomey. The application of numerical filtering to the solution of integral equations encountered in indirect sensing measurements. *J. Franklin Inst.*, 279:95–109, 1965.

[218] P.M. van den Berg, M.G. Coté, and R.E. Kleinman. "Blind" shape reconstruction from experimental data. *IEEE Trans. Ant. Prop.*, 43:1389–1396, 1995.

[219] J.M. Varah. Pitfalls in the numerical solution of linear ill-posed problems. *SIAM J. Sci. Stat. Comput.*, 4:164–176, 1983.

[220] W. Wendland. On Galerkin collocation methods for integral equations of elliptic boundary value problems. In: *Numerical Treatment of Integral Equations* (R. Leis, editor). ISNM 53, Birkhäuser–Verlag, Basel, pages 244–275, 1979.

[221] J. Werner. *Optimization Theory and Applications.* Vieweg–Verlag, Braunschweig, Wiesbaden, 1984.

[222] G.M. Wing. *A Primer on Integral Equation of the First Kind. The Problem of Deconvolution and Unfolding.* SIAM Publications, Philadelphia, 1992.

[223] J. Wloka. *Funktionalanalysis und Anwendungen.* de Gruyter, Berlin, New York, 1971.

[224] X.G. Xia and M.Z. Nashed. The Backus–Gilbert method for signals in reproducing kernel Hilbert spaces and wavelet subspaces. *Inverse Problems*, 10:785–804, 1994.

[225] X.G. Xia and Z. Zhang. A note on 'the Backus–Gilbert inversion method and the processing of sampled data'. Preprint, 1992.

[226] Y. Yan and I.H. Sloan. On integral equations of the first kind with logarithmic kernels. *J. Integral Equations Appl.*, 1:549–579, 1988.

[227] Y. Yan and I.H. Sloan. Mesh grading for integral equations of the first kind with logarithmic kernel. *SIAM J. Numer. Anal.*, 26:574–587, 1989.

[228] T. Yosida. *Lectures on Differential and Integral Equations*. Wiley Interscience, New York, 1960.

[229] D. Zidarov. *Inverse Gravimetric Problems in Geoprospecting and Geodesy*. Elsevier, Amsterdam, 1980.

Index

Applied Mathematical Sciences

(continued from page ii)